CHEMICAL FUNGAL TAXONOMY

CHEMICAL FUNGAL TAXONOMY

edited by

Jens C. Frisvad

Technical University of Denmark
Lyngby, Denmark

Paul D. Bridge

International Mycological Institute
Egham, Surrey, United Kingdom

Dilip K. Arora

Banaras Hindu University
Varanasi, India

MARCEL DEKKER, INC. NEW YORK • BASEL • HONG KONG

Library of Congress Cataloging-in-Publication Data

Chemical fungal taxonomy / edited by Jens C. Frisvad, Paul D. Bridge, Dilip K. Arora.
p. cm.
Includes bibliographical references and index.
ISBN 0-8247-0069-4 (alk. paper)
1. Fungi--Chemotaxonomy. I. Frisvad, Jens C. II. Bridge, Paul D. III. Arora, Dilip K.
QK603.2.C48 1998
579.5'01'2--dc21

98-8044
CIP

This book is printed on acid-free paper.

Headquarters
Marcel Dekker, Inc.
270 Madison Avenue, New York, NY 10016
tel: 212-696-9000; fax: 212-685-4540

Eastern Hemisphere Distribution
Marcel Dekker AG
Hutgasse 4, Postfach 812, CH-4001 Basel, Switzerland
tel: 44-61-261-8482; fax: 44-61-261-8896

World Wide Web
http://www.dekker.com

The publisher offers discounts on this book when ordered in bulk quantities. For more information, write to Special Sales/Professional Marketing at the headquarters address above.

Current printing (last digit):
10 9 8 7 6 5 4 3 2 1

PRINTED IN THE UNITED STATES OF AMERICA

Preface

A large volume of literature is available on basic and applied mycology, covering aspects such as taxonomy and phylogeny, physiology, biochemistry, comparative morphology, genetics and molecular biology, pathology, and fungal metabolites including mycotoxins and ecology. Mycology is also often a significant component in studies of foods, biodeterioration, and health publications. A stable, reliable taxonomy for the fungi is a prerequisite for workers in all of these disciplines, to facilitate communication and to allow comparisons to be made between different subject areas. But until now, the various aspects of chemical fungal taxonomy had never been covered in one book.

Historically, fungal taxonomy has developed from classical botanical approaches, and most taxonomic characters are based on morphological attributes or associations with particular hosts. However, chemical methods have played an important role in the taxonomy of yeasts, where there are few morphological features, and in the taxonomy of lichens and some lichen-associated fungi. Chemical taxonomy, in its broadest sense, has often been limited to a small number of filamentous microfungi, although the application of some chemical and physiological characters in wider taxonomic schemes is becoming more common. For example, immunological methods are widely used for the diagnosis of fungal diseases in animals and are now considered for the diagnosis of a wide range of plant pathogenic fungi.

Recent developments in molecular biology have resulted in new techniques, such as restriction site mapping, becoming widely available in biological laboratories. The incorporation of such methods in systematics has revolutionized many taxonomic schemes, allowing the characterization of individual isolates while also permitting phylogenetic and other evolutionary inferences to be made.

One disadvantage to the adoption of new techniques is that they have often been undertaken by specialists. As a result, chemical and molecular biology approaches in systematics have been applied independently, so comparisons between such schemes, and with those based on classical characters, are difficult to make. Similarly, important publications in these fields have been published in a wide range of *specialist* journals and books.

It is therefore timely and important to present a broad range of information on these techniques, and their applications in systematic mycology, in a single volume. We have asked the authors of individual chapters to give a full and critical account of the use of particular methods in fungal systematics and hope that this volume will then serve as a valuable reference source. The book will be of particular value for university and industrial scientists actively involved in many applied mycological disciplines, but especially human and plant pathology, food quality, industrial mycology including screening for strains, applications or production of stereospecific transformations, fermentations of enzymes, and production of pharmacological products. It will be of particular value for interdisciplinary scientists and for mycologists, botanists, soil microbiologists, molecular biologists, microbial ecologists, biotechnologists, agriculturalists, and graduate school students in microbiology, mycology, botany, plant pathology, microbial ecology, and biotechnology.

We would like to thank the authors for their contributions and for all their effort. We would also like to thank Ms. Sandra Beberman for her dedicated editorial assistance.

Jens C. Frisvad
Paul D. Bridge
Dilip K. Arora

Contents

Contributors

Dilip K. Arora, M.Sc. Professor of Microbiology and Mycology, Centre for Advanced Study in Botany, Banaras Hindu University, Varanasi, India

Søren Banke, M.Sc. Research Assistant, Department of Mycology, Botanical Institute, University of Copenhagen, Copenhagen, Denmark

Manuel Bernabé, Ph.D. Senior Research Scientist, Departamento de Química Orgánica Biológica, Instituto de Química Orgánica, Grupo de Carbohidratos, Centro de Investigaciones Biológicas (CSIC), Madrid, Spain

A. Botha, Ph.D. Department of Microbiology and Biochemistry, University of the Orange Free State, Bloemfontein, South Africa

Paul D. Bridge, B.Sc., Ph.D. International Mycological Institute, Egham, Surrey, United Kingdom

M. A. Cousin, Ph.D. Professor, Department of Food Science, Purdue University, West Lafayette, Indiana

G. A. De Ruiter, Ph.D. Hercules European Research Center, Barneveld, The Netherlands

Véronique Edel, Ph.D. Researcher, Department of Plant Pathology, Institut National de la Recherche Agronomique, Dijon, France

Ole Filtenborg, M.Sc. Associate Professor, Department of Biotechnology, Technical University of Denmark, Lyngby, Denmark

J. M. Frank, Ph.D. Research Fellow, School of Biological Sciences, University of Surrey, Guildford, Surrey, United Kingdom

Jens C. Frisvad, Ph.D., M.Sc. Associate Professor of Food Mycology, Department of Biotechnology, Technical University of Denmark, Lyngby, Denmark

Grégoire L. Hennebert, Ph.D. Professor Emeritus, Université Catholique de Louvain, Louvain-la-Neuve, Belgium

J. L. F. Kock, Ph.D. Professor, Department of Microbiology and Biochemistry, University of the Orange Free State, Bloemfontein, South Africa

Thomas Ostenfeld Larsen, Ph.D. Assistant Professor, Department of Biotechnology, Technical University of Denmark, Lyngby, Denmark

J. Antonio Leal, Ph.D. Research Scientist, Department of Molecular Microbiology, Centro de Investigaciones Biológicas (CSIC), Madrid, Spain

Helge Thorsten Lumbsch Botanical Institute, University of Essen, Essen, Germany

S. H. W. Notermans, Ph.D. National Institute of Public Health and the Environment, Bilthoven, The Netherlands

R. Russell Monteith Paterson, Ph.D. Senior Biochemist, Department of Environmental and Industrial Development, International Mycological Institute, Egham, Surrey, United Kingdom

Gaby E. Pfyffer, Ph.D. Head, Swiss National Center for Mycobacteria, Department of Medical Microbiology, University of Zurich, Zurich, Switzerland

F. M. Rombouts, Ph.D. Professor, Department of Food Science, Wageningen Agricultural University, Wageningen, The Netherlands

Søren Rosendahl, Ph.D. Associate Professor, Department of Mycology, Botanical Institute, University of Copenhagen, Copenhagen, Denmark

Gerard S. Saddler, Ph.D. Department of Biosystematics and Biochemistry, International Mycological Institute, Egham, Surrey, United Kingdom

Ulf Thrane, Ph.D. Associate Professor, Department of Biotechnology, Technical University of Denmark, Lyngby, Denmark

Marc Vancanneyt, Ph.D. Laboratorium voor Microbiologie, Universiteit Gent, Ghent, Belgium

1

Chemical Fungal Taxonomy: An Overview

Jens C. Frisvad
Technical University of Denmark, Lyngby, Denmark

Paul D. Bridge
International Mycological Institute, Egham, Surrey, United Kingdom

Dilip K. Arora
Banaras Hindu University, Varanasi, India

I. INTRODUCTION

Chemotaxonomy or biochemical systematics is based on chemical characteristics, and has been an important part of taxonomy of many organisms for several decades. The development of analytical and molecular biology methods and the associated instrumentation has led to a large number of different approaches to chemotaxonomy. Chemotaxonomy has been used extensively in the lichenized fungi and yeasts, and more recently in many other fungal groups. Chemotaxonomy is concerned mainly with on DNA, RNA, or proteins (1,2), and to a lesser extent with lipids, carbohydrates, proteins, and cell wall membrane components (3–7). Other researchers regard secondary metabolites as the main field of chemotaxonomy (8–17). The significance of these biochemical characteristics may depend on the taxonomic level (species, genus, family, etc.) and the systematic school adhered to, whether neo-Darwinistic, numerical phenetic, or cladistic (18). In the broadest sense, chemotaxonomy includes both the evaluation of chemical constituents of the organisms (primary and secondary metabolites) and the organisms' chemical and physiological activities (19–21). Physiological tests have been used extensively in the taxonomy of bacteria (22,23) and yeasts (24), but have not been used extensively in the classification of common filamentous fungi (25–28). Chemical tests have long been established in the taxonomy of lichenized fungi (29), and although the use of chemotaxonomy in lichenology has increased in recent years, the use of chemical characters is not common in the classification and identification of fungi (30). Very few descriptions of new fungal species contain chemical data; however, taxonomic revisions and descriptions of lichenized fungi would probably be considered incomplete if chemical data were not included (31). Chemical and molecular biological methods have developed rapidly, and in recent years a large number of chemotaxonomic studies have been undertaken (32–36). Molecular studies with lichenized fungi have been slower to develop (37–42), mainly because of difficulties in separating the phycobiont and mycobiont and the slow growth rate of mycobionts in pure culture (42). This situation has, however, been greatly improved by the introduction of PCR technology (40).

Chemotaxonomic approaches with the ascomycetous yeasts have been mainly based on physiological and chemical tests (24,43), cell wall biopolymers (7), quantitative profiles of sterols and total fatty acids, and molecular methods (7,44–49). Nakase (50) pointed out that a polyphasic approach to the systematics of basidiomycetous yeast based on molecular data, monosaccharide composition, and ubiquinones gave more reliable criteria than gross morphological and physiological characters. Chemotaxonomy has rarely been used with the Chytridiomycota (51). Recent work has chemically characterized species of *Zygomycota* based on fatty-acid profiles, antigens, polyols, and DNA sequences (52–55). However, more data are needed if phylogenetic conclusions are to be drawn (56).

In higher basidiomycetes, indirect chemical characters such as color, smell, and taste of the basidiocarp have been used for many years. Since the 1960s many basidiomycetes have been analyzed chemically (57,58), especially in respect to their secondary-metabolite profiles. However, although structures have been elucidated for a large number of secondary metabolites, several isolates of the same species have rarely been compared in a systematic manner. Molecular (DNA) characters have been considered quite extensively in recent years for basidiomycetes; however, fewer data are available in basidiomycete DNA systematics than in ascomycetes (59–68). Chemotaxonomy and molecular methods have an important role in the identification of mycorrhizal fungi which are difficult to isolate and culture (68,69), and methods that have been used for identification and classification of these include immunology, isozymes, pyrolysis, of cells, phospholipid fatty acids, and DNA sequences (69).

A considerable number of chemotaxonomic studies have been carried out within the mitosporic ascomycetes (70–72). The genera *Aspergillus*, *Penicillium*, and *Fusarium* and their teleomorphs have been extensively examined using different chemotaxonomic and morphological methods (73–78). Many other ascomycete genera have also been examined extensively, including *Trichoderma*, *Verticillium* (79) and *Monascus* (80).

II. CHEMICAL METHODS

Two different approaches have been used for the analysis of primary and secondary metabolites: specific compounds are separated and characterized (6,16), or mixtures of compounds are directly characterized without any prior separation, often resulting in a polyphenic trace or a compound spectrum (81). The most notable example of the latter approach is pyrolysis gas chromatography (82–103). In general, early studies in this area included only a few strains and rather inefficient multivariate statistical techniques for the evaluation of the pyrograms (96,100), but later studies have included Principal Component Analysis for the evaluation of gas chromatographic traces and Discriminant Analysis for pyrolysis mass spectra (100,102). These studies have grouped isolates of different species using these methods. Although pyrolysis gas chromatography has been quite effective in chemotaxonomic studies of fungi, it has been used less frequently in recent years due to a desire to obtain a more accurate elucidation of structures of the polymers which give rise to the pyrolysis products—i.e., carbohydrates (104,105).

The indirect characterization of surface biopolymers of fungi may be based on differences in hydrophobicity and electrokinetic properties (106,197). These properties can be measured from cross-partition data based on aqueous two-phase systems (108–110). Blomquist and co-workers (111–113) used this technique in combination with colony diameters and Disjoint Principal Component Analysis (SIMCA pattern recognition) to separate closely related species of *Penicillium*. Only a few isolates (one to three of each species) were studied, and it is therefore difficult to conclude whether this method will be of significant value in chemotaxonomy in the future.

In the future, rapid methods for identification may be based on spectroscopy of fungi or fungal extracts. Methods already in use in the food industry for detection of microorganisms include UV, NIR, Raman, and fluorescence spectrometry combined with multivariate calibration (114).

III. MOLECULAR AND IMMUNOLOGICAL METHODS

Before the advent of modern molecular methods many fungi were studied using total electrophoretic protein profiles and isozyme profiles, and these methods are still of great value (105,115–118). Methods based on the electrophoretic separation of isozymes are among the most effective in investigations of genetic phenomena at the broad molecular level, but are also of great value in taxonomic studies (115). These methods are treated in detail in Chapters 4 and 5.

Molecular methods play a major role in modern taxonomy (1). They are based on the analysis of chromosomes, genes, and their translation products (proteins). Early studies were based on DNA reassociation (119), and this technique has been used in several mycological studies (120). This method has its weaknesses in that the individual characters, the nucleotides, are not identified (121). Orthogonal field gel electrophoresis (or pulsed-field gel electrophoresis) of chromosomes (karyotyping) is another technique that has been used in mycology (122–125). Other methods are based on a further characterization of certain genes or defined regions of DNA, through either restriction enzyme patterns (126) or direct sequencing (127). The restriction fragment length polymorphism (RFLP) method can be applied to small or large parts of the genome, and are most commonly obtained from the internally transcribed spacer (ITS) or intragenic spacer (IGS) regions of the ribosomal DNA (128–138).

Direct sequencing of rDNA has been used widely in fungal phylogenetic studies (45,139–147). These studies have provided mycologists with a much better understanding of fungal phylogeny, rDNA is possibly one of the oldest and most conserved regions of DNA available. rDNA is only a minor part of the total genome (61), so genes other than rDNA have been suggested as molecular markers, especially the β-tubulin gene (134).

The PCR reaction has been a major step forward in the study of phylogeny, classification, and identification of fungi. This technique was quickly taken up by mycologists (148,149) and has been used in combination with other methods in experimental mycology (68,151–153).

The RAPD (random amplification of polymorphic DNA) method uses PCR to amplify genomic DNA segments with nonspecific oligonucleotide primers (154,155). RAPDs have been used particularly in race, pathotype, and population studies (156–167). An alternative method for fingerprinting fungi below the species level is AFLP (amplified fragment length polymorphism). This method is based on the selective PCR amplification of restriction fragments from a total digest of genomic DNA (168,169) and has been used in fungal studies (170). The uses of many of these molecular techniques are detailed by Edel in Chapter 3 of this volume.

Although serological tests have been used in bacterial identification since the 1920s, immunological techniques have only become powerful tools in fungal identification and detection in the past three decades. Serology has played an important role in the identification and detection of fungi from soil, plants, animals, and humans. Furthermore, serology has helped in the understanding of phylogenetic relationships, the detection and management strategies in disease control, the detection of in vitro translation products of nucleic acids, and the detection of intimate structural changes within the cells as a result of fungal invasion (171,172). Although polyclonal antibodies (PAbs) are the simplest to produce and have been used in many cases for the detection and identification of fungi, their general cross-reactivity has limited their usefulness. The advent of hybridoma techniques brought about a revolution in this area with the production of monoclonal antibodies (MAbs). The use of monoclonal antibodies was, however, initially only widely exploited in medical sciences, especially for the diagnosis of human and animal pathogens. MAbs have, however, since been introduced for the detection and identification of fungi in agricultural and food sciences (172). Recent work has shown that it is possible to raise genus-specific and pathovar or isolate-specific MAbs to identify, detect and quantify particular fungal isolates (173–177). ELISA-based assays have been developed for the detection and identification of *Rhizoctonia*, *Verticillium*, *Pythium*, *Fusarium*, *Trichoderma*, *Botrytis*, *Phytophthora*, and many other common fungi (173,175,176,178–180).

IV. STATISTICAL ANALYSIS AND COMPUTER-ASSISTED IDENTIFICATION

Chemotaxonomic methods can provide complex data sets and multivariate statistical techniques such as cluster analysis or ordination methods have been used to analyze such data (81,181–183). Phylogenetic analyses can also be undertaken with chemical data and numerical methods (184–186).

In mycology, most chemotaxonomic data have been analyzed by cluster analysis (187–191), although Principal Component Analysis (81,192,193) and Correspondence Analysis (189,194) have also been used. Molecular sequence data have generally been analyzed with the neighbor joining technique (essentially a single-linkage cluster technique) or parsimony analysis (195). Once a good classification is achieved, it is possible to construct computer-based identification schemes. The schemes may be based on data bases, probabilistic keys, synoptic keys, or multivariate ordination principles. Bridge et al. (197,198) describe a identification system for the terverticillate penicillia based on a probabilistic concept. Some of the characters they used were chemical, as was the synoptic key for the terverticillate penicillia proposed by Frisvad and Filtenborg (199). Few other keys for filamentous fungi are based predominantly on chemical characters, but new principles in chemometrics have been used for proposing keys. Smedsgaard and Frisvad (196) suggested that obtaining electrospray mass profiles of fungal extracts and the use of mass spectrum search routines could be applied for identifications.

The SIMCA (soft independent modeling of class analogy) principle is based on Disjoint Principal Component Analysis for each class that is recognized. New isolates can be judged as belonging to any one of these disjoint classes if they are within a confidence cylinder or box based on approximative F-tests (200–202). The principle is well suited for identification using quantitative chemical data, but has not been used formally yet for a large group of fungi.

V. CONCLUSION

Classification methods based on proteins, carbohydrates, and lipids may be of great value on many taxonomic levels, and can often be automated. Some characteristics, such as secondary-metabolite profiles may be species-specific. DNA-based methods are playing an increasing role in the elucidation of the phylogeny and classification of fungi and can be routinely used for fungal identification.

Chemotaxonomy, if used on its own, is not without problems. A polyphasic approach to taxonomy is needed to be sure that a natural classification is obtained. Where a consensus classification has been reached, or if fundamentally different methods give the same classifications, simplified identification schemes based on a combination of morphology and chemistry may be proposed. On the other hand, few mycological laboratories have routine access to advanced chromatographs or spectrometers, and therefore most mycologist prefer mainly morphologically based keys for everyday identification. Another problem in chemotaxonomy, and also in transformed cladistics and numerical phenetics, is the availability of a sufficient number and spread of isolates of each taxon.

The many advanced chromatographic and spectrometric methods combined with multivariate statistics that have been developed will probably become more common in classification. Useful rapid methods for identification can be developed based on these advances in technology. Furthermore, understanding about biosynthetic pathways and their genetic regulation and the ecological/functional relevance of the chemical compounds will also be important in the fungal chemotaxonomy of the future.

REFERENCES

1. Hillis DM, Moritz C, eds. Molecular Systematics. Sunderland, MS: Sinauer Associates, 1990.
2. Kohn L. Developing new characters for fungal systematics: an experimental approach for determining the rank of resolution. Mycologia 1992; 84:139–153.
3. Bartnicki-Garcia S. Cell wall chemistry, morphogenesis and taxonomy of fungi. Ann Rev Microbiol 1968; 22:87–108.
4. Gorin PAJ, Spencer JFT. Proton magnetic resonance spectroscopy—an aid in identification and chemotaxonomy of yeast. Adv Appl Microbiol 1970; 13:25–89.
5. Bisby FA, Vaughan JG, Wright CA, eds. Chemosystematics: Principles and Practice. London: Academic Press, 1980.
6. Fox A, Morgan SL, Larsson L, Odham G, eds. Analytical Microbiological Methods. Chromatography and Mass Spectrometry. New York: Plenum Press, 1990.
7. De Hoog GS, Smith MT, Weijman ACM, eds. The Expanding Realm of Yeast-like Fungi. Amsterdam: Elsevier, 1987.
8. Benedict RC. Chemotaxonomic relationships among basidiomycetes. Adv Appl Microbiol 1970; 13:1–23.

9. Heywood VH. Chemosystematics—an artificial discipline. In: Bendz G, Santeson J, eds. Chemistry in Botanical Classification. New York: Academic Press, 1973: 41–53.
10. Harborne JB, Turner BL. Biochemical Systematics. London: Academic Press, 1984.
11. Hegnauer R. Accumulation of secondary products and its significance for biological systematics. Nova Acta Leopoldina Suppl 1977; 7:45–75.
12. Hegnauer R. Phytochemistry and plant taxonomy—an essay on the chemotaxonomy of higher plants. Phytochemistry 1986; 25:1519–1535.
13. Aparecida M, Cagnin H, Gomes CMR, et al. Biochemical systematics: methods and principles. Plant Syst Evol Suppl 1977; 1:53–76.
14. Dettner K. Chemosystematics and evolution of beetle chemical defenses. Annu Rev Entomol 1987; 32:17–48.
15. Whalley AJS, Edwards R. Xylariaceous fungi: use of secondary metabolites. In: Rayner ADM, Brasier CM, Moore D, eds. The Evolutionary Biology of Fungi. Cambridge: Cambridge University Press, 1987:423–434.
16. Whalley AJS, Edwards R. Secondary metabolites and systematic arrangement within the Xylariaceae. Can J Bot 1995; 73:S802–S810.
17. Culberson WL, Culberson CF. Secondary metabolites as a tool in ascomycete systematics: lichenized fungi. In: Hawksworth DL, ed. Ascomycete Systematics. Problems and Perspectives in the Nineties. New York: Plenum Press, 1994:155–163.
18. Hawksworth DI, ed. Prospects in Systematics. Oxford: Clarendon Press, 1988.
19. Goodfellow M, O'Donnell AG, eds. Chemical Methods in Procaryotic Systematics. Chichester: John Wiley, 1993.
20. Hawksworth DL, ed. The Identification and Characterization of Pest Organisms. Wallingford: CAB International, 1994.
21. Hawksworth DL, Kirk PM, Sutton BC, Pegler DN. Ainsworth & Bisby's Dictionary of the Fungi. Wallingford: CAB International, 1995.
22. Barrow GI, Feltham RKA, eds. Cowan and Steel's Manual for the Identification of Medical Bacteria. 3rd ed. Cambridge: Cambridge University Press, 1993.
23. Goodfellow M, James AL. Rapid enzyme tests in the characterization and identification of microorganisms. In: Hawksworth DL, ed. The Identification and Characterization of Pest Organisms. Wallingford: CAB International, 1994:289–300.
24. Barnett JA, Payne RW, Yarrow D. Yeasts: Characteristics and Identification. 2nd ed. Cambridge: Cambridge University Press, 1990.
25. Murray IG. Some aspects of the biochemical differentiation of pathogenic fungi: a review. J Gen Microbiol 1968; 52:213–221.
26. Pitt JI. An appraisal of identification methods for *Penicillium* species: novel taxonomic criteria based on temperature and water relations. Mycologia 1973; 65:1135–1157.
27. Frisvad JC. Physiological criteria and mycotoxin production as aids in identification of common asymmetric penicillia. Appl Environ Microbiol 1981; 41:568–579.
28. Paterson RRM, Bridge PD. Biochemical Techniques for Filamentous Fungi. IMI Technical Handbooks No. 1. Wallingford: CAB International, 1994.
29. Nylander W. Circa novum in studio lichenum criterium chemicum. Flora (Jena) 1866; 49:198–201.
30. Carlille MJ, Watkinson SC. The Fungi. London: Academic Press, 1994.

31. Hawksworth DL. Lichen chemotaxonomy. In Brown DH, Hawksworth DL, Bailey RH, eds. Lichenology: Progress and Problems. London: Academic Press, 1976: 139–184.
32. Hawksworth DL. Problems and perspectives in the systematics of the Ascomycotina. Proc Indian Acad Sci Plant Sci 1985; 94:319–339.
33. Culberson CF, Elix JA. Lichen substances. Methods Plant Biochem 1989; 1: 509–535.
34. Rogers RW. Chemical variation and the species concept in lichenized ascomycetes. Bot J Linn Soc 1989; 101:229–239.
35. Culberson WL, Culberson CF. Secondary metabolites as a tool in ascomycete systematics: lichenized fungi. In: Hawksworth DL, ed. Ascomycete Systematics: Problems and Perspectives in the Nineties. New York: Plenum Press, 1994:155–163.
36. Elix JA. Biochemistry and secondary metabolites. In: Nash TH III, ed. Lichen Biology. Cambridge: Cambridge University Press, 1996:181.
37. Hafellner J. Principles in classification and main taxonomic groups. In: Galun M, ed. CRC Handbook of Lichenology 3. Boca Raton: CRC Press, 1988:41–52.
38. Bruns TD, White TJ, Taylor JW. Fungal molecular systematics. Annu Rev Ecol Syst 1991; 22:525–564.
39. Armelo D, Clerc P. Lichen chimeras: DNA analysis suggests that one fungus forms two morphotypes. Exp Mycol 1991; 15:1–10.
40. Gargas A, Taylor JW. Polymerase chain reaction (PCR) primers for amplifying and sequencing nuclear 18S rDNA from lichenized fungi. Mycologia 1992; 84:589–592.
41. DePriest PT, Been MD. Numerous group I introns in the lichen fungus *Cladonia*. J Mol Biol 1992; 228:315–321.
42. Grube M, Depriest PT, Gargas A, Hafellner J. DNA isolation from lichen ascomata. Myc Res 1995; 99:1321–1324.
43. Kreger–van Rij NJW, ed. The Yeasts, a Taxonomic Study. Amsterdam: Elsevier, 1984.
44. Kurtzman CP. The systematics of ascomycetous yeasts defined from ribosomal RNA sequence divergence: theoretical and practical considerations. In: Reynolds DR, Taylor JW, eds. The Fungal Holomorph: Mitotic, Meiotic and Plesiomorphic Speciation in Fungal Systematics. Wallingford: CAB International, 1993:271–279.
45. Kurtzman CP, Robnett CJ. Molecular relationships among hyphal ascomycetous yeasts and yeastlike taxa. Can J Bot 1995; 73:S824–S830.
46. Boekhout T, Kurtzman CP. Principles and methods used in yeast classification, and an overview of currently accepted yeast genera. In: Wolf K, ed. Nonconventional Yeasts in Biotechnology. Berlin: Springer, 1996:1–81.
47. Van Eijk GW, Roeymans HJ, Weijman ACM. Biochemical characteristics: volatiles, carotenoids, sterols and fatty acids. Stud Mycol (Baarn) 1982; 22:39–49.
48. O'Donnell AG. Quantitative and qualitative analysis of fatty acids in the classification and identification of microorganisms. In: Hawksworth DL, ed. The Identification and Characterization of Pest Organisms. Wallingford: CAB International, 1994:323–335.
49. Augustyn OPH, Kock JLF, Ferraira D. Differentiation between yeast species, and strains within a species, by cellular fatty acid analysis. 5. A feasible technique? Syst Appl Microbiol 1992; 15:105–115.

50. Nakase T, Hamamoto M, Sugiyama J. Recent progress in the systematics of basidiomycetous yeast. Jpn J Med Mycol 1991; 32(suppl 1):21–30.
51. Powell MJ. Looking at mycology with a Janus face: a glimpse at Chytridiomycetes active in the environment. Mycologia 1993; 85:1–20.
52. Amano N, Shinmen Y, Akimoto K, Kawashima H, Amachi T. Chemotaxonomic significance of fatty acid composition in the genus *Mortierella* (Zygomycetes, Mortierellaceae). Mycotaxon 1992; 44:257–265.
53. De Ruiter GA, van Bruggen–van der Lugt AW, Rombouts FM. Approaches to the classification of *Mortierella isabellina* group: antigenic extracellular polysaccharides. Mycol Res 1993; 97:690–696.
54. Pfyffer GE, Pfyffer BU, Rast DM. The polyol pattern, chemotaxonomy and phylogeny of fungi. Sydowia 1986; 39:160–201.
55. O'Donnell, Cigelnik E. Phylogeny of the Zygomycota. Abstracts of the Fifth International Mycological Congress, 14–21 Aug., 1994, Vancouver, B.C., 1994:160.
56. Benny GL. Classical morphology in zygomycete taxonomy. Can J Bot 1995; 73: S725–S730.
57. Tyrrell D. Biochemical systematics and fungi. Bot Rev 1969; 35:305–316.
58. Moser M. The relevance of chemical characters for the taxonomy of Agaricales. Proc Indian Acad Sci Plant Sci 1985; 94:381–386.
59. Bruns TD, Szaro TM. Rate and mode differences between nuclear and mitochondrial small-subunit rRNA genes in mushrooms. Mol Biol Evol 1992; 9:836–855.
60. Hibbett DS. Ribosomal RNA and fungal systematics. Trans Mycol Soc Jpn 1992; 33:533–556.
61. Seifert KA, Wingfield BD, Wingfield MJ. A critique of DNA analysis in the taxonomy of filamentous Ascomycetes and ascomycetous anamorphs. Can J Bot 1995; 73:S760–S767.
62. Swann E, Taylor JW. Higher taxa of basidiomycetes: an 18S rRNA gene perspective. Mycologia 1993; 85:923–936.
63. Barbee ML, Taylor JW. From 18S ribosomal sequence data to evolution of morphology among fungi. Can J Bot 1995; 73:S677–S683.
64. McLaughlin DJ, Berres ME, Szabo LJ. Molecules and morphology in basidiomycete phylogeny. Can J Bot 1995; 73:S684–S692.
65. Hibbett DS, Donoghue MJ. Progress toward a phylogenetic classification of the Polyporaceae through parsimony analysis of mitochondrial ribosomal DNA sequences. Can J Bot 1995; 73:S853–S861.
66. Parmasto E. Corticioid fungi: a cladistic study of a paraphyletic group. Can J Bot 1995; 73:S843–S852.
67. Swan E, Taylor JW. Phylogenetic perspectives on basidiomycete systematics: evidence from the 18S rRNA gene. Can J Bot 1995; 73:S862–868.
68. Bruns TD. Identification of ecomycorrhizal fungi using a combination of PCR-based approaches. In: Rossen L, Rubio V, Dawson MT, Frisvad JC, eds. Fungal Identification Techniques. Brussels: European Commission, 1996:116–121.
69. Söderström B. Identification of mycorrhizal fungi. In: Rossen L, Rubio V, Dawson MT, Frisvad JC, eds. Fungal Identification Techniques. Brussels: European Commission, 1996:110–114.
70. Reynolds DR, Taylor JW, eds. The Fungal Holomorph: Mitotic, Meiotic and

Pleomorphic Speciation in Fungal Systematics. Wallingford: CAB International, 1994.
71. Hawksworth DI, ed. Ascomycete Systematics. Problems and Perspectives in the Nineties. New York: Plenum Press, 1994.
72. Rossen L, Rubio V, Dawson MT, Frisvad JC, eds. Fungal Identification Techniques. Brussels: European Commission, 1996.
73. Samson RA, Pitt JI, eds. Advances in *Penicillium* and *Aspergillus* Systematics. New York: Plenum Press, 1985.
74. Samson RA, Pitt JI, eds. Modern Concepts in *Penicillium* and *Aspergillus* Classification. New York: Plenum Press, 1990.
75. Pitt JI, Samson RA. Approaches to *Penicillium* and *Aspergillus* systematics. Stud Mycol (Baarn) 1990; 32:77–90.
76. Sugiyama J, Rahayu ES, Chang J, Oyaizu H. Chemotaxonomy of *Aspergillus* and associated teleomorphs. Jpn J Med Mycol 1991; 32(suppl 1):39–60.
77. Chelkowski J, ed. *Fusarium*. Mycotoxins, Taxonomy and Pathogenicity. Amsterdam: Elsevier, 1989.
78. Logrieco A, Peterson SW, Bottalico A. Phylogenetic relationship within *Fusarium sambucinum* Fuckel sensu lato determined from ribosomal RNA sequences. Mycopathologia 1995; 129:153–158.
79. Jun Y, Bridge PD, Evans HC. An integrated approach to the taxonomy of the genus *Verticillium*. J Gen Microbiol 1991; 137:1437–1444.
80. Bridge PD, Hawksworth DL. Biochemical tests as an aid to the identification of *Monascus* species. Lett Appl Microbiol 1985; 1:25–29.
81. Vogt NB. Soft modelling and chemosystematics. Chemom Intell Lab Syst 1987; 1:213–231.
82. Vincent PG, Kulik MM. Pyrolysis-gas-liquid chromatography of fungi: differentiation of species and strains of several members of the *Aspergillus flavus* group. Appl Microbiol 1970; 20:957–963.
83. Sekhon AS, Carmichael JW. Classification of some Gymnoascaceae by pyrolysis-gas-liquid chromatography using added marker compounds. Sabouraudia 1971; 13: 83–88.
84. Sekhon AS, Carmichael JW. Pyrolysis-gas-liquid chromatography of some dermatophytes. Can J Microbiol 1972; 18:1593–1601.
85. Sekhon AS, Carmichael JW. Column variation affecting a pyrolysis-gas-liquid chromatographic study of strain variation in two species of *Nannizzia*. Can J Microbiol 1973; 19:409–411.
86. Carmichael JW, Sekhon AS, Sigler L. Classification of some dermatophytes by pyrolysis-gas-liquid chromatography. Can J Microbiol 1973; 19:403–407.
87. Kulik MM, Vincent PG. Pyrolysis-gas-liquid chromatography of fungi: observations on variability among nine *Penicillium* species of the section Asymmetrica, subsection Fasciculata. Mycopathol Mycol Appl 1973; 51:1–18.
88. Vincent PG, Kulik MM. Pyrolysis-gas-liquid chromatography of fungi: numerical characterization of species variation among members of the *Aspergillus glaucus* group. Mycopathol Mycol Appl 1973; 51:251–265.
89. Seviour RJ, Chilvers GA, Crow WD. Characterization of eucalypt mycorrhizas by pyrolysis-gas chromatography. New Phytol 1974; 73:321–332.

90. Burns DT, Stretton RJ, Jaytilake SDAK. Pyrolysis gas chromatography as an aid to the identification of *Penicillium* species. J Chromatogr 1976; 116:107–115.
91. Stretton RJ, Campbell M, Burns DT. Pyrolysis gas chromatography as an aid to the identification of *Aspergillus* species. J Chromatogr 1976; 129:321–328.
92. De Hoog GS. *Rhinocladiella* and allied genera. Stud Mycol (Baarn) 1977; 15:1–140.
93. Brosseau JD, Carmichael JW. Pyrolysis gas-liquid chromatography applied to a study of variation in *Arthroderma tuberculatum*. Mycopathologia 1978; 63: 67–79.
94. Weijman ACM, Meuzelaar HLC. Biochemical contributions to the taxonomic status of the Endogonaceae. Can J Bot 1978; 57:284–291.
95. Gunasekaran M, Weber DJ, Hess WM. Differentiation of races of *Tilletia caries* and *T. foetida* by pyrolysis-gas-liquid chromatography. Mycologia 1979; 71:1066–1071.
96. Blomquist C, Johansson E, Söderström B, Wold S. Data analysis of pyrolysis chromatograms by means of SIMCA pattern recognition. J Anal Appl Pyrol 1979; 1:53–65.
97. Blomquist C, Johansson E, Söderström B, Wold S. Classification of fungi by means of pyrolysis-gas chromatography-pattern recognition. J Chromatogr 1979; 173: 19–32.
98. Blomquist C, Johansson E, Söderström B, Wold S. Reproducibility of pyrolysis-gas chromatograms of the mould *Penicillium brevicompactum*. J Chromatogr 1979; 173: 7–17.
99. Windig W, de Hoog GS, Haverkamp J. Factor analysis of pyrolysis mass spectra for chemical characterization of yeasts and yeast-like fungi. J Anal Appl Pyrol 1981; 3: 213–220.
100. Söderström B, Wold S, Blomquist G. Pyrolysis-gas chromatography combined with SIMCA pattern recognition for classification of fruit-bodies of some ectomycorrhizal *Suillus* species. J Gen Microbiol 1982; 128:1773–1784.
101. Windig W, Heverkamp J. Pyrolysis mass spectrometry. I. *Rhodosporidium*. Stud Mycol (Baarn) 1982; 22:50–59.
102. Windig W, de Hoog GS. Pyrolysis mass spectrometry. II. *Spodidiobolus* and related taxa. Stud Mycol (Baarn) 1982; 22:60–64.
103. Söderström B, Frisvad JC. Separation of closely related asymmetric penicillia by pyrolysis gas chromatography and mycotoxin production. Mycologia 1984; 76: 408–419.
104. Weijman ACM, Golubev WI. Carbohydrate patterns and taxonomy of yeasts and yeast-like fungi. Stud Mycol (Baarn) 1987; 30:361–371.
105. De Hoog GS, Guého E, Boekhout T. Experimental fungal taxonomy. In: Arora DK, Ajello L, Mukerji KG, eds. Handbook in Applied Mycology. Vol. 2. Humans, Animals and Plants. New York: Marcel Dekker, 1991:369–394.
106. Mozes N, Rouxet PG. Methods for measuring hydrophobicity of microorganisms. J Microbiol Meth 1987; 6:99–112.
107. Fisher DJ, Richmond DV. The electrokinetic properties of some fungal spores. J Gen Microbiol 1969; 57:51–60.
108. Blomquist GK, Ström GB, Söderström B. Separation of fungal propagules by partition in aqueous polymer two-phase systems. Appl Environ Microbiol 1984; 47: 1316–1318.

109. Ström GB, Blomquist GK. Selective partitioning of conidia of some *Penicillium* and *Aspergillus* species. Appl Environ Microbiol 1986; 52:723–726.
110. Ström GB, Palmgren U, Blomquist GK. Separation of organic dust from microorganism suspensions by partitioning in aqueous polymer two-phase systems. Appl Environ Microbiol 1987; 53:860–863.
111. Ström GB. Cross-point determination of *Penicillium* conidia—characterization of closely related fungi. J Appl Bacteriol 1986; 60:557–561.
112. Ström GB, Blomquist GK, Söderström B. A chemotaxonomic method for classification of asymmetric penicillia by means of cross-partition in aqueous two-phase systems combined with SIMCA pattern recognition analysis. J Appl Bacteriol 1989; 66:461–467.
113. Blomquist GK, Ström GB. Classification of *Penicillium* fungi by cross-partition analysis. Meth Enzymol 1994; 228:354–362.
114. Martens H, Naes T. Multivariate Calibration. Chichester: John Wiley and Sons, 1989.
115. Murphy RW, Sites JW, Buth DG Jr, Haufler CH. Proteins I: isozyme electrophoresis. In: Hillis DM, Moritz C, eds. Molecular Systematics. Sunderland: Sinauer Associates, 1990:45–126.
116. Micales JA, Bonde MR, Peterson GL. The use of isozyme analysis in fungal taxonomy and genetics. Mycotaxon 1986; 27:405–449.
117. Bonde MR, Micales JA, Peterson GL. The use of isozyme analysis for identification of plant-pathogenic fungi. Plant Dis 1993; 77:961–968.
118. Rosendahl S, Sen R. Isozyme analysis of mycorrhizal fungi and their mycorrhizas. In: Varma AK, Read DJ, Norris JR, eds. Methods in Microbiology. Vol. 24. Experiments with Mycorrhizae. London: Academic Press, 1992:169–194.
119. Bak AL, Stenderup A. Deoxyribonucleic acid homology in yeasts. Genetic relatedness within the genus *Candida*. J Gen Microbiol 1969; 59:21–30.
120. Kurtzman CP. Impact of nucleic acid comparisons on systematic mycology. Bot J Linn Soc 1989; 99:3–10.
121. Werman SD, Springer MS, Britten RJ. Nucleic acids I: DNA-DNA hybridization. In: Hillis DM, Moritz C, eds. Molecular Systematics. Sunderland: Sinauer Associates, 1990:204–249.
122. Carle GF, Olson MV. Separation of chromosomal DNA molecules from yeast by orthogonal-field-alternation gel electrophoresis. Nucl Acids Res 1984; 12:5647–5664.
123. Schwartz DC, Cantor CR. Separation of yeast chromosome sized DNA's by pulsed field gradient gel electrophoresis. Cell 1984; 37:67–75.
124. Boekhout T, van Gool J, van den Bogert H, Jille T. Karyotyping and G+C composition as taxonomic criteria applied to the systematics of *Tilletiopsis* and related taxa. Mycol Res 1992; 95:331–342.
125. Zimmerman M, Fournier P. Electrophoretic karyotyping of yeast. In: Wolf K, ed. Nonconventional Yeasts in Biotechnology. Berlin: Springer-Verlag, 1996:101–116.
126. Dowling TE, Moritz C, Palmer JD. Nucleic acids II: restriction site analysis. In: Hillis DM, Moritz C, eds. Molecular Systematics. Sunderland: Sinauer Associates, 1990:250–317.

127. Hillis DM, Larson A, Davis SK, Zimmer EA. Nucleic acids III: sequencing. In: Hillis DM, Moritz C, eds. Molecular Systematics. Sunderland: Sinauer Associates, 1990:318–370.
128. Magee BB, D'Souza TM, Magee PT. Strain and species identification by restriction fragment length polymorphisms in the ribosomal DNA repeat of *Candida* species. J Bacteriol 1987; 169:1639–1643.
129. Vilgalys R, Gonzales D. Organization of ribosomal DNA in the basidiomycete *Thanathephorus praticola*. Curr Gen 1990; 18:277–280.
130. Cubeta MA, Echandi E, Abernathy T, Vilgalys R. Characterization of anastomosis groups of binucleate *Rhizoctonia* species using restriction analysis of an amplified ribosomal RNA gene. Phytopathology 1991; 81:1395–1400.
131. Hibbett DS, Vilgalys R. Evolutionary relationships of *Lentinus* to the Polyporaceae: evidence from restriction analysis of enzymatically amplified ribosomal DNA. Mycologia 1991; 83:425–439.
132. Ward E, Akrofi AY. Identification of fungi in the *Gaeumannomyces-Phialophora* complex by RFLPs of PCR-amplified ribosomal DNAs. Mycol Res 1994; 98: 219–224.
133. Bunyard BA, Nicholson MS, Royse DJ. A systematic assessment of *Morchella* using RFLP analysis of the 28S ribosomal RNA gene. Mycologia 1994; 86:762–772.
134. O'Donnell K. Progress towards a phylogenetic classification of *Fusarium*. Sydowia 1996; 48:57–70.
135. Croft JH. Application of RFLPs in systematics and population genetics of aspergilli. In: Powell KA, Renwick A, Peberdy JF, eds. The Genus *Aspergillus*. From Taxonomy and Genetics to Industrial Application. New York: Plenum Press, 1994: 277–289.
136. Erland S, Henrion B, Martin F, Glower LA, Alexander IJ. Identification of the ectomycorrhizal basidiomycete *Tylospora fibrillosa* Donk by RFLP analysis of the PCR amplified ITS and IGS regions of ribosomal DNA. New Phytol 1994; 126: 525–532.
137. Persson Y, Erland S, Jansson H-B. Identification of nematode-trapping fungi using RFLP analysis of the PCR-amplified ITS region of ribosomal DNA. Mycol Res 1996; 100:531–534.
138. Bunyard BA, Chaichuchote S, Nicholson MS, Royse DJ. Ribosomal DNA analysis for resolution of genotypic classes of *Pleurotus*. Mycol Res 1996; 100:143–150.
139. Lutzoni F, Vilgalys R. Integration of morphological and molecular data sets in estimating fungal phylogenies. Can J Bot 1995; 73:S649–S659.
140. Tehler A. Morphological data, molecular data, and total evidence in phylogenetic analysis. Can J Bot 1995; 73:S667–S676.
141. Burbee ML, Taylor JW. From 18S ribosomal sequence data to evolution of morphology among fungi. Can J Bot 1995; 73:S677–S683.
142. Maclaughlin DJ, Berres ME, Szabo LJ. Molecules and morphology in basidiomycete phylogeny. Can J Bot 1995; 73:S684–S692.
143. Eriksson OE. DNA and ascomycete systematics. Can J Bot 1995; 73:S784–S789.
144. Spatafora JW. Ascomal evolution of filamentous ascomycetes: evidence from molecular data. Can J Bot 1995; 73:S811–S815.

145. Rehner SA, Samuels GJ. Molecular systematics of the Hypocreales: a teleomorph gene phylogeny and the status of their anamorphs. Can J Bot 1995; 73:S816–S823.
146. Hibbett DS, Donoghue MJ. Progress toward a phylogenetic classification of the Polyporaceae through parsimony analysis of mitichondrial ribosomal DNA sequences. Can J Bot 1995; 73:S853–S861.
147. Swann EC, Taylor JW. Phylogenetic perspectives on basidiomycete systematics: evidence from the 18S rRNA gene. Can J Bot 1995; 73:S862–S868.
148. White TJ, Bruns T, Lee S, Taylor JW. Amplification and direct sequencing of fungal ribosomal RNA genes for phylogenetics. In: Innis MA, Gelfand DH, Sninsky JJ, White TJ, eds. PCR Protocols. A Guide to Methods and Applications. San Diego: Academic Press, 1990:315–322.
149. Bruns TD, White TJ, Taylor JW. Fungal molecular systematics. Annu Rev Ecol Syst 1992; 22:525–561.
150. Foster LM, Kozak KR, Loftus MG, Stevens JJ, Ross IK. The polymerase chain reaction and its application to filamentous fungi. Mycol Res 1993; 97:769–781.
151. Gardes M, White TJ, Fortin JA, Bruns TD, Taylor JW. Identification of indigenous and introduced symbiotic fungi in ectomycorrhizae by amplification of nuclear and mitochondrial ribosomal DNA. Can J Bot 1991; 69:180–190.
152. Rosewich UL, McDonald BA. DNA fingerprinting in fungi. Meth Mol Cell Biol 1994; 5:41–48.
153. McDonald BA. Using RFLPs to elucidate the population genetics of fungi. In: Rossen L, Rubio V, Dawson MT, Frisvad JC, eds. Fungal Identification Techniques. Brussels: European Commission, 1996:62–67.
154. Welsh J, McClelland M. Fingerprinting genomes using PCR with arbitrary primers. Nucl Acids Res 1990; 18:7213–7218.
155. Williams JGK, Kubelik AR, Livak KJ, Rafalski JA, Tingey SV. DNA polymorphisms amplified by arbitrary primers are useful as genetic markers. Nucl Acids Res 1990; 18:6531–6535.
156. Crowhurst RN, Hawthorn BT, Rikkerink EH, Templeton MD. Differentiation of *Fusarium solani* f. sp. *curcubitae* races 1 and 2 by random amplification polymorphic DNA. Curr Gen 1991; 20:931–936.
157. Guthrie PAI, Magill CW, Frederiksen RA, Odvody GN. Random amplified polymorphic DNA markers: a system for identifying and differentiating isolates of *Colletotrichum graminicola*. Phytopathology 1992; 82:832–835.
158. Mills PE, Sreenivasprasad S, Brown AE. Detection and differentiation of *Colletotrichum gloeosporioides* isolates using PCR. FEMS Microbiol Lett 1992; 98: 137–144.
159. Hamelin RC, Oullette GB, Bernier L. Identification of *Gremmeniella abientina* races with random amplified polymorphic DNA markers. Appl Environ Microbiol 1993; 59:1752–1755.
160. Nicholson P, Rezanoor HN. The use of random amplified polymorphic DNA to identify pathotype and detect variation in *Pseudocercosporella herpotrichoides*. Mycol Res 1994; 98:13–21.
161. Garbelotto M, Bruns TD, Cobb FW, Otrosina WJ. Differentiation of intersterily groups and geographic provenances among isolates of *Heterobasidion annosum*

detected by random amplified polymorphic DNA assay. Can J Bot 1993; 71: 565–569.

162. Meijer G, Megnegneau B, Linders EGA. Variability for isozyme, vegetative compatibility and RAPD markers in natural populations of *Phomopsis subordinaria*. Mycol Res 1994; 98:267–276.
163. Buscot F, Wipf D, Di Battista C, Munch J-C, Botton B, Martin F. DNA polymorphism in morels: PCR/RFLP analysis of the ribosomal DNA spacers and microsatellite-primed PCR. Mycol Res 1996; 100:63–71.
164. Cooke DEL, Kennedy DM, Guy DC, Russell J, Unkles SE, Duncan JM. Relatedness of group I species of *Phytophthora* as assessed by randomly amplified polymorphic DNA (RAPDs) and sequences of ribosomal DNA. Mycol Res 1996; 100:297–303.
165. Gosselin L, Jobidon R, Bernier L. Assessment of genetic variation within *Chondrostereum purpureum* from Quebec by random amplified polymorphic DNA analysis. Mycol Res 1996; 100:151–158.
166. Duncan S, Barton JE, O'Brian PA. Analysis of variation in isolates of *Rhizoctonia solani* by random amplified polymorphic DNA assay. Mycol Res 1993; 97:1076–1082.
167. Manulis S, Kogan N, Reuven M, Ben-Yephet Y. Use of RAPD technique for identification of *Fusarium oxysporum* f. sp. *dianthi* from carnation. Phytopathology 1994; 84:98–101.
168. Vos P, Hogers R, Bleeler M, et al. AFLP: a new technique for DNA fingerprinting. Nucl Acids Res 1995; 23:4407–4414.
169. Lin J-J, Kuo J. AFLP: a novel PCR-based assay for plant and bacterial DNA fingerprinting. Focus 1995; 17:66–70.
170. Mueller DG, Lipari SE, Milgroom MG. Amplified fragment length polymorphism (AFLP) fingerprinting of symbiotic fungi cultured by the fungus-growing ant *Cyphomyrmex minutus*. Mol Ecol 1996; 5:119–122.
171. Polonelli L, Fischiaro P, Bertolotti D, et al. Immunological identification of pathogenic fungi. In: Rossen L, Rubio V, Dawson MT, Frisvad JC, eds. Fungal Identification Techniques. Brussels: European Commission, 1996:31–34.
172. Banks JN, Barker I, Turner JA, Northway BJ, Rizvi RH, Rahman S. Immunochemical techniques for the identification of fungi with particular reference to Fusarium species. In: Rossen L, Rubio V, Dawson MT, Frisvad JC, eds. Fungal Identification Techniques. Brussels: European Commission, 1996:138–152.
173. Dewey FM. Development of diagnosis assays for fungal plant pathogens. In: Brighton Crop Protection Conference, Pest and Diseases. Vol. 2. Thornton Heath: BCPC Registered Office, 1988:777–786.
174. Dewey FM. Detection of plant invading fungi by monoclonal antibodies. In: Duncan JM, Torrance L, eds. Techniques for Rapid Detection of Plant Pathogens. Oxford: Blackwell Science Publications, 1992:47–62.
175. Dewey FM, Thronton CR. Fungal immunodiagnostics in plant agriculture. In: Skerrit J, Apels R, eds. New Diagnostics in Crop Sciences. Wallingford: CAB International, 1994:47–62.
176. Kaufman L, Standard PG, Padhye AA. Exoantigen tests for the immunoidentification of fungal cultures. Mycopathologia 1980; 82:3–12.

177. Schots A, Gommers FG, Egberts E. Modern Assays for Plant Pathogenic Fungi: Identification, Detection and Quantification. Wallingford: CAB International, 1994.
178. Sekhon AS, Padhye AA. Exoantigens in the identification of medically important fungi. In: Arora DK, Ajello L, Mukerji KG, eds. Handbook of Applied Mycology. Vol. II. Animals, Humans and Insects. New York: Marcel Dekker, 1991:757–764.
179. Samson RA, Hocking AD, Pitt JI, King AD, eds. Modern Methods in Food Mycology. Amsterdam: Elsevier, 1992.
180. Torrance L. Use of monoclonal antibodies in plant pathology. Eur J Pl Pathol 1995; 101:351–363.
181. Sneath PHA, Sokal RR. Numerical Taxonomy. San Fransisco: Freeman, 1973.
182. Harris JA, Bisby FA. Classification from chemical data. In: Bisby FA, Vaughan JG, Wright CA, eds. Chemosystematics: Principles and Practice. London: Academic Press, 1980:305–327.
183. Podani J. Multivariate Data Analysis in Ecology and Systematics. Ecological Computations Series. Vol. 6. The Hague: SPB Academic Publishing, 1994.
184. Sneath PHA. The history and future potential of numerical concepts in systematics: the contribution of H.G. Gyllenberg. Binary 1995; 7:32–36.
185. Sneath PHA. Thirty years of numerical taxonomy. Syst Biol 1995; 44:281–298.
186. Pankhurst RJ. Some problems in the methodology of cladistics. Binary 1995; 7: 37–41.
187. Sheard JW. The taxonomy of the *Ramalina siliquosa* species aggregate (lichenized Ascomycetes). Can J Bot 1978; 56:916–938.
188. Bridge PD, Hawksworth DL, Kozakiewicz Z, et al. A reappraisal of the terverticillate penicillia using biochemical, physiological and morphological features. I. Numerical taxonomy. J Gen Microbiol 1989; 135:2941–2966.
189. Frisvad JC. Chemometrics and chemotaxonomy: a comparison of multivariate statistical methods for the evaluation of binary fungal secondary metabolite data. Chemom Intell Lab Syst 1992; 14:253–269.
190. Svendsen A, Frisvad JC. A chemotaxonomic study of the terverticillate penicillia based on high performance liquid chromatography of secondary metabolites. Mycol Res 1994; 98:1317–1328.
191. Larsen TO, Frisvad JC. Chemosystematics of *Penicillium* based on profiles of volatile metabolites. Mycol Res 1995; 99:1167–1174.
192. Thrane U. Grouping *Fusarium* section Discolor isolates by statistical analysis of quantitative high performance liquid chromatography data on secondary metabolites production. J Microbiol Meth 1990; 12:23–39.
193. Blomquist G, Andersson B, Andersson K, Brondz I. Analysis of fatty acids. A new method for characterization of moulds. J Microbiol Meth 1992; 16:59–68.
194. Frisvad JC. Correspondence, principal coordinate, and redundancy analysis used on mixed chemotaxonomical qualitative and quantitative data. Chemom Intell Lab Syst 1994; 23:213–229.
195. Swofford DL, Olson GJ. Phylogeny reconstruction. In: Hillis DM, Moritz C, eds. Molecular Systematics. Sunderland: Sinauer Associates, 1990:411–501.
196. Smedsgaard J, Frisvad JC. Using direct electrospray mass spectrometry in taxonomy and secondary metabolite profiling of crude fungal extracts. J Microbiol Meth 1996; 25:5–17.

197. Bridge PD, Hawksworth DL, Kozakiewicz Z, Onions AHS, Paterson RRM, Sackin MJ. A reappraisal of the terverticillate penicllia using biochemical, physiological and morphological features. II. identification. J Gen Microbiol 1989; 135:2967–2978.
198. Bridge PD, Kozakiewicz Z, Paterson RRM. PENIMAT: a computer assisted identification scheeme for the terverticillate *Penicillium* isolates. Mycol Pap 1992; 165: 1–59.
199. Frisvad JC, Filtenborg O. Secondary metabolites as consistent criteria in *Penicillium* taxonomy and a synoptic key to *Penicillium* subgenus *Penicillium*. In: Samson RA, Pitt JI, eds. Modern Concepts in *Penicillium* and *Aspergillus* Classification. New York: Plenum Press, 1990:373–384.
200. Wold S, Sjöström M. SIMCA: a method for analyzing chemical data in terms of similarity and analogy. In: Kowalski BR, ed. Chemometrics: Theory and Application. ACS Symposium Series 52. Washington D.C.: American Chemical Society, 1977:254–282.
201. Albano C, Dunn W III, Edlund U, et al. Four levels of pattern recognition. Anal Chim Acta 1978; 103:429–443.
202. Ståhle L, Wold S. Multivariate analysis of variance (MANOVA). Chemom Intell Lab Syst 1990; 9:127–141.

2

Numerical Analysis of Fungal Chemotaxonomic Data

Paul D. Bridge and Gerald S. Saddler
International Mycological Institute, Egham, Surrey, United Kingdom

I. INTRODUCTION

One of the major features of biochemical and molecular taxonomic approaches is that they often generate a considerable amount of data, and in many cases these data must be analyzed in some way before any interpretations can be made. The procedures most commonly used for analyzing taxonomic data have often had very mixed histories; some have originated from, and been developed through, specific biochemical and molecular methods while others are based on the application of established techniques from other areas of science. The implementation of numerical methods has revolutionized taxonomy in that they enable some measure of precision to be given to taxonomic interpretations. This measure may be a single value such as a likelihood for an identification or a similarity value for relatedness, or it may be some graphical representation based on a mathematical model, such as an ordination based on measures of resemblance. In this chapter we will attempt to deal with numerical procedures that have commonly been used in chemotaxonomic and molecular studies of filamentous fungi. In common with many of the approaches in this volume, numerical techniques were widely used in other areas of biology before their application to filamentous fungi. Numerical methods have been used in a wide variety of areas, and at a variety of taxonomic levels; it is therefore not appropriate to refer only to "organisms" when discussing the methods. The term "operational taxonomic unit" (OTU) (1) is used to describe the lowest-ranking taxa in a study; these will commonly be isolates or strains, but may be higher groups such as special forms or species. Most of the key references and procedures have their origins in bacteriology, zoology, botany, and population studies (1–4).

Excluding the statistical techniques associated with good laboratory practice and experimental design, such as confidence levels, the major applications of numerical methods with fungal data have been in grouping organisms, identifying organisms, and determining phylogenies. While there are some common features to these approaches, each approach can be considered in isolation. It may be useful in this chapter to point out that while chemical and molecular data are well suited to numerical analyses, morphological features can also be analyzed in the same way. Historically, the first applications of numerical methods with fungi were in fact based on morphological features (5). The use of morphological features requires some form of coding strategy to be developed, and this is discussed in more detail later. The numerical procedures that we will describe in this chapter are all essentially computer-based, and a wide selection of suitable software is available, either from commercial sources or as share-ware. Consequently we have not attempted to review individual packages, and those mentioned in the text are only those packages of which we have had practical experience. As with any approach to data analysis, final interpretations are generally made by the scientist, and it must be remembered that dendrograms, ordinations cladograms, etc. are representations of the data used. Several numerical "cutoff" levels have been suggested for determining clusters and identification; these are only guidelines, however, and fluctuate considerably depending on the type and quality of the original data.

II. NUMERICAL METHODS FOR GROUPING ORGANISMS

A. Data Format

In order to use a numerical algorithm to group organisms, it is necessary to first calculate a measure of resemblance between the organisms. These measures of resemblance fall broadly into three categories, distance, association, and shape or pattern (1,6,7). These categories are not entirely exclusive, and there may be simple relationships and equivalences between individual measures in different categories.

The first step in calculating the resemblances among a group of organisms is to arrange the data into a matrix. This is a table showing each test or character for each organism, and is generally arranged with each row representing an organism, and each column representing the results of each individual test or character. This is termed a $t \times n$ table, where t stands for taxa and n for characters or tests (1). So a 45×50 matrix would consist of 50 test results or characters for 45 organisms.

To determine groups based on resemblance it is necessary to calculate the resemblance of each organism to each other organism. The number of calculations necessary to achieve this is described by $tx(t-1)/2$ where t is the number of organisms. The number of calculations necessary in any study therefore rises significantly with each organism added; 45 calculations are needed for a 10-organism data matrix, but 4950 are needed for a 100-organism data base. Numerical systematics is therefore not a practical proposition without suitable computers and software. A wide range of software such as SYSTAT, CLUSTAN, GENSTAT, SPSS/PC+, and MVSP is available for the "PC-type" computers.

B. Character Coding

The construction of the data matrix is an important part of grouping the organisms, as this is the form in which the experimental data will be analyzed (1,3). Some characters, such as those shown in Table 1, are easy to code into a matrix and they can be represented in a binary form without any loss of information. Other characters, however, may be less easy to enter into a matrix and may be qualitative or quantitative. In these cases the method chosen to code the data must be appropriate to the information contained in that character, the information contained in other characters, and the type of resemblance coefficient chosen.

Data presented for numerical analysis do not have to be in a binary form, and quantitative and qualitative values may be used directly. This may not always be appropriate, however. Most resemblance calculations give equal weight to each character, although some individual characters may convey considerably more information than others. One way in which this can be dealt with is to split significant quantitative characters into a number of binary characters. For example, the single character amylase production could be split into two characters as "amylase production $<$ 10U/ml" and "amylase production $>$ 10U/ml." An alternative would be to use the quantitative character as a multistate character so production of amylase could be coded as 0, 1, or 2 for "no", "weak," and "strong" production. In practice, class limits can be applied to split up most quantitative characters.

In the early numerical taxonomic studies with fungi, considerable attention was given to the weighting of characters, which in those cases were morphological (5,8). Methods investigated then included the designation of primary and second-

Table 1 Extract of a $t \times n$ Table

Organisms	Growth at 37°C	Growth on NO_2	Production of cyclopenol	Amylase activity
Penicillium expansum	—	—	+	+
P. citrinum	+	+	—	+
P. roquefortii	—	+	—	—

ary characters and the use of weighted resemblance coefficients. More recently, Bridge and Sackin (9) have considered character coding, particularly the effect of splitting single multistate characters into a series of related binary characters. In practice, with a large data set this does not appear to have a significant effect. What can be more significant, however, is the coding of missing or inappropriate characters. An example of this is production of a metabolite or enzyme, if the culture is unable to grow under the conditions used. In this situation it is incorrect to code those characters as negative, as the true result is "not known"; this should be taken into account when constructing the data matrix.

C. Resemblance Coefficients

As mentioned earlier, a wide range of resemblance coefficients are available for numerical taxonomic studies; only a small selection will be discussed here. There has in general been no recent study on the suitability of particular coefficients with different types of fungal data, but a large number of coefficients have been evaluated for bacterial data (6).

1. Association Coefficients

Probably the most commonly used resemblance coefficients are the association coefficients. These coefficients have been fully described in a number of places, and their full derivations will not be considered here (1,3,4,6). In general, association coefficients are used with qualitative data sets, although some, such as Gower's coefficient, may be used with qualitative and quantitative data. The different association coefficients calculate a measure of similarity on a scale of 0 to 1 or 0 to 100 and are often described as similarity coefficients. Within the similarity coefficients some, such as the Simple Matching coefficient, are only available for binary data, while others, such as Jaccard's coefficient, discount similarities due only to matching negative characters. Some characteristics of association coefficients are listed in Table 2.

2. Distance Coefficients

There is a wide selection of distance coefficients in use with fungal taxonomic data. Many distance coefficients are based on measures of the distances between organisms in character or attribute space (*A* space). *A*-space is the space defined when each dimension is a character or attribute. For a data matrix of two characters this is analogous to plotting the characters at right angles, as in the axis of a conventional graph. If all the characters are plotted onto straight line axes at right angles to each other (e.g., orthogonally), this then describes n-dimensional space, where n is the number of characters. In a situation where all of the character axes are orthogonal and of the same lengths, then distances between organisms can be calculated by an extension of Pythagoras' theorem. This situation describes Euclidean space, and the measures are Euclidean distances (1,16).

Table 2 Characteristics of Some Association Coefficients

Coefficient	Binary data	Matching negatives	Reference
Simple Matching (S_{SM})	+	+	Sokal and Michener (10)
Jaccard (S_J)	+	—	Sneath and Sokal (1)
Dice[a] (S_D)	+	—	Dice (11)
Gower (S_G)	—	+	Gower (12)
Yule (S_Y)	+	+	Brisbane and Rovira (13)
Total difference[b] (D_T)	+	+	Sneath (14)

[a]Dice's coefficient is the same as that described by Sorensen (15). Dice's coefficient is similar to Jaccard's coefficient but gives greater weight to character matches.
[b]Total difference is directly related to the Simple Matching coefficient as $D_T = 1-S_{SM}$.

Figure 1 shows a two-dimensional situation where the Euclidean distance c can be calculated from $c^2 = a^2 + b^2$, where the distances a and b are defined as the difference in the character states between the two organisms for each character. Euclidean distance for an n-dimensional situation is therefore defined as $D_{AB} = [\Sigma(X_A-X_B)^2]^{1/2}$. As can be seen, Euclidean distance will increase with the number of characters, so it is more usual to use a standardized value defined as $d_{AB} = (D^2/n)^{1/2}$, termed Average distance by Sneath and Sokal (1).

Not all distance measures are Euclidean; one of the more commonly used is termed "Manhattan distance." In this case the distance between the two organisms A and B in Figure 1 is measured not by the hypotenuse of the triangle (c) but by the distance along each of the other two sides (a and b) (see also 3,4). This type of coefficient has particular advantages in cladistic and phylogenetic studies, as an easily quantifiable part of the measure can be attributed to particular characters.

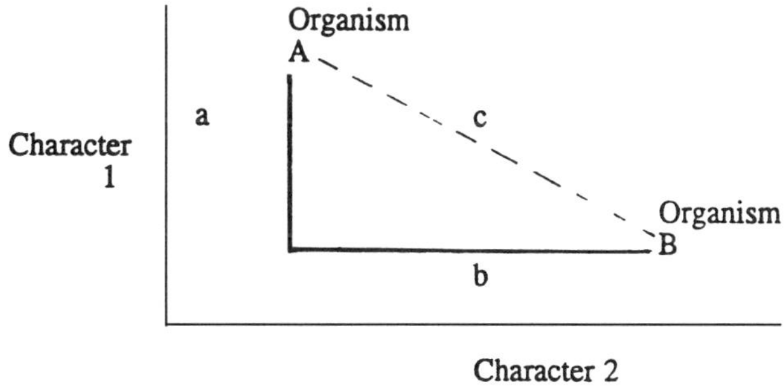

Figure 1 Two-dimensional representation of taxonomic distances.

As with Euclidean distance there is a standardized form, which is called Mean Character Difference (17).

3. Shape or Pattern Coefficients

There are a number of resemblance coefficients that do not fit comfortably within the previous categories. Three that have been used in fungal chemotaxonomy are Pattern difference, the Correlation coefficient, and Cosine theta.

The Pattern coefficient, although sometimes considered a shape coefficient, can also be considered as a specialized association coefficient. Essentially, the Pattern difference (D_P) is an association coefficient that takes into account the vigor of the organisms compared. This is particularly important in data sets where organisms may have significantly different growth rates or levels of activity. In such a situation, similar organisms may appear different due to growth rate differences, and the Pattern difference may alleviate this. Pattern difference is calculated from total difference and differences in vigor (14,18); this is detailed later.

The Correlation coefficient and Cosine theta have both been suggested for comparing microbial chemotaxonomic data that are generated in the form of a trace, such as from a densitometer or a high-performance liquid chromatography (HPLC) trace (19). Both of these coefficients are shape coefficients in that they are more affected by the overall pattern of the traces than by individual peaks. Both these coefficients are less sensitive to concentration differences than association and distance coefficients. It must be noted, however, that although the Correlation coefficient ranges from 0 to 1 and Cosine theta from -1 to 1, a maximum score does not imply identity, so these two coefficients should not in the strictest sense be termed "similarity" coefficients.

4. Relationships Between Coefficients

As mentioned earlier, the different types of resemblance coefficients are not mutually exclusive; many can be directly related to each other. In some cases, one coefficient may have been described independently by different workers, such as Dice's coefficient and Sorensen's coefficient. In other cases special circumstances may result in equivalences—e.g., calculation of Gower's coefficient with binary data gives identical results to the Simple Matching coefficient. Other coefficients may be related by simple equations—e.g., average distance where $d = (1-S_{SM})^{1/2}$, or Pattern difference where $D_P^2 + D_V^2 = D_T^2$, D_T being defined as $1-S_{SM}$ and D_V is the difference in vigor (14).

D. Clustering Similar Organisms

Once the resemblance values have been calculated for every possible pair of organisms, they need to be used to arrange the most similar organisms into groups. Although not all coefficients give similarities (see above), such as distance measures, for the purposes of clustering the highest values are considered similar for similarity measures, and the lowest values are considered the most similar for distance and difference measures. There are a number of ways in which this may be achieved, and in this section we will consider the methods that gradually build up groups or clusters in a hierarchic classification. These methods have been termed as sequential, agglomerative, hierarchic clustering (SAHN) by Sneath and Sokal (1). Nonhierarchic methods will be considered later. The two most commonly encountered forms of SAHN clustering are Single linkage and Average linkage clustering. Each will be briefly described here.

The final output from clustering regimens is a dendrogram, which is a representation of the similarities or differences among the organisms. This is drawn in the form of a graph with a scale for similarity/difference and vertical links at the appropriate points to represent relationships (Fig. 2). It must be remembered that a dendrogram is only a representation of the information contained in the original similarity/difference matrix. Testing this will be discussed later.

1. Single Linkage

In Single linkage clustering, organisms are grouped on the basis of their single highest similarity level. With the SAHN procedure, clustering occurs in steps, starting with the single highest value. This then continues until the single highest values that link all of the organisms together have been selected. This is illustrated in Table 3 which shows the similarity values obtained for a set of five organisms.

The first values to be selected would be the 0.8 similarities between organisms A and C and between B and E. After this the next highest values are the 0.6 between D and A and between D and E. These values then link all five organisms, and no others are required (Fig. 2).

2. Average Linkage Clustering

While Single linkage clustering allows organisms to be grouped into cluster, it does not use all of the data available and therefore may not give an accurate

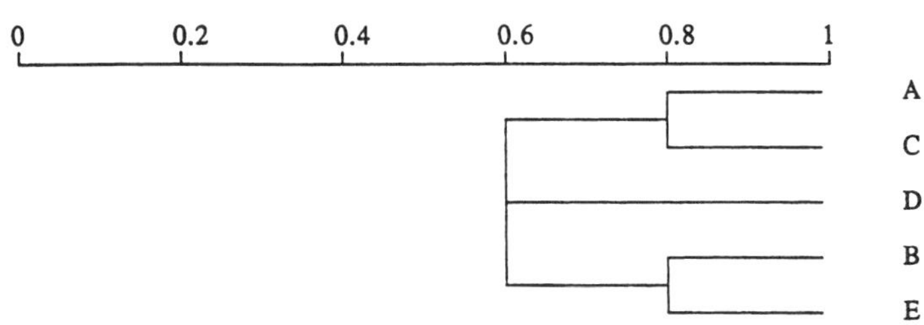

Figure 2 Single linkage clustering of data in Table 3.

Table 3 Similarity Matrix for Five Organisms

Organism						
	A	X				
	B	0.4	X			
	C	0.8	0.4	X		
	D	0.6	0.4	0.2	X	
	E	0.2	0.8	0.4	0.6	X
		A	B	C	D	E
				Organism		

representation of the overall similarities. For example, in the situation shown above, the implied similarity, from the dendrogram, of organisms A and E is 0.6, although the true value in the matrix is 0.2. One way in which this can be considered is to calculate the arithmetic average among all relevant values when constructing the dendrogram, which is also known as Group average clustering. Average linkage clustering in one form or another is the most commonly encountered form of clustering. Much has been written on the merits and disadvantages of these methods, and these are reviewed in Abbott et al. (3).

There are two general forms of Average linkage clustering—weighted and unweighted. In Weighted clustering, the sizes of the groups involved is considered when branch points are calculated. This leads to an unequal weighting being applied to each individual organism. This technique has rarely been used with fungal data and will not be discussed further. The most commonly encountered form of Average linkage clustering is the Unweighted Pair Group Method using Arithmetic Averages (UPGMA). The UPGMA clustering of the data in Table 3 would start in the same way as for Single linkage, as the highest value for pairs of individuals must be the only relevant figures. However, after the two 0.8 pairs, the position of D is calculated as the average of A and D and C and D. This is 0.3 and lower than other values in the matrix; it is therefore not the next link. The next link must be either D to B and E or A and C to B and E; the average values for these two cases are 0.5 and 0.35. The highest value is selected and so D is linked to B and E. The final link on the dendrogram is the average of A and C to B, E, and D, which is 0.37. The Average linkage dendrogram is illustrated in Figure 3.

A further form of Average linkage clustering is Centroid clustering. Centroid clustering computes the position of each organism from the theoretical center of the clusters and may be used in weighted and unweighted forms. Centroid methods are generally used with Euclidean distance measures; they have not been used widely with fungal data.

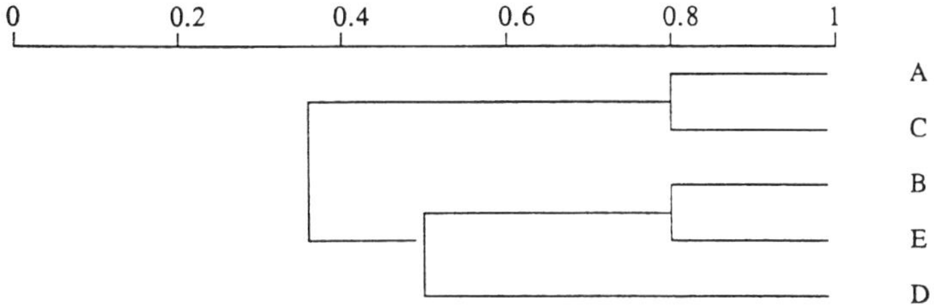

Figure 3 Average linkage dendrogram of data in Table 3.

3. Assessing Dendrograms

As stated earlier, a final dendrogram is a representation of the similarities between organisms, based on the original similarity/difference matrix. It must be remembered that the object of any clustering regimen is to group organisms, so a dendrogram can always be produced. How faithfully a dendrogram represents the true similarity values will vary, however, and obviously an inaccurate dendrogram can be worse than no dendrogram at all. One way of assessing how faithfully a dendrogram represents the original data is to calculate a correlation between the dendrogram and the original similarity difference matrix. To do this it is first necessary to generate a matrix of the similarity difference values as represented by the dendrogram. Taking Figure 3 as an example, the link levels give the matrix shown in Table 4. The correlation coefficient (r) between the dendrogram values and the original values in Table 3 can then be calculated. This is often termed the Cophenetic Correlation; as a general rule it would be expected to be > 0.75 for a reasonably faithful representation, and > 0.9 for a good representation (20,21).

Some work has been carried out on the direct comparison of different dendrograms, although there is no universally agreed procedure for this (22,23).

Table 4 Matrix of Similarity Values as Shown by Figure 3

Organism		A	B	C	D	E
	A	X				
	B	0.37	X			
	C	0.8	0.37	X		
	D	0.37	0.5	0.37	X	
	E	0.37	0.8	0.37	0.5	X

(Columns: Organism)

One approach that has been used with fungal data is to calculate the significance of differences among the cophenetic values (9).

There have been a number of attempts to develop general rules and algorithms to determine groups within a dendrogram; some such methods are listed by Milligan and Cooper (24). One commonly used criteria is to test the distinctness of potential groups (25,26). This can be undertaken by determining the observed overlap between groups, and then testing whether this is significantly greater than that expected for different group types. One criticism of this method is that comparisons among groups of equal sizes may be very useful, but comparisons among groups of very unequal sizes may not be accurate. For a full description of this approach, see Sneath (25).

E. Applications

The earliest applications of numerical methods in the hierarchical grouping of filamentous fungi probably date from the 1960s, when workers were most concerned with morphological features. The methods for coding morphological features were also scrutinized at this time (5,8,27).

The use of biochemical and physiological data in numerical schemes was not applied to filamentous fungi until the 1970s, and in these studies there was generally a mixture of both biochemical and morphological characters (28–31). This integrated approach of analyzing biochemical and morphological features together is commonly encountered today, two examples being the terverticillate penicillia (32) and isolates of *Beauveria* sp. from insects (33). However, some studies were still based entirely on morphological characters (34,35). Another major area where numerical taxonomic methods have been used with filamentous fungi is from entirely chemotaxonomic approaches. Examples include fatty-acid profiles (36), carbon source utilizations (37), and isoenzyme patterns (38,39).

The most recent examples of fungal numerical taxonomies have been based on molecular data. It is important to stress at this point that the majority of molecular studies are interpreted in terms of phylogeny, and the trees produced by such methods are not necessarily comparable to those described here. This will be discussed in further detail later. However, some molecular studies have used the methods described here, and one example involved selected species of *Pythium* (40).

III. NONHIERARCHICAL METHODS

Nonhierarchical methods, sometimes referred to as ordination techniques or multidimensional scaling methods, have also been used to group or classify fungi. Although the use of these techniques is not as widespread as the more established hierarchical methods, and they are seldom relied on alone, they are being used increasingly to confirm or to add a new perspective to clusters or cluster groups uncovered using hierarchical methods. The examples outlined here deal mainly with the most widely used nonhierarchic methods and do not represent an exhaustive list. A more complete analysis of these methods, largely using bacterial examples, is featured in an excellent review by Alderson (41) and, in a more general sense, by Dunn and Everitt (4), Manly (42), and Digby and Kempton (43). The general principle behind nonhierarchic methods is to attempt to represent the variation contained within a data matrix (*txn*) or a distance matrix in some sort of summarized format. Results are generally presented as a two- or three-dimensional scatter diagram in which each OTU is represented by a point. In such diagrams OTUs that are closely related are depicted as clouds or swarms and will appear distant or distinct from other groupings. The axes in ordination plots reflect the variation within the data, the first of which (generally the x-axis) represents the direction of greatest possible variation, the second the maximum possible residual variation. This "window" into the data or distance matrix allows the user to visualize relationships among OTUs. Clusters are not formed as they are with hierarchical methods; they are detected by eye.

A. Principal Components Analysis

Principal components analysis (PCA), which was first described by Pearson (44), is by far the most widely used nonhierarchic technique in microbial systematics. PCA is generally performed on quantitative data sets, though binary data can also be studied. Typically the types of data analyzed are taken from morphological measurements, or complex chemical data such as fatty acid profiles, pyrograms etc., in which there are common characters found in all OTUs which vary in relative amount. These data can be represented in multidimensional space, each character defined by an axis, the relative amount of which determines the coordinates of the OTU. PCA works by detecting a dimension or principal component, also known as an eigenvector, which expresses the greatest spread or scatter and arrays the OTUs along this dimension. The technique can be explained more simply if we take an example in which three measurements (a, b, c) are made from a group of strains. Data can then be presented graphically by plotting the measurements of a, b, and c for each strain, as shown in Figure 4. PCA attempts to express the variation in these data by projecting the strains (points) onto a line which maximizes the variation. This process is analogous to calculating a line of best fit between points in the least-squares sense. That is the line that minimizes the sum of squares between the points and itself (Fig. 5). The analysis is sequential, after detecting the dimension of maximum variation the direction of maximum residual variation is defined (Fig. 6). The power of this method is that it can work with data sets with large numbers of variables—e.g., whole-organism fingerprinting by pyrolysis mass spectrometery generates data on 150 variables (45). PCA can

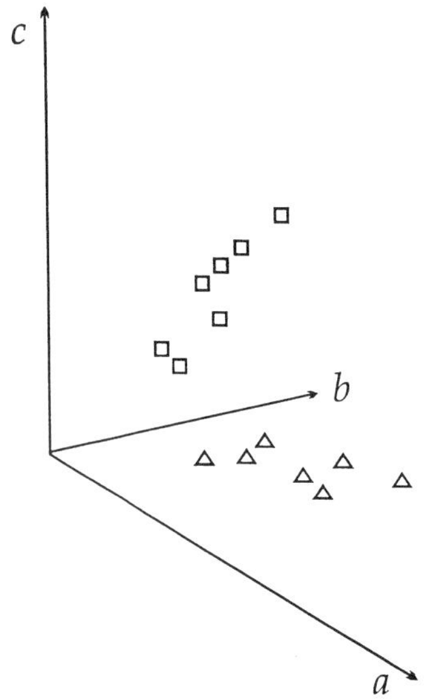

Figure 4 Graphical representation of quantitative data obtained by measuring characters a, b, and c for two populations of strains.

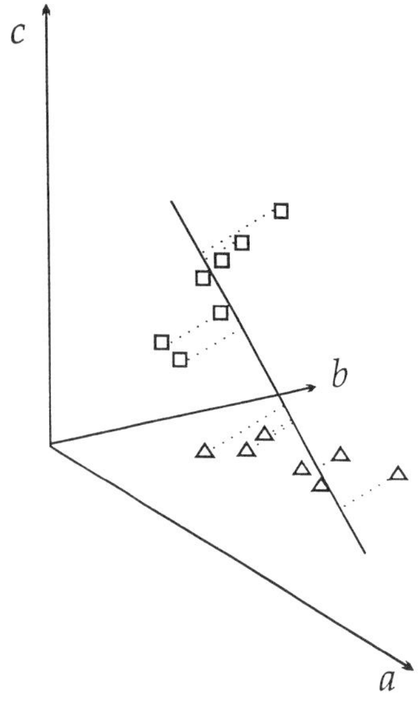

Figure 5 Graphical representation of quantitative data obtained by measuring characters a, b, and c for two populations of strains. The first principal component is shown to demonstrate the direction of maximum variation between the two populations.

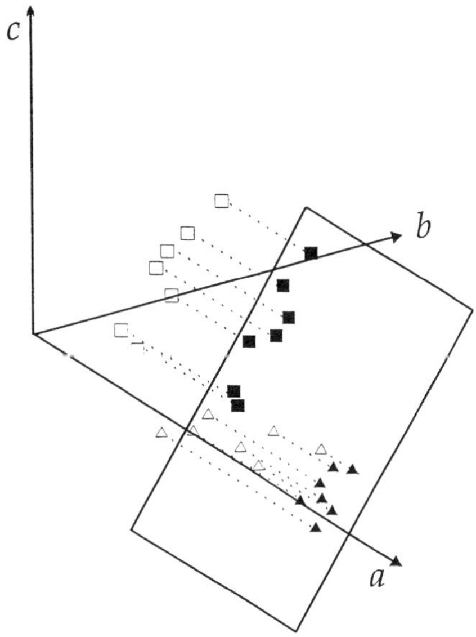

Figure 6 Graphical representation of quantitative data obtained by measuring characters a, b, and c for two populations of strains. The first and second principal components are shown to demonstrate how PCA can be used to "summarize" the information in multidimensional space into a smaller number of dimensions—in this case two.

reduce the information contained in multidimensional space, in this case 150-dimensional space, down to two or three principal components.

The interpretation of results must be performed with care. As stated previously ordination techniques such as PCA do not provide a classification per se; groupings must be detected by the user on the basis of prior knowledge or in conjunction with alternative techniques such as hierarchical clustering. It is very dangerous to form groupings subjectively on the basis of one PCA plot, as points or OTUs that appear close to each other in two dimensions may in reality be quite distant in the third or subsequent dimensions. PCA is most effective when the variation in the data can be well represented by the first three principal components, primarily because it is difficult to visualize or represent on paper or a computer screen if additional principal components are needed to describe the data. In addition, it should be borne in mind that it is impossible to represent a large number of variables by a smaller number of principal components if the original variables are uncorrelated. PCA, inherently, attempts to represent the maximum variation within a data set, uncovering gross differences between OTUs or outliers. To this end, a distorted perspective can often occur, particularly between closely related OTUs.

There have been a number of recent studies that have employed PCA to good effect. Johnk and Jones (46) were able to differentiate populations of *Rhizoctonia solani* AG-2-2 by analysis of cellular fatty acids. Data were analyzed by hierarchical clustering of Euclidean distance and by PCA. By and large, the results were comparable showing clear differences between isolates from corn, turf, sugarbeet, and mat rush. Although within-group relationships lacked definition with PCA, group membership was well defined and identical to the clusters produced by hierarchical methods. Similarly, in a chemotaxonomic study of *Beauveria* (33), reasonably good correlation was found between results from hierarchical clustering of percentage similarity values calculated using Gower's coefficient and PCA. In this study it was possible to differentiate among strains using API ZYM and isoenzyme analysis on the basis of host and geographic origin.

B. Principal Coordinates Analysis

Unlike principal components analysis (PCA), principal coordinates analysis (PCO) (47) is not dependent on an *txn* matrix. PCO works on a symmetric matrix which can represent associations between the OTUs. The association matrix can be derived from the *txn* matrix by calculating a measure of association. For example, by converting a matrix of similarities calculated by the S_{SM} coefficient to dissimilarity or total distance (see Sect. IIC). Once the raw data are transformed in this way, the analysis and representation of results are largely similar to PCA; indeed, if Euclidean distance is used as an association measure, then the results of PCO are equivalent to PCA. As with PCA the effectiveness of the method lies in the fact that the dominant patterns in the data are reflected in the first few dimensions. The advantages of PCO over PCA are listed by Sneath and Sokal (1). In summary, PCO can be used directly from association/distance matrices, which are generally published in the literature, unlike the raw data. It is also thought to be more robust than PCA when dealing with missing data.

PCO analysis has not been used widely in fungal taxonomy; however, a recent study demonstrates its usefulness (48). To characterize endophytic fungi from Mycorrhizae, restriction fragment length polymorphism (RFLP) data were determined for a number of isolates. The presence or absence of particular restriction fragments were scored for each strain, and these binary data were analyzed using the coefficient of Nei and Li (49). The subsequent distance matrix was then studied by UPGMA hierarchical clustering and by PCO. Broadly, results of hierarchical clustering showed that strains could be assigned to two clusters—the first of which encompassed members of the genus *Phialocephala*, and the second, strains of *Hymenoscyphus ericae* and *Phialophora finlandia* and some other isolates. Ordination by PCO showed good congruence with the hierarchical method by revealing two groupings with identical membership to the hierarchic clusters. However, within-group relationships did not show much similarity to those found within the hierarchic clusters. Using two techniques in this way it is possible to have greater confidence in the groupings/clustering of strains as similar results are obtained with different methods. Further examples of PCO in comparisons of nonhierachical methods for fungal data include studies on the carbohydrate and secondary metabolite profiles of some species of *Penicillium* (50,51).

C. Canonical Variate Analysis

Canonical variates analysis (CVA) is a discriminant technique which attempts to derive axes that maximize differences among populations. Similar to PCA and PCO, CVA allows the determination of relationships between OTUs in multidimensional space. However, unlike PCO and PCA, CVA requires an established or a priori knowledge of group structure. Again, like PCA, transformed axes are sought from the data matrix, but these are based on group means and not on individual values as they are in PCA. In this way the first canonical variate axis is in the direction of greatest variability between the means of different taxa. The second canonical variate is calculated from the residual variation after the first, and so on. In this way CVA can be used to statistically interpret the differences between two or more populations. Typically the type of data analyzed by this method are obtained from strains of the same taxon or replicate analyses of the same strain. In addition, there is a general assumption that there is a common within-group dispersion matrix.

Typically, generalized distance measures (52) are derived from the data matrix; these are commonly referred to as Mahalanobis distances. This coefficient is dependent on quantitative or continuous data and maximizes the variance between pairs of means for characters which have maximal variance between groups relative to pooled variance within groups. This measure is independent of scale and takes into account correlated variables—i.e., variables which essentially measure the same thing. Thus OTUs can be arranged along a line that maximizes differences, the canonical variate. The canonical variates can then be plotted against each other much in the same way as found in PCA.

An excellent example of how CVA can be used and a concise description of the method is given in MacFie et al. (53). In this study the authors were able to differentiate among a number of bacterial taxa by pyrolysis gas chromatography (PyGC). Replicate analyses were performed, and the mean of 24 separate peak heights was determined. From these data Mahalanobis distances were calculated for each pair of groups, and the resultant CVA plot was produced. Each group was represented by a point, and a confidence limit could be imposed on top of this to determine whether there was any cluster overlap. This technique has also been used to study of RFLP data obtained from 153 isolates of *Phythium infestans* obtained from 14 fields (54). In this particular case the data from each field were taken as a priori groupings to determine the genetic diversity of this pathogen within fields of potatoes and tomatoes in the Netherlands. In essence the population of *P. infestans* is similar in fields of commercial potatoes regardless of geographic distribution, but quite distinct from the populations found within tomato fields. This would suggest that variation in population structure is evident and can be correlated to the types of crops grown. In a study of 80 isolates representing 40 species of the genus *Phythium*, CVA was used to analyze measurements and derived indices from 20 individual oogonia taken from each of the isolates (55). Each set of measurements of the oogonia are treated as the individuals, and the isolates as the groups. In this study isolates from the same species tended to group together; in addition, isolates that had been doubtfully assigned to a particular species were found to be more scattered on the canonical variate plot, as in the case of *P. vexans*.

D. SIMCA

Soft Independent Modeling of Class Analogy (SIMCA) (56) is a multivariate technique based on PCA which can be used to test out hypotheses on group membership or as a template for identification.

In general, once groups have been detected using PCA, the group members are defined and modeled using disjoint principal components (57). The technique describes a group of similar objects, a class, by an empirical model. Each class is represented by a distinct principal component model so that the within-group variation is modeled independently and is not assumed to be equal. This procedure offers considerable advantages in systematics, as taxa rarely show equal intraspecific homogeneity. From the scatter of objects within the model, the tolerance interval for any given probability level, usually 95%, can be calculated creating a cylinder of "SIMCA can" in multidimensional space (Fig. 7).

The number of statistically significant components necessary to describe each class model is determined by cross-validation (58). Cross-validation randomly divides the class into subgroups. Each subgroup represents elements from all rows and columns in the data matrix. One of the subgroups is then held out while the principal components are recalculated and predictions are calculated for holes in the matrix. These predictions are then compared with the actual data. In this way the procedure provides a useful estimate of how much of the variance within a class is systematic and how much is random noise. The statistical stability and therefore the predictive potential of each class model can then be determined.

The models can then be used as templates for identification, as each OTU can be tested against each model using linear multiple regression. An approximate F-test is used to compare the OTU standard deviation (SD) with the typical SD of the class. An observed F-value larger than the critical F value on a given level of significance, generally $P = .05$, within the calculated degrees of freedom indicates

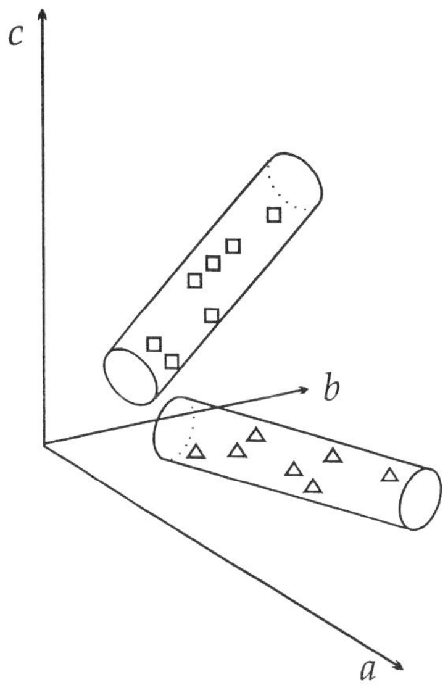

Figure 7 Graphical representation of how a SIMCA model or "can" is calculated for distinct populations. Each model is delineated independently by the scatter of objects within it at given confidence limit.

the object to be an outlier hence it is not assigned to a class. In this way SIMCA can be used as an identification procedure. A graphical interpretation of the procedure illustrating the class distance of a test strain to two models plotted at right angles to each other is given in Figure 8. Point coordinates can be determined by the object SD of each strain when challenged against each model.

An example of how SIMCA can be used is demonstrated by the classification of Streptomycetes by quantitative fatty acid analysis (59). In this particular study PCA analysis revealed that members of the species *Streptomyces cyaneus* were diverse and appeared to form two groupings. This finding was confirmed by disjoint principal components supporting the view that this species contains a wide diversity and encompasses many atypical strains. The use of quantitative fatty-acid analysis followed by SIMCA on the resultant data has been demonstrated for fungi in a study of various members of the genera *Aspergillus*, *Mucor*, and *Penicillium* (60). Statistically valid models of all three genera could be produced; in addition, it was possible to separate species within the same genus—e.g., *A. flavus* and *A. oryzae*. This statistical technique has also been used to group members of the genus *Fusarium* on the basis of HPLC analysis of secondary metabolites (61). An examination of the raw data revealed no qualitative differences or obvious quantitative differences, yet hierarchical clustering and PCA

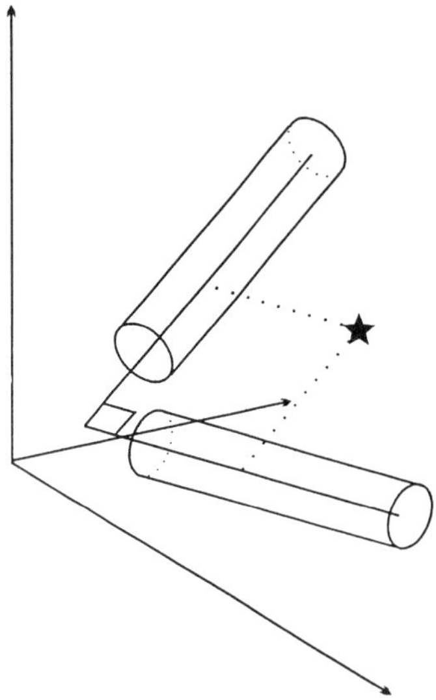

Figure 8 Use of SIMCA models for identification of an unknown. Models are plotted at right angles to each other, and the class distance of the unknown to each model is calculated. In this instance the unknown falls outside the confidence limits of the models and is therefore not identified.

were clearly able to distinguish among *F. culmorum*, *F. graminearum*, and *F. crookwellense*. SIMCA has also been used with binary fungal metabolite data (50) and for the separation of closely related species of *Penicillium* on the basis of pyrolysis gas chromatography (62). The use of SIMCA as an identification tool is demonstrated in a study designed to recognize new strains of the species *Streptomyces tsukabaensis* using the technique of pyrolysis mass spectrometry (PyMS) (63). By using disjoint principal components to define a model of the species it was possible to add fresh PyMS data and then determine whether the unknowns were identified—i.e., if they fell within the confidence limits of the model.

E. Minimum Spanning Trees

Minimum spanning trees (MST), sometimes referred to as shortest spanning trees, are a useful adjunct to ordination techniques. The main use of this technique is to counteract losses when displaying ordinations by imposing an MST onto the plot (64). In this technique, single-linkage clustering (see Sect. IID) can be calculated from a resemblance matrix, and the points or OTUs can be linked subsequent to production of an ordination plot. This is used primarily because relationships among close neighbors are frequently distorted in a two- or three-dimensional representation of OTUs as produced in PCA and PCO. A pair of OTUs that appear close together in two dimensions may be far apart in higher dimensions. This is particularly likely when the first two dimensions fail to describe a large percentage of the total variation between OTUs.

F. Merits of Hierarchical and Nonhierarchical Methods

Nonhierarchical methods possess a number of advantages and disadvantages over the more widely used hierarchic methods (41). It is generally believed that ordination methods are inappropriate for uncovering distances between closely related strains, as these can be obscured, and that these are better represented by hierarchical methods. However, it is frequently the case that the converse applies, in that ordination techniques are far better at depicting variations between distantly related taxa than hierarchical methods. A valid criticism of some hierarchical methods is that they set out by assuming clusters are present (65); nonhierarchic methods avoid this assumption. Ordination merely presents the OTUs as points in space. It is for the user to decide the significance of the groupings. Nonhierarchic methods can be viewed as data reduction techniques, and it is therefore inevitable that some relationships between OTUs will be obscured as the variation evident in multidimensional space is reduced down to two or three axes. To some extent these problems can be overcome by the judicious use of MST or techniques such as SIMCA, which test out the validity of groupings.

IV. PHYLOGENY AND EVOLUTION

Many studies have used numerical methods to construct phylogenetic representations from chemical and molecular data. These representations are generally in the form of a "tree," which can be either "rooted" or "unrooted." In an unrooted tree, the earliest point or ancestor to the phylogeny is not identified, whereas in a rooted tree it is either identified or postulated. Such studies with filamentous fungi have almost exclusively used DNA sequence data, although a number of different numerical approaches have been considered. The current interest in this area has already started to have a significant effect on fungal systematics, although some caution may still be required in the interpretation of results. There is a considerable amount of recent literature on phylogeny reconstruction already available, and so in this section we will only consider some of the more common techniques. The two general approaches for phylogenetic work are either from some form of distance or similarity calculation, or from the direct construction of optimal trees from the character data (66,67).

A. Distance and Similarities

Distances between pairs of organisms may be used in phylogenetic analysis along with an appropriate clustering strategy. In general, similarities and matches are transformed into distance measures, although some measures such as the Manhattan distance have been used directly. The three most common forms of transformation are listed by Swofford and Olsen (67) as: $d = 1 - S$, $d = -\ln S$ and $d = 1/S - 1$, where S is the fractional similarity between organisms.

The two most commonly encountered types of data for filamentous fungi are either band patterns, as in isoenzyme and RFLP analyses, and DNA sequence data. These two types of data sets may be treated in similar ways. Sequences of DNA consist of the four bases adenine, thymine, guanine, and cytosine. Therefore a DNA sequence of 100 bases can be considered as a set of 100 characters, each of which is a particular base. Each base is independent and there is no quantitative sequence between them, so the data set consists of multistate qualitative characters. With band patterns, each band may be treated as a single binary character. The simplest way of deriving a similarity in these cases is to calculate the fraction of bases in common positions or the fraction of common bands. This is analogous to calculating a Simple Matching coefficient. Some of the coefficients that have been used for these procedures are identical to previously described association coefficients such as Nei and Li's coefficient (49), which is the same as Dice's coefficient (see Sect. IIC). The similarity may then be converted to a direct distance by one of the transformations given above. The individual characters can be given independent weights, so that different parts of a sequence can be given greater significance in the final calculation (68).

The above is a very basic example. In reality the comparison of isoenzyme, restriction fragment, and sequence data can become much more complicated; for example, some types of differences in sequences are more common than others, as a result a number of calculations and transformations have been suggested for dealing with transition and transversion substitutions, deletions, and insertions (67). The use of restriction fragment data is further complicated by the strict definition of the characters. Because each fragment size may be affected by differences in two restriction sites, it is therefore more correct to use restriction site data than fragments (49). In isoenzyme data the number of loci and alleles may be considered and more complex coefficients generated; one example of this is the calculation of genetic diversity at a locus described by Selander et al. (69).

Distance values may be clustered by any of the methods described earlier. There has been considerable discussion about the relative merits of different clustering strategies; Sneath and Sokal (1) have suggested that unweighted clustering may give the most theoretically correct trees. However, most tree-forming methods in the literature, while not strictly cluster analysis, are based on forms of analysis analogous to single linkage such as neighbor joining (70). These types of trees are termed additive trees, as there is an assumption that distances on individual branches can be summed to give a meaningful quantity related to the evolutionary pathway (67). Other techniques that fit into this category are Fitch-Margoliash and Wagner distance methods. These are described in detail in a number of texts including Swofford and Olsen (67), Sneath and Sokal (1), and Abbott et al. (3).

B. Parsimony

One of the guiding principles of phylogeny reconstruction is that evolutionary pathways have occurred through the minimum number of steps. In this way a tree that depicts lineages involving the least number of changes in characters will be the most parsimonious (71,72).

The maximum parsimony contained within any tree may be calculated directly from the character data without the need to calculate distances or similarities. Many trees may be constructed from a data set, and the larger the number of organisms, the larger this set of trees will be. The final number will depend on the type of tree constructed and whether both the tips of the tree (the living organisms) and the branch points (hypothetical ancestors) are included. Felsenstein (73) has suggested formulae for calculating the number of possible trees, but in the worst case there will be n^{n-1}, where n is the number of known organisms—i.e., 100^{99} trees where there are 100 known organisms (65). In parsimony methods it is possible that there may be more than one tree that are equally parsimonous. Thus a definite answer may not be reached.

Trees produced from parsimony analysis may be rooted or unrooted, and a common way of rooting trees is to include an outgroup. An outgroup is generally accepted as a taxon or group of taxa that have the same ancestor as the organisms being studied (ingroup), but that are not closely related to them (65). More than one outgroup may be included in an analysis, but the points at which they link to the ingroup may not be significant. This type of analysis assumes that the ingroup is monophyletic—i.e., derived from a single line of descent (67).

C. Errors and Confidence in Phylogenies

In attempting to fit data into distinct evolutionary lines there will be some degree of distortion or uncertainty included in any tree. This can be estimated, and apart from measuring the confidence of a final representation, these measures can also suggest alternative ancestries. In the case of trees produced by cluster analysis of distance or similarity data, an estimate of the reliability of the tree can be taken from the Cophenetic Correlation Coefficient (see Sect II.D) (1). Where trees have been derived directly from the character data, as in parsimony methods, measuring confidence is harder as there is no matrix of absolute values. Felsenstein (74) has reviewed some of the methods available in this situation; the only one we will consider here is bootstrapping.

A general premise of any scientific study is that the larger the database, the more accurate the results taken from it. However, the amount of data available is limited by the scope of an individual study, although estimates of reliability may be obtained by repeated resampling the original data set. In bootstrapping the original data are resampled randomly to produce a matrix of equal size. As the resampling is random, some data will be repeated and not all data will be included. This process is repeated many times, and data are scored as to whether they are present or absent in the individual resamplings. From this, characters and character combinations can be weighted according to their frequency of occurrence, and confidence limits can be applied to particular branches in the final tree (67). However, it should be borne in mind that the final confidence of the bootstrap will depend on the critical values used, and it has been reported that critical values below 95% may lead to bootstrap values that are underestimates of the confidence level and that this underestimate will increase with the number of competing topologies (75).

A further type of error that should be considered in phylogenetic analysis is sampling error. Where individual sequences are being compared, a number of assumptions must be made regarding the evolution and commonality of that sequence. Errors or misinterpretations in these assumptions can give rise to a significant reduction of confidence in a final tree. There are already considerable data available in this area; the reader is directed to Felsenstein (74) and Sneath (76).

V. NUMERICAL METHODS FOR IDENTIFYING ORGANISMS

Just as numerical systems have been developed for grouping organisms, other numerical systems have been developed for identifying organisms. Numerical identification systems have not been widely used with filamentous fungi, however, and those that have must be considered experimental (77). Numerical identification methods can be considered in four groups—computer generated/assisted keys, profile matching, probabilistic/distance measures, and neural networks. Most work with filamentous fungi has been restricted to the first two categories (78), and such schemes are generally not numerically based in the strictest sense. Computer-generated/assisted keys and profile matching are nearly always computer-based, although final diagnostic keys or profiles can be produced as "hard copy." Probabilistic distance measures and neural networks are reliant on access to a computer for every identification.

A. Computer-Generated/Assisted Keys

Computer keys start in the same way as other keys with some form of data matrix. In most cases this consists of a series of positive or negative responses by taxa to particular characters, usually morphological. There are a number of mechanisms that can be employed to generate a key (65); only one general approach will be considered here.

Once a matrix of responses has been produced, then one of a variety of indices can be calculated to consider the information content of each character or a separation function. That is, the usefulness of each character or branch in the key can be mathematically assessed, and a computer program can then select and offer the most useful character as the first step in the key. In an interactive key, once a character has been selected and a response noted, the indices can be recalculated on the basis of the reduced number of characters and possible identifications, and a further character selected. This will continue until only one possible organism is left. Alternatively, for a standalone key for "hard copy" use, information content coefficients can be calculated to plan the most efficient route through the data matrix. There are a wide range of indices and coefficients that have been proposed for these tasks (1,79), and, just as with numerical classification, it is important to select a method that is suitable for both the data used and the type of key required.

In constructing an interactive key, the usual calculation is one that produces a separation index—that is, a measure of how well a character separates the different taxa. Separation indices can be grouped into two main types—those that give high scores for "key" characters, and those that give high scores for "diagnostic" characters. This is an important distinction, as the keys produced with each type of character will be quite different (80).

The term "key" character here refers to characters that give the best separation of the organisms into two groups. For example, if there are eight organisms in a key, then a good "key" character is that which is positive for four and negative for four. Thus a key of "key" characters may consist of characters that split the eight organisms into two groups of four, followed by characters which split the groups of four into groups of two, and so on. This type of character is generally regarded as the most useful to construct traditional dichotomous keys. The term "diagnostic" character can be used to refer to characters where only one organism differs from the others. Thus a key constructed from "diagnostic" characters will consist of one less character than there are organisms. This type of character is generally considered undesirable in a key but can be extremely useful in some circumstances, particularly as a confirmatory character.

A number of separation indices and information content formulas have been suggested for identification schemes. Two that have been used with microbial data are Gyllenberg's Sum of C and Sneath's Consistency and Strain Potential (CSP), as described in Sneath (79). These formulas are an example of the situation described above, as the Sum of C will give a high score with "diagnostic" characters and the CSP will give high scores with "key" characters.

B. Profile Matching

As mentioned earlier, profile matching is not a mathematical method in the strictest sense, although mathematical elements may be included and the final schemes are often computer-based. Profile matching involves comparing a set of characters from one organism (the unknown) against a table of characters for all the reference organisms. The best match is then selected from the reference organisms, either mathematically by calculating an association coefficient or, more usually, empirically. One example of profile matching in fungal identification is the computer program for yeast identification provided by Barnett et al. (81). Profile matching has often been associated with some form of data reduction, so the results from a standard set of characters can be represented as a series of numbers or characters (82). In fact this form of data reduction was considered in relation to multiple entry keys for smut fungi as long ago as 1941 (83).

C. Probabilistic and Distance Methods for Identification

Probabilistic and distance-based schemes for identification have been used widely in bacteriology (84) but rarely in mycology. These methods are similar to profile matching in that a set of characters for an unknown are compared to a table of characters for the reference organisms. A "best match" is then calculated. The best match is either the probability that the unknown belongs to the reference taxon or a measure of the distance of the unknown from the reference taxon in *A* space (see Sect. II.C).

Probability and distance methods commonly use frequency matrices as the reference databases. A frequency matrix is one where the response for each character is given as the frequency of occurrence in the taxon, rather than plus/minus. As the frequency of occurrence of each character in each reference taxon is listed, then the probability of an individual set of characters occurring in (i.e., belonging to) a particular taxon can be calculated (see Table 5). In these matrices, although the observed frequency of occurrence may be 100% or 0%, there is always a margin of error and the entire population is rarely sampled, so the terms "always" and "never" are recorded as 99% and 1%.

The most commonly used probability calculations are those based on Bayes' theorem, and of these the most commonly used in microbiology is the Willcox likelihood (85). Working from the frequency matrix (Table 5), the likelihood that an unknown with the characteristics +,+,+,− belongs to each of the reference taxa can be calculated by determining the product of the individual likelihoods. This is acheived by dividing the frequency matrix values by 100, and

Table 5 Frequency Matrix for Four Characters and Three Reference Taxa

Reference taxa	Character 1	2	3	4
A	95%	99%	75%	1%
B	50%	75%	50%	1%
C	1%	99%	99%	50%

multiplying either these values where a character is positive, or 1 minus these values where a character is negative:

Taxon A	$0.95 \times 0.99 \times 0.75 \times (1\ \ 0.01) = 0.6983$
Taxon B	$0.50 \times 0.75 \times 0.50 \times (1-0.01) = 0.1856$
Taxon C	$0.01 \times 0.99 \times 0.99 \times (1-0.50) = 0.0049$

As can be seen, it is most likely that the unknown belongs to taxon A; however, the absolute probability decreases with the number of characters used. One solution to this is to use a normalized probability, based on the sum of all of the taxa. In this case this would be 0.6983/(0.6983 + 0.1856 + 0.0049), which would then give a score of 0.7857. In reality with microbial systems normalized scores higher than this are expected, and various "cutoff" levels for good identifications have been suggested between 0.85 and 0.99 (80,85,86).

The example given above is a very simplified form of Bayes' theorem, which in its complete form allows for prior probabilities to be set so that the effects of very rare or particularly important characters may also be included.

Calculating distance measures for identification is a little different. In this technique it is assumed that the individual members of a taxon show roughly equivalent variation in characters. When represented in *A* space the individual members would then form a group around a central point which would represent the most typical member of the taxon. In a three-dimensional Euclidean space model this would form a sphere, and in multidimensional space it would form a hypersphere. The distances of an unknown from the centers or edges of these hyperspheres can be calculated, so an unknown can be identified as being closest to, or in, a particular taxon. Interpretation of distance-based identifications is much harder than for probability systems, as uneven variation between taxa makes it difficult to define general "cutoff" levels (1,79).

D. Mixed Identification Systems

It must be remembered with a probability system based on Bayes' theorem that the identification assumes that the unknown belongs to one of the reference taxa. If the unknown is not represented in the database, for example an *Aspergillus* in a *Penicillium* scheme, it is still possible to get a high normalized identification score (80). However, in these circumstances a distance measure could be expected to be very large. However, a single, very discrepant character can give an increased distance measure in a single dimension, so it is useful in an identification program to be able to list discrepant characters. The different types of measures and discrepant character lists should be combined in a single program to enable unequivocal identifications to be made. In such an identification program links can be made between the different features, and the overall performance of the system can be considerably enhanced (80,87,88).

E. Neural Networks

The most recent developments in computing for identification are neural networks, so called because they have been constructed to try to simulate the neural activities within a human brain. Neural networks can be broken down into a number of specific types such as back-propagation networks, self-organizing maps, learning vector quantization, and probabilistic networks. All of these have a common feature in that identifications are based on a learn/train system, rather than specific rules.

The neural network consists of layers of nodes, with the input to the first nodes being the pattern of characters, and the output from the last layer being the identification. The network can then be trained by presenting it with a series of known patterns and taxa. This is termed supervised learning, and the network can be used to assign new patterns to the correct taxa. An alternative is unsupervised learning, where the network is presented with unassigned patterns which it then groups according to similarities between these. Unsupervised learning can therefore be used as a classification aid; however, it must be remembered that, as the network is not strictly rule-based, it is not possible to determine why particular patterns have been grouped together.

To train a network to identify organisms it is usually necessary to present character patterns many times, as each initial attempt gets closer to the correct identification. In reality this may mean many thousand times, and although this is not a problem with modern computers, considerable data may be required in setting up a neural network system.

As mentioned above, there are several different types of artificial neural network; the most commonly encountered in identification systems is learning vector quantization. In learning vector quantization there is an input layer of one node per character, which connects to a Kohonen layer with a variable number of nodes which in turn are divided into groups that connect to a single output node. A Kohonen layer acts as a nearest-neighbor classifier, all nodes competing with each other on the basis of Euclidean distances. The node with the minimum distance to the weights of the input data then adjusts its own weighting criteria so as to respond more strongly if the input data are repeated (89).

F. Applications

Despite their relatively long history, there are few examples of entirely computer-based or numerical identification schemes for filamentous fungi. This contrasts widely with bacteriology, where such schemes have been widely accepted for some years (84). As mentioned earlier, numerical profile data reduction schemes were proposed for fungi as long ago as 1941 (83). While such systems have been proposed from time to time, there has been little further development in such identification schemes for fungi, except for yeasts (see Sect. V.B). Computer-generated keys have been accepted for yeasts but are not widely available for filamentous fungi; one example, however, is the key to *Penicillium* species produced by Pitt (78). Similarly, probabilistic schemes have not been adopted with fungal data, although there are a few individual examples, one of which is the scheme for terverticillate penicillia produced by Bridge et al. (80). Neural networks are the most recent innovation in this area of identification, and these have been used, although in somewhat experimental forms, in a small number of cases (90).

As numerical schemes have not been used widely with filamentous fungi, it is not possible to accurately assess their performance or full potential. However, the relative performances of a computer-generated multiple entry key, a probabilistic identification system, and a back-propagation neural network with data from species of *Penicillium* have been compared by Bridge et al. (77). This study used a small experimental database and found that the neural network and the probabilistic scheme performed equally as well as and better than the computer-generated key. This is perhaps as expected, but further trials with larger databases are required.

REFERENCES

1. Sneath PHA, Sokal RR. Numerical Taxonomy. San Francisco: W. H. Freeman, 1973.
2. Sneath PHA. The application of computers to taxonomy. J Gen Microbiol 1957; 17: 201–226.
3. Abbott LA, Bisby FA, Rogers DJ. Taxonomic Analysis in Biology. New York: Columbia University Press, 1985.
4. Dunn G, Everitt BS. An Introduction to Mathematical Taxonomy. Cambridge: Cambridge University Press, 1982.
5. Kendrick WB, Proctor JR. Computer taxonomy in the fungi imperfecti. Can J Bot 1964; 42:65–88.
6. Austin B, Colwell RR. Evaluation of some coefficients for use in numerical taxonomy of microorganisms. Int J Syst Bacteriol 1977; 27:204–210.
7. Bridge PD. Classification. In: Fry JC, ed. Biological Data Analysis. Oxford: IRL Press, 1992:219–242.
8. Proctor JR, Kendrick WB. Unequal weighting in numerical taxonomy. Nature (Lond) 1963; 197:716–717.
9. Bridge PD, Sackin MJ. Stability of classification of filamentous fungi under changes in character coding strategy. Mycopathologia 1991; 115:105–111.
10. Sokal RR, Michener CD. A statistical method for evaluating systematic relationships. Univ Kans Sci Bull 1958; 38:1409–1438.
11. Dice LR. Measures of the amount of ecologic association between species. Ecology 1945; 26:297–302.
12. Gower JC. A general coefficient of similarity and some of its properties. Biometrics 1971; 27:857–871.

13. Brisbane PG, Rovira AD. A comparison of methods for classifying rhizosphere bacteria. J Gen Microbiol 1961; 26:379–392.
14. Sneath PHA. Vigour and pattern in taxonomy. J Gen Microbiol 1968; 54:1–11.
15. Sorensen T. A method of establishing groups of equal amplitude in plant sociology based on similarity of species content and its application to analyses of the vegetation on Danish commons. Biol Skrift 1948; 5:1–34.
16. Cowan ST. In: Hill LR, ed. A Dictionary of Microbial Taxonomy. Cambridge: Cambridge University Press, 1978.
17. Cain AJ, Harrison GA. An analysis of the taxonomist's judgement of affinity. Proc Zool Soc Lond 1958; 131:85–98.
18. Sackin MJ. Vigour and pattern as applied to multistate quantitative characters in taxonomy. J Gen Microbiol 1981; 122:247–254.
19. Jackman PJH, Feltham RKA, Sneath PHA. A program in BASIC for numerical taxonomy of micro-organisms based on electrophoretic protein patterns. Microbios Lett 1983; 23:87–98.
20. Sokal RR, Rohlf FJ. The comparison of dendrograms by objective methods. Taxon 1962; 11:33–40.
21. Jones D, Sackin MJ. Numerical methods in the classification and identification of bacteria with especial reference to the Enterobacteriaceae. In: Goodfellow M, Board RG, eds. Microbiological Classification and Identification. London: Academic Press, 1980:73–106.
22. Podani J, Dickinson TA. Comparison of dendrograms: a multivariate approach. Can J Bot 1984; 62:2765–2778.
23. Sackin MJ. Comparisons of classifications. In: Goodfellow M, Jones D, Priest FG, eds. Computer-Assisted Bacterial Systematics. London: Academic Press, 1985: 21–36.
24. Milligan GW, Cooper MC. An examination of procedures for determining the number of clusters in a data set. Psychometrika 1985; 50:159–179.
25. Sneath PHA. A method for testing the distinctness of clusters: a test of the disjunction of two clusters in Euclidean space as measured by their overlap. Mathemat Geol 1977; 9:123–143.
26. Perruchet C. Significance tests for clusters: overview and comments. In: Felsenstein J, ed. Numerical Taxonomy. Berlin: Springer Verlag, 1983:199–208.
27. Kendrick WB, Weresub LK. Attempting neo-adansonian computer taxonomy at the ordinal level in the basidiomycetes. Syst Zool 1966; 15:307–329.
28. Whalley AJS, Greenhalgh GN. Numerical taxonomy of *Hypoxylon*. I. Comparison of classifications of the cultural and the perfect states. Trans Br Mycol Soc 1973; 61: 435–454.
29. King DS. Systematics of *Conidiobolus* (Entomophthorales) using numerical taxonomy. I. Biology and cluster analysis. Can J Bot 1976; 54:45–65.
30. Dabinett PE, Wellman AM. Numerical taxonomy of certain genera of fungi imperfecti and Ascomycotina. Can J Bot 1978; 56:2031–2049.
31. Sheard JW. The taxonomy of the *Ramalina siliquosa* species aggregate (lichenized Ascomycetes). Can J Bot 1978; 56:916–938.
32. Bridge PD, Hawksworth DL, Kozakiewicz Z, et al. A reappraisal of terverticillate

penicillia using biochemical, physiological and morphological features. I. Numerical taxonomy. J Gen Microbiol 1989; 135:2941–2966.

33. Mugnai L, Bridge PD, Evans HC. A chemotaxonomic evaluation of the genus *Beauveria*. Mycol Res 1989; 92:109–209.
34. Zambettakis C, Joly P. Application de traitements numeriques a la systematique des ustilaginales. III. Le genre *Thecaphora*. Bull Soc Mycol Fr 1975; 91:71–88.
35. Mueller GM. Numerical taxonomic analyses on *Laccaria* (Agaricales). Mycologia 1985; 77:121–129.
36. Dart RK, Stretton RJ, Lee JD. Relationships of *Penicillium* species based on their long-chain fatty acids. Trans Br Mycol Soc 1976; 66:525–529.
37. Manczinger L, Polner G. Cluster analysis of carbon source utilization patterns of *Trichoderma* isolates. Syst Appl Microbiol 1985; 9:214–217.
38. Yamazaki M, Goto S. An electrophoretic comparison of enzymes in the genera *Lipomyces* and *Myxozyma*. J Gen Appl Microbiol 1989; 31:313–321.
39. Jun Y, Bridge PD, Evans HC. An integrated approach to the taxonomy of the genus *Verticillium*. J Gen Microbiol 1991; 137:1437–1444.
40. Belkhiri A, Dick MW. Comparative studies on the DNA of *Pythium* species and some possibly related taxa. J Gen Microbiol 1988; 134:2673–2683.
41. Alderson G. The application and relevance of nonhierarchic methods in bacterial taxonomy. In: Goodfellow M, Jones D, Priest FG, eds. Computer-Assisted Bacterial Systematics. London: Academic Press, 1985:227–263.
42. Manly BFJ. Multivariate Statistical Methods. A Primer. London: Chapman & Hall, 1986.
43. Digby PGN, Kempton RA. Multivariate Analysis of Ecological Communities. London: Chapman & Hall, 1987.
44. Pearson K. On lines and planes of closest fit to a system of points in space. Philos Magazine 1901; 2:557–572.
45. Magee J. Whole-organism fingerprinting. In: Goodfellow M, O'Donnell AG, eds. Handbook of New Bacterial Systematics. London: Academic Press, 1993:383–427.
46. Johnk JS, Jones RK. Differentiation of populations of AG-2-2 of *Rhizoctonia solani* by analysis of cellular fatty acids. Phytopathology 1993; 83:278–283.
47. Gower JC. Some distance properties of latent root and vector methods used in multivariate analysis. Biometrika 1966; 53:325–338.
48. Stoyke G, Egger KN, Currah RS. Characterization of streile endophytic fungi from the mycorrhizae of subalpine plants. Can J Bot 1992; 70:2009–2016.
49. Nei M, Li WH. Mathematical model for studying genetic variation in terms of restriction endonucleases. Proc Natl Acad Sci USA 1979; 76:5269–5273.
50. Frisvad JC. Chemometrics and chemotaxonomy: a comparison of multivariate statistical methods for the evaluation of binary fungal secondary metabolite data. Chemometrics Intelligent Lab Syst 1992; 14:253–269.
51. Frisvad JC. Correspondence, principal coordinate, and redundancy analysis used on mixed chemotaxonomical qualitative and quantitative data. Chemometrics Intelligent Lab Syst 1994; 23:213–229.
52. Mahalanobis PC. On the generalised distance in statistics. Proc Natl Inst Sci India 1936; 2:49–55.

53. MacFie HJH, Gutteridge CS, Norris JR. Use of canonical variates analysis in differentiation of bacteria by pyrolysis gas-liquid chromatography. J Gen Microbiol 1978; 104:67–74.
54. Drenth A, Goodwin SB, Fry WE, Davidse LC. Genotypic diversity of *Phytophthora infestans* in the Netherlands revealed by DNA polymorphisms. Phytopathology 1993; 83:1087–1092.
55. Shazhad S, Coe R, Dick MW. Biometry of oospores and oogonia of *Pythium* (Oomycetes): the independant taxonomic value of calculated ratios. Bot J Linn Soc 1992; 108:143–165.
56. Wold S, Sjostrom MJ. SIMCA: a method for analysing chemical data in terms of similarity and analogy. In: Kowalski B, ed. Chemometrics: Theory and Application. ACS Symposium Series No. 52. Washington DC: American Chemical Society, 1977: 243–282.
57. Wold S. Pattern recognition by means of disjoint principal components models. Pattern Recogn 1976; 8:127–139.
58. Wold S. Cross validatory estimation of the number of components in factor and principal components analysis. Technometrics 1978; 20:397–406.
59. Saddler GS, O'Donnell AG, Goodfellow M, Minnikin DE. SIMCA pattern recogntion in the analysis of streptomycete fatty acids. J Gen Microbiol 1987; 133:1137–1147.
60. Blomquist G, Andersson B, Andersson K, Brondz I. Analysis of fatty acids. A new method for characterization of moulds. J Microbiol Methods 1992; 16:59–68.
61. Thrane U. Grouping *Fusarium* section Discolor isolates by statistical analysis of quantitative high performance liquid chromatographic data on secondary metabolite production. J Microbiol Methods 1990; 12:23–39.
62. Söderström B, Frisvad JC. Separation of closely related asymmetric penicillia by pyrolysis gas chromatography and mycotoxin production. Mycologia 1984; 76: 408–419.
63. Sanglier J-J, Whitehead D, Saddler GS, Ferguson EV, Goodfellow M. Pyrolysis mass spectrometry as a method for the classification, identification and selection of actinomycetes. Gene 1992; 115:235–242.
64. Sackin MJ, Jones D. Computer-assisted classification. In: Goodfellow M, O'Donnell AG, eds. Handbook of New Bacterial Systematics. London: Academic Press, 1993: 281–313.
65. Pankhurst RJ. Practical Taxonomic Computing. Cambridge: Cambridge University Press, 1991.
66. Felsenstein J. Evolutionary trees from gene frequencies and quantitative characters: finding maximum likelihood estimates. Evolution 1981; 35:1229–1242.
67. Swofford DL, Olsen GJ. Phylogeny reconstruction. In: Hillis DM, Moritz C, eds. Molecular Systematics. Sinauer Associates, Sunderland USA, 1990:411–501.
68. Olsen GJ. Phylogenetic analysis using ribosomal RNA. Methods Enzymol 1988; 164:793–838.
69. Selander RK, Caugant DA, Ochman H, Musser JM, Gilmour MN, Whittam TS. Methods of multilocus enzyme electrophoresis for bacterial population genetics and systematics. Appl Environ Microbiol 1986; 51:873–884.
70. Saitou N, Nei M. The neighbor-joining method: a new method for reconstructing phylogenetic trees. Mol Biol Evol 1987; 4:406–425.

71. Camin JH, Sokal RR. A method for deducing branching sequences in phylogeny. Evolution 1965; 19:311–326.
72. Edwards AWF, Cavalli-Sforza LL. reconstruction of evolutionary trees. In: Heywood VH, McNeill J, eds. Phenetic and Phylogenetic Classification. London: Systematics Association, 1964:67–76.
73. Felsenstein J. Estimation of hominoid phylogeny from a DNA hybridization data set. J Mol Evol 1987; 26:123–131.
74. Felsenstein J. Phylogenies from molecular sequences; inference and reliability. Annu Rev Genet 1988; 22:521–565.
75. Zharkikh A, Li W-H. Estimation of confidence in phylogeny: the complete- and partial-bootstrap technique. Mol Phyl Evol 1995; 4:44–63.
76. Sneath PHA. Analysis and interpretation of sequence data for bacterial systematics: the view of a numerical taxonomist. Syst Appl Microbiol 1987; 12:15–31.
77. Bridge PD, Boddy L, Morris CW. Information resources for pest identification: an overview of computer-aided approaches. In: Hawksworth DL, ed. The Identification and Characterization of Pest Organisms. Wallingford: CAB International, 1994: 153–167.
78. Pitt JI. PENNAME. A Computer Key to Common Penicillium Species. North Ryde, Australia: CSIRO Division of Food Processing, 1991.
79. Sneath, PHA. BASIC program for character separation indices from an identification matrix of percent positive characters. Comput Geosci 1979; 5:349–357.
80. Bridge PD, Kozakiewicz Z, Paterson RRM. PENIMAT: A Computer Assisted Identification Scheme for the Terverticillate *Penicillium* isolates. Mycol Papers 1992; 165.
81. Barnett JA, Payne RW, Yarrow D. Yeast Identification PC Program. Norwich: Barnett, 1990.
82. Beech FW, Davenport RR, Goswell RW, Burnett JK. Two simplified schemes for identifying yeast cultures. In: Gibbs BM, Shapton DA, eds. London: Academic Press, 1968:151–175.
83. Ainsworth GC. A method for characterizing smut fungi exemplified by some British species. Trans Br Mycol Soc 1941; 25:141–147.
84. Willcox WR, Lapage SP, Holmes B. A review of numerical methods in bacterial identification. Antonie van Leeuwenhoek 1980; 46:233–299.
85. Willcox WR, Lapage SP, Bascomb S, Curtis MA. Identification of bacteria by computer: theory and programming. J Gen Microbiol 1973; 77:317–330.
86. Williams ST, Goodfellow M, Wellington EMH, et al. A probability matrix for identification of some streptomycetes. J Gen Microbiol 1983; 129:1815–1830.
87. Sneath PHA. BASIC program for identification of an unknown with presence-absence data against an identification matrix of percent positive characters. Comput Geosci 1979; 5:195–213.
88. Priest FG, Alexander B. A frequency matrix for probabilistic identification of some bacilli. J Gen Microbiol 1988; 134:3011–3018.
89. Boddy L, Morris CW, Wimpenny JWT. Introduction to neural networks. Binary Comput Microbiol 1990; 2:179–185.
90. Morris CW, Boddy L, Allman R. Identification of basidiomycete spores by neural network analysis of flow cytometry data. Mycol Res 1992; 96:697–701.

3

Use of PCR and RFLP in Fungal Systematics

Véronique Edel
Institut National de la Recherche Agronomique, Dijon, France

I. INTRODUCTION

The development of molecular nucleic acid techniques has revolutionized fungal systematics. As with bacterial taxonomy, fungal classification has traditionally been based on observations of various morphological features and physiological properties. In many cases, these traits are variable, and they generally yield insufficient information for comparative taxonomic studies. DNA techniques now permit the analysis of genetic markers to establish the identity of individuals, as well as taxonomic and phylogenetic relationships between individuals.

Direct analysis of DNA polymorphisms is now a general approach to identify and compare fungi at the intraspecific, species, genus, or higher level. As with other organisms, DNA-DNA hybridization (1) and sequencing of ribosomal RNA (2,3) were among the first molecular methods employed for the classification of fungi. Simple molecular tools such as restriction enzymes, gel electrophoresis, and Southern hybridization (4) have led to the development of restriction fragment length polymorphism (RFLP) analysis. This procedure involves the comparison of restriction patterns of genomic, ribosomal, or mitochondrial DNA by gel electrophoresis or after hybridization with a DNA probe. Depending on the level of discrimination required, different kinds of probes can be used: ribosomal or mitochondrial DNA, randomly cloned DNA fragments, repeated sequences, and minisatellite probes. RFLP analyses have already been used extensively to type fungi and to investigate their evolutionary relationships.

More recently, the emergence of the polymerase chain reaction (PCR) (5–7) has led to the development of new approaches in molecular systematics. This powerful technique allows exponential amplification of specific DNA sequences by in vitro DNA synthesis, and it provides a simple and rapid way to generate microgram quantities of target DNA. Using PCR, it is now possible to proceed to RFLP analysis without any need for Southern hybridization since the PCR amplification product can be used directly as the substrate for restriction enzyme analysis. Thus, direct sequencing and restriction analysis of amplified DNA are two new approaches which have become widely used in fungal taxonomy. In classical RFLP analysis with hybridization, the specificity of the analysis is given by the probe, whereas in PCR-based methods, the specificity depends on the oligonucleotide primers flanking the region of DNA to be amplified. In this way, the technique used can change while the target DNA sequences chosen could still be the same, defined either by the primers in one method, or by the probe in the other one.

Regions most commonly used for taxonomic and phylogenetic studies are ribosomal DNA regions because many sequence data are now available and because they contain both variable and conserved domains, allowing discrimination at the genus, species, or subspecies level. Universal primer sequences are now available and provide access to different ribosomal DNA regions for all the fungi.

Other PCR-based methods can also be used for discrimination among fungi, particularly at the intraspecific level. DNA amplification by PCR with primers specific for repeated elements, arbitrarily chosen or defined, is another strategy, which can be called interrepeat-PCR or PCR fingerprinting (8). The analysis of random amplified polymorphic DNA (RAPD) (9) has been proposed to resolve genetic variations between fungal strains.

Molecular methods for fungal systematics have been reviewed previously (10–12). This chapter will discuss the use of PCR and RFLP in fungal systematics, with an emphasis on current methods. Because they are now the most currently used, PCR-based methods will be considered first. There are two particularly important approaches available for the use of PCR in systematic studies: analysis after PCR amplification of a defined gene or DNA region, where target DNA sequences are often ribosomal DNA sequences; and PCR fingerprinting, which is based on PCR amplification of multiple loci. Despite the emergence of these PCR-based methods, conventional RFLP analysis can offer other possibilities in fungal systematics and remains one molecular technique in use for fungal identification and classification. The origin, principle, and applications in systematics of each method will be described and illustrated by some recent examples of use for various fungi. From a practical point of view, wherever possible, some references for information or protocols which can be considered as universal for the fungi will be given. Finally, advantages and disadvantages of the methods will be discussed, as well as their taxonomic level of discrimination.

II. PRINCIPLES OF PCR AMPLIFICATION AND PRODUCT ANALYSIS

The PCR is a relatively new molecular procedure, first described in 1986 by Mullis et al. (5,6). This method is based on the use of a thermostable DNA polymerase isolated from *Thermus aquaticus*, called *Taq* polymerase (7). PCR amplification requires a DNA template containing the region to be amplified and two oligonucleotides homologous to the DNA at the ends of the region to be amplified. The procedure consists of a repetition of the three following steps: high temperature denaturation of the double-stranded DNA template; annealing of the oligonucleotide primers to the two single-stranded templates; and an extension step in which new DNA strands are synthesized from the primers. These steps are repeated in 30 to 40 successive cycles, each one doubling the amount of target DNA in the reaction. The PCR reaction produces sufficient DNA to be directly sequenced or visualized in agarose gels following staining with ethidium bromide. White et al. (13) reviewed the protocol for fungal DNA amplification by PCR; they described a standard procedure with all the information required for reaction conditions and cycling parameters.

Numerous protocols for extraction of fungal DNA are available (14,15), including some plant DNA extraction protocols that are also useful for fungi (16). Since the DNA template for the PCR need not be very concentrated or very pure, simplified and rapid DNA extraction procedures are often adequate (17,18). The extraction of fungal DNA directly from soil for subsequent PCR amplification has also been described (19). Various ribosomal DNA (rDNA) sequences are useful in fungal systematics (13); the choice of the target region for PCR amplification will be detailed further on.

The most precise method for detecting polymorphisms between PCR products is the determination and the comparison of their nucleotide sequences. PCR products were initially sequenced after cloning (20), but misincorporation of bases, which can occur during PCR amplification (21), is not detected in the sequencing of a single clone. Even if the fidelity of the PCR can be improved by changing reaction conditions and DNA polymerase (22), incorporation errors can still occur at a low frequency. To avoid these problems and also to save time, PCR products can be sequenced directly without cloning. Several strategies have been developed for the direct sequencing of PCR-amplified DNA (23–25), and these have been reviewed recently (26,27).

Should sequencing be not required or not feasible, other approaches are available for the analysis of PCR products, and these could be more suitable—for example, in large-scale characterization studies. The PCR products obtained from different strains can be digested with restriction endonucleases and the profiles compared. A restriction endonuclease recognizes and cleaves at a specific site, generally 4 or 6 nucleotides in length, in double-stranded DNA. This restriction analysis of PCR products, also called PCR/RFLP analysis, is generally performed with several different enzymes and thus allows the comparative study of several polymorphic restriction sites. From RFLP data, it is then possible to estimate the genetic divergence between two organisms (28), either from the proportion of common restriction fragments or from the changes in restriction sites if the map location of the sites can be inferred from the different restriction patterns. A disadvantage of the first possibility over the second is that the simple comparison of common fragments can lead to misinterpretations since comigrating fragments of the same size do not necessarily correspond to the same DNA sequence. This can be confirmed by excision of a DNA band from the gel and hybridization with other fragments or subsequent digestion with different restriction endonucleases. To avoid such verifications, the strategy commonly adopted is to "dilute" these potential errors by using several restriction enzymes, enabling comparison of numerous restriction fragments. The methods used to estimate genetic relationships between strains from PCR/RFLP data and to construct trees will be described in the next section.

There are many restriction endonucleases available. Those chosen should have enough sites in the target DNA to yield profiles of sufficient complexity to reveal polymorphisms among the taxa considered. Strains can be compared and contrasted through similarities and differences in their restriction profiles due to differences in the number and location of the enzyme recognition sites. Enzymes can be chosen empirically or, if the sequence of the target area is known, by computer analysis using software that locates recognition sites. Furthermore, if the sequence is known for several strains, it is then possible to compare their restriction maps and to select enzymes with polymorphic sites directly without prior testing. One strategy in taxonomic studies of a large collection can be to sequence the PCR products for one or a few strains to determine the most suitable enzymes, and then to characterize other individuals by PCR/RFLP.

The restriction fragments can be separated by electrophoresis in agarose or acrylamide gels, using simple and inexpensive equipment and well-established protocols (29). Examples of electrophoretic patterns are shown in Figure 1. Other approaches for studying large samples have been described, such as the use of capillary electrophoresis for the automation of PCR product analysis (30). In denaturing gradient gel electrophoresis (DGGE), DNA fragments of the same length but with different nucleotide sequences can be separated (31). The DNA fragments migrate until the denaturing conditions induce DNA melting. Sequence variation results in a difference in the melting temperature of the DNA fragments and leads to different migration distances. DGGE can be used for direct analysis of PCR-amplified DNA and is especially useful for the detection of mutations. Furthermore, a G+C-rich sequence, called a GC clamp, can be incorporated into one of the PCR primers to improve the detection of single-base differences (32). Another electrophoretic method, in which amplified products are subjected to

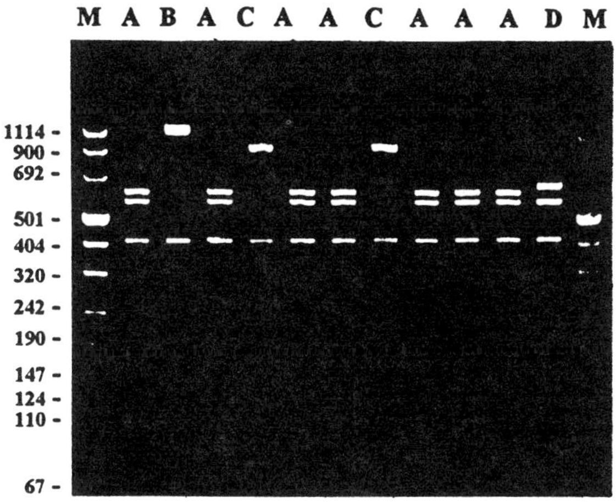

Figure 1 Examples of electrophoretic patterns obtained by restriction analysis of PCR-amplified ribosomal DNA. A DNA fragment of 1700 bp was amplified and digested with the restriction enzyme *Rsa*I. Similar restriction patterns were grouped and assigned the same letter (A through D). Lane M, molecular weight marker. (From 65).

single-strand conformation polymorphism (SSCP) analysis, can be used to detect sequence variations as small as single-base substitutions (33). In contrast to the DGGE technique, for SSCP analysis the electrophoresis is performed in a neutral acrylamide gel, but the DNA is denaturated before migration. Sequence variations in single-stranded DNA, which result in different conformations of the renaturated strands, are detected by changes in mobility. Few papers describe the use of DGGE or SSCP analysis for fungal studies. As an example, Simon et al. (34) combined PCR with SSCP analysis to characterize endomycorrhizal fungi.

III. FROM PCR/RFLP DATA TO TREES

Electrophoretic patterns generated by PCR/RFLP analysis can be compared to quantify the variations between strains, to estimate their relationships, and to represent them in a dendrogram or tree. Two different approaches can be used: a phenetic analysis, or a cladistic analysis. In a phenetic analysis, a dendrogram is constructed on the basis of phenotypic similarities without considering evolutionary pathways. In a cladistic approach, the data are analyzed by considering the evolutionary history, and a phylogenetic tree is constructed (28).

A phenetic analysis yields results in the form of pairwise similarities. The degree of relationships between two DNA sequences analyzed by PCR/RFLP is correlated with their proportion of shared restriction fragments. For each restriction enzyme, comigrating bands observed among restriction patterns are considered common restriction fragments. Comparison of strains can be made by statistical analyses commonly used for phenotypic characters. For each strain, presence or absence of each character (here, each restriction fragment) is coded by 1 or 0, respectively (Table 1), and these data are used to calculate a similarity coefficient for each pairwise comparison, generating a similarity matrix. The simple matching coefficient and the Jaccard coefficient (35) are examples of similarity coefficients frequently used. Similarity is often expressed as a function of the fraction F of common fragments: $F = 2N_{xy}/(N_x + N_y)$, where N_{xy} is the number of fragments common to strain x and strain y considering all enzymes, and N_x and N_y are the total number of fragments in strain x and strain y, respectively. This measure of similarity can be converted to a value of genetic distance (D) between pairs of strains. As an example, the mathematical model defined by Nei and Li (36) is frequently used to estimate D from RFLP data. D, also called sequence divergence, is interpreted as the number of nucleotide substitutions per site. Distance values obtained for all pairs of strains are presented in a pairwise distance matrix.

The similarity matrix or distance matrix is displayed as a dendrogram representing the relationships among the strains. Several algorithms are available to construct dendrograms from matrix, such as the unweighted-pair group method with arithmetic mean (UPGMA) (35), the neighbor-joining method (37), and the Fitch-Margoliash method (38). Examples of computer programs that can be used to calculate similarity coefficients or distances and to construct dendrograms are: NTSYS-pc (39), RESTSITE (40), and PHYLIP (41).

Estimation of genetic distances from proportions of common restriction fragments can lead to errors because comigrating fragments can correspond to different DNA sequences and because small fragments are often undetected in electrophoretic patterns. Moreover, estimation of similarities from restriction fragments implies the absence of length differences between the PCR products. Otherwise, a mapping approach can be adopted to analyze PCR/RFLP data. The

Table 1 Data Matrix Showing Presence (1) or Absence (0) of Restriction Fragments in Restriction Patterns Presented in Figure 1

	Restriction fragments in base pairs							
Pattern	1200	900	650	610	560	400	275	90
A	0	0	0	1	1	1	0	1
B	1	0	0	0	0	1	0	1
C	0	1	0	0	0	1	1	1
D	0	0	1	0	1	1	0	1

map of restriction sites is deduced from the restriction patterns, and instead of the polymorphism restriction fragments, the polymorphism of restriction sites is analyzed. Though time-consuming, mapping is more precise because it explains the physical relationships of restriction fragments. Several methods can be used to map the sites. When the nucleotide sequence of the PCR product is known for a few strains, the map location of restriction sites found for all the strains can be inferred from known sequences and patterns (Fig. 2). Other strategies can be used: double digestions, partial digestions, and PCR mapping (42). In this latter, different DNA fragments successively larger but with a common end are amplified by PCR and digested with restriction enzymes. Presence or absence of restriction sites is coded as 1 or 0, respectively (Table 2). Data matrix and dendrograms can be constructed from similarities in restriction sites with the same methods and programs as those described above for restriction fragment analysis.

Finally, a phylogenetic approach can be adopted. Some authors recommend against using restriction fragment data for phylogenetic analysis (43). Thus, restriction sites need to be mapped. But even with restriction site data, special precautions are required in phylogenetic studies because of the asymmetry in the probabilities of gaining and losing restriction sites (43). Phylogenetic analysis are frequently based on the principle of parsimony, which is to accept the shortest tree

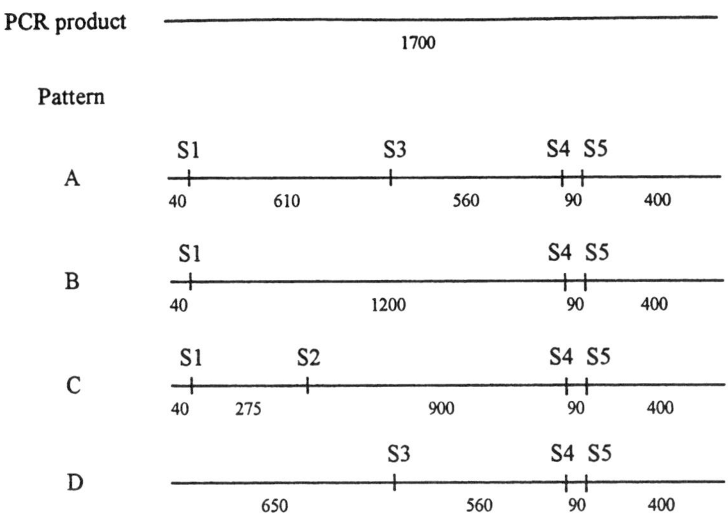

Figure 2 Map location of restriction sites in the PCR product inferred from restriction patterns shown in Figure 1 and from the known nucleotide sequence of one strain presenting pattern A. S1 to S5 correspond to the restriction sites. The sizes (in base pairs) of the restriction fragments are given below the lines.

Table 2 Data Matrix Showing Presence (1) or Absence (0) of Restriction Sites S1 to S5 Defined in Figure 2.

	Restriction sites				
Pattern	S1	S2	S3	S4	S5
A	1	0	1	1	1
B	1	0	0	1	1
C	1	1	0	1	1
D	0	0	1	1	1

constructed from the data as the best estimate of the evolutionary tree. Phylogenetic trees can be constructed with the computer programs PAUP (phylogenetic analysis using parsimony) (44) or PHYLIP (41).

IV. ANALYSIS OF RIBOSOMAL DNA BY PCR

Ribosomal DNA sequences are often chosen for taxonomic and phylogenetic studies because this DNA region is found universally in living cells and corresponds to an important function in the cell, and so its evolution might reflect the evolution of the whole organism. This region contains some highly conserved sequences but also some variable sequences, allowing the comparison of organisms at different taxonomic levels.

In fungi and other eukaryotes, there are two locations for rDNA: the nuclear genome, and the mitochondrial genome. The latter contains two genes coding for rRNA—the small and the large mitochondrial rRNA genes. Nuclear rDNA in fungi is generally organized in a nuclear unit, which is tandemly repeated. A rDNA unit, illustrated in Figure 3, includes three rRNA genes: the small nuclear (18S-like) rRNA, the 5.8S rRNA, and the large nuclear (28S-like) rRNA genes. In one unit, the genes are separated by two internal transcribed spacers (ITS1 and ITS2), and two rDNA units are separated by the intergenic spacer (IGS). The last rRNA gene (5S) may or may not be within the repeated unit. Numerous sequence data are now available and allow the determination of primer sequences for the PCR amplification of different parts of the nuclear and mitochondrial rDNAs (13). These primers, considered as universal primers for the fungal kingdom, are located in conserved regions, allowing the amplification of the fragment they flank in most fungi.

Among rRNA genes, the smallest one (5S) was first used in taxonomic studies. It is particularly useful for studying relationships between distantly re-

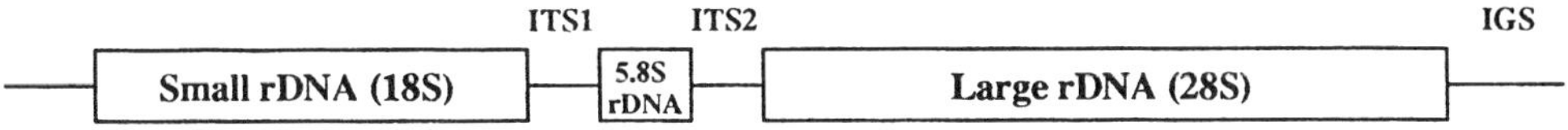

Figure 3 Diagram of a nuclear ribosomal DNA repeat unit.

lated organisms. As an example, Hori and Osawa (2) aligned and compared several 5S rRNA sequences to construct a phylogenetic tree which included most of the major groups of organisms. However, the 5S rRNA is too short and too conserved to be used for studying closely related organisms.

The 18S-like rRNA sequences are particularly useful for resolving taxonomic and evolutionary questions at the level of orders or higher. 18S rDNA were used to determine the phylogenetic relationships within the fungal class of Chytridiomycetes (45,46) and among the three major classes—Basidiomycetes, Ascomycetes, and Chytridiomycetes (47). These studies clarified the taxonomic position of the Chytridiomycetes. The 28S-like rRNA gene contains both slowly and rapidly evolving domains. These latter, termed domains D1 and D2, are highly variable within a fungal genus. For example, Guadet et al. (48) evaluated the divergence between eight species of *Fusarium* by comparing their D1 and D2 sequences.

Before the emergence of the PCR, many of these phylogenetic studies were performed by RNA extraction and sequencing. More recently, phylogenetic analysis based on sequence comparisons of PCR-amplified rDNA regions have been used in preference. For example, Cubeta et al. (49) characterized the anastomosis groups of binucleate *Rhizoctonia* species by restriction analysis of an amplified portion of the 28S-like rDNA. Similarly, the phylogenetic relationships of the basidiomycete *Lentinus* were inferred from sequence comparison of PCR-amplified large nuclear rRNA gene (50), and most of the sequence variation was observed in the regions that correspond to eukaryote-specific divergent domains D1 and D2 at the 5′ end of 28S rDNA. These domains were also helpful for taxonomic studies at the species level within the genus *Colletotrichum* (51) or *Gliocladium* (52).

With the exception of some variable domains of the rRNA genes, the coding regions are highly conserved among organisms, thus allowing comparisons between distantly related fungi. In contrast, because they evolve rapidly, noncoding regions have more variability than coding regions. The noncoding internal transcribed spacers (ITS1 and ITS2) can be used to discriminate between closely related species within a fungal genus. The ITS region, including the ITS1, the 5.8S rRNA gene, and the ITS2, is about 600 to 1000 base pairs and can be amplified either fully or partly, using "universal" primers described by White et al. (13). For example, sequencing of PCR-amplified ITSs has been used to infer taxonomic and phylogenetic relationships among rust species (53). *Phytophthora* species (54), *Colletotrichum* species (55), *Sclerotinia* species (56), and *Penicillium* species (57). Instead of sequencing, restriction analysis of PCR-amplified ITSs can be used to investigate interspecific variability. A useful strategy is to sequence the DNA portion in a few species belonging to the same genus, and to use these sequence data to develop a PCR/RFLP method. This approach is suitable when large numbers of isolates are needed. Vilgalys et al. (42) described a rapid identification method of *Cryptococcus* species by restriction typing of PCR-amplified ITS and other rDNA regions. Bernier et al. (58) used PCR to amplify and compare ITS and 18S-like rDNA in *Gremmeniella* isolates to investigate interspecific variability within this genus and to reassess its taxonomic position. RFLP analysis of PCR-amplified ITSs allowed the discrimination of *Tuber* species (59) and the identification of species within the *Gaeumannomyces-Phialophora* complex (60). The method was also used for the molecular comparison of *Pythium* species (61,62), and revealed that morphologically related *Pythium* species were genetically distinct (62). Moreover, since the primers designed by White et al. (13) have been shown to enable the amplification of ITSs from many genera of fungi, the RFLP analysis of PCR-amplified ITS region will certainly be applied to the identification of species in other fungal genera. The ITS region is therefore a powerful taxonomic indicator at the species level. On the other hand, variability in the ITS sequences is generally very low or undetected at the intraspecific level (52,54), with the exception of the species *Fusarium sambucinum*, in which a high level of divergence was reported for ITS sequences (63).

Nontranscribed sequences such as the intergenic spacer (IGS) that separates rDNA repeat units have a lower degree of conservation than the ITS and can be useful for intraspecific characterization. Restriction analysis of PCR-amplified IGS DNA has allowed discrimination of closely related fungi, including *Laccaria* (64) and *Fusarium oxysporum* (65) strains. However, this intergenic region may also resolve genetic variations at the interspecific level. For example, PCR amplification and direct sequencing of IGS regions were used to infer phylogenetic relationships among *Armillaria* species (66).

In addition, length mutations are common in the internal and intergenic spacers; these differences in length may also be used as molecular criteria in taxonomic studies (64,67). On the other hand, a PCR/RFLP analysis can be biased by the presence of deletions or insertions, and these length variations must be identified in the PCR products before starting the restriction analysis; otherwise they may introduce misinterpretations in the restriction site polymorphism analysis.

Consensus primers were also proposed for the PCR amplification of mitochondrial small rDNA (mtSrDNA) and large rDNA (mtLrDNA) (13). These primers were tested with many fungal genera and found to amplify mitochondrial DNA correctly. Bruns et al. (68) used them as early as 1990 to amplify and sequence mtDNA from dried fungal herbarium specimens. Lobuglio et al. (57) used the PCR amplification and sequencing of mtLrDNA in addition to the ITS region for their phylogenetic analysis within the genus *Penicillium*, and separate analyses of both ribosomal DNA regions gave topologically similar phylogenetic trees. Combined with the ITS, the mtLrDNA was also useful in separating three heterothallic *Pythium* species (61).

Until now, most taxonomic studies have been based on PCR amplification and analysis of rRNA genes or rDNA spacer, but other strategies based on the amplification of a particular gene can be developed. For example, primer sets were designed to amplify conserved genes such as histones and β-tubulin from filamentous ascomycetes (69), and these DNA regions were used in PCR/RFLP analysis in addition to ITSs for differentiating *Fusarium* species (70). Mehman et al. (71) described the potential use of chitin synthase genes in taxonomy of ectomycorrhizal fungi. Because of the presence of variable introns in these genes, their PCR amplification generated polymorphisms in both the number and length of DNA fragments among the strains. Resulting DNA patterns permitted the identification of basidiomycetous ectomycorrhizal species. Similarly, PCR amplification of a particular gene could be helpful for the discrimination of pathotypes within a pathogenic species. As an example, PCR amplification of a gene involved in the production of a host-specific toxin was useful for differentiating races of *Cochliobolus carbonum* (72). This strategy could be applied to other studies at the pathotype level, but it supposes the previous characterization of a particular gene.

V. RAPD AND OTHER PCR FINGERPRINTING

Characteristic fingerprints can be generated by amplifying genomic DNA at low stringency with short (5 to 10 bp) primers (9). This method is based on PCR amplification of random DNA fragments using a single primer with an arbitrary nucleotide sequence. PCR products are separated electrophoretically in agarose or acrylamide gel. The complexity of the profiles will vary according to the number and distribution of sites in the genome to which the primers hybridize. This technique has variously been called random amplified polymorphic DNA (RAPD) (9), arbitrarily primed polymerase chain reaction (AP PCR) (73), DNA amplification fingerprinting (DAF) (74), or amplification fragment length polymorphism (AFLP). RAPD fingerprints can be used to establish identity at various taxonomic levels (75), including species (76), according to the primers chosen. Primers for intra- and interspecies polymorphism analyses have to be chosen empirically and tested experimentally.

In fungi, most RAPD analyses concern intraspecific studies. The RAPD method has been successfully used to differentiate isolates within fungal species belonging to different genera such as *Aspergillus* (77), *Colletotrichum* (78), *Fusarium* (79,80), and *Rhizoctonia* (81), or arbuscular-mycorrhizal fungi (82).

Several authors tried to correlate RAPD characters with other properties, including host specificity of pathogenic fungi. For example, in *Fusarium oxysporum*, RAPD markers were used to assess genetic relationships between races of the formae speciales *vasinfectum* (79), *pisi* (83), or *dianthi* (84). Although the RAPD method established DNA fingerprints useful for race characterization in *F. oxysporum* f. sp. *vasinfectum*, generalized race-specific patterns were not found in *F. oxysporum* f. sp. *pisi* or *dianthi*. Similarly, a low association was found between virulence and RAPD patterns in *Puccinia striiformis* (85). However, RAPD markers allowed the identification of a pathotype within *Pseudocercosporella herpotrichoides* (86).

While RAPD fingerprints are generated by arbitrarily chosen primers, defined PCR primers specific for repeated DNA motifs can also be used for PCR fingerprinting (8,87). This strategy is also called interrepeat-PCR. Tandemly repeated elements, such as microsatellites (2 to 10 bp) or minisatellites (15 to 30 bp) can be amplified using specific primers (88). The number and size of amplified fragments will vary according to the number and distribution of the repeated elements. Electrophoretic separation of the PCR products will yield characteristic patterns. Meyer et al. (88) used this technique to characterize isolates of *Cryptococcus neoformans* and closely related species, and suggested that this technique has a potential application in epidemiological studies. Intraspecific genetic variation in *Heterobasidion annosum* was revealed by amplification of minisatellite DNA (89), and the comparison of the PCR products allowed intersterility group classification in this species. Interrepeat-PCR with primers directed against various repetitive sequences was described for the typing of *Aspergillus* strains (90). PCR with eukaryotic repeat motif primers allowed discrimination of *Aspergillus* species, but no intraspecific polymorphism was observed. On the other hand, prokaryotic repeat motif primers derived from enterobacterial repetitive intergenic consensus (ERIC) elements permitted the individual typing of *Aspergillus fumigatus* isolates. Discrimination of closely related strains of *Fusarium oxysporum* was also achieved by ERIC-PCR fingerprinting (65). Other fingerprints were produced by PCR amplification using consensus tRNA gene primers with the object of identifying bacterial, plant, or animal species (91), and this may also be useful for revealing variations among fungal genomes.

Some limitations of these PCR fingerprinting methods must be taken into consideration. In some cases, the complex banding patterns generated can be difficult to analyze to quantify variations. In addition to the major PCR products, there are often some minor weak bands on gels, because of less specific or nonspecific amplifications, or because of variations in DNA conformation. These minor products can also explain the problems of reproducibility that can occur. These PCR fingerprinting procedures must be well standardized because they are more sensitive to reaction conditions than conventional PCR amplifications. The brand of *Taq* DNA polymerase and the thermal cycler employed are factors that can affect the reproducibility of RAPD (92). The technique is particularly sensitive to Mg^{2+} concentration and the number and stringency of temperature cycles. Effects of primer and template DNA concentrations have also been investigated and shown to be important (93). Finally, another limitation of PCR fingerprinting is that two distinct DNA fragments that comigrate on gels because of similar size could be considered identical fragments. To resolve this kind of problem, a PCR fragment can be eluted from the gel and used as an hybridization probe in Southern analysis against the other products. However, these PCR-based DNA fingerprinting methods have several advantages and give a pattern of multiple bands allowing analysis of multiple loci, without any need for target sequence information. These rapid and simple tools are useful for resolving genetic variations among closely related organisms, generally at the species or intraspecific level.

VI. USE OF PCR FOR MONITORING FUNGI FROM THE NATURAL ENVIRONMENT

The development of specific oligonucleotide primers for selective PCR amplification of a given taxon provides powerful tools for the monitoring of microorganisms in their natural environment, such as water, soil, plant, or clinical materials, without the need to isolate them by cultivation. Fungal-specific primers are especially useful to characterize obligate parasites and symbionts, allowing their monitoring from their host without contamination by host DNA. For example, specific amplification of mycorrhizal fungal DNA was obtained from colonized root (94–96). Similarly, Gardes and Bruns (67) have described specific primer sets for the identification of mycorrhizae and rusts in plant tissues. The first step consists of recovering DNA from environmental samples. DNA can be easily extracted from plant material such as leaves or roots (16,17). Bonito et al. (96) were able to amplify arbuscular mycorrhizal fungal DNA in roots without DNA extraction by simply boiling root tissues in a buffer. DNA extraction from complex medium such as soils is more difficult to perform because PCR can be inhibited by impurities in soil-derived DNA samples (97). Different protocols for direct extraction of DNA from soil, which produced DNA capable of being amplified, have been described (19,97–99).

PCR assays for examining fungi in natural substrates require specific primers. Any DNA sequence can be used for the selective amplification of an organism by PCR provided that differences in sequences enable the designation of specific primers for this particular organism. As RAPD analyses do not need any sequence information, it is an attractive method for identifying fungi from their specific RAPD patterns. However, it supposes the previous isolation of fungi from their environmental matrix. As random primers are not specific, they can generally not be used for the detection of an organism in natural samples. Whatever the DNA region used, specific primers are defined from sequence alignments and comparisons. Ribosomal DNA genes and spacers are ideal targets for specific PCR amplification because of the growing database and because they are highly repeated in the fungal genome. Thus, choosing rDNA as target can increase the sensitivity for amplifying fungi that are few in number in natural samples. The use of primers based on sequence differences in the ITS region is a common approach for species-specific amplification. This strategy has been used to design primer sets that specifically amplify *Verticillium dahliae*, *V. albo-atrum*, or *V. tricorpus* (100,101). Species-specific primers have also been defined for other phytopathogenic fungi such as *Phytophthora* spp. (102), *Septoria* spp. (103), and *Ophiosphaerella* spp. (104), allowing their DNA amplification from infected plant tissues. Ribosomal DNA has also been selected as target DNA for detection of *Candida* spp. (105) and *Aspergillus* spp. (106) in clinical samples. At the intraspecific level, some variety-specific or pathotype-specific primers were also defined on the basis of differences in ITS sequences, and were used to differentiate pathogenic isolates in plant material (107,108).

Amplified rDNA from environmental samples can be digested with restriction enzymes to characterize the diversity of microbial communities in these samples (98). Johnston and Aust (109) have compared *Phanerochaete* species directly in soil by PCR/RFLP analysis of ITS regions after DNA extraction from soil. Thus, taxon-selective primers can enable the identification and characterization of fungi in natural samples, and have several applications in environmental microbiology for ecological studies of natural populations, quantitation of pathogens, and indicator populations (95,110); disease diagnosis (111); detection of specific genes, and measure of gene expression (112,113).

VII. CONVENTIONAL RFLP

RFLP analysis can be performed either without hybridization or after hybridization with a probe. In the first case, it consists of a digestion of the DNA with a restriction enzyme, a separation of the resulting DNA fragments according to their size by gel electrophoresis, and a comparison of the restriction patterns obtained. In the second case, this procedure is followed by Southern blotting and hybridization (4). The DNA is transferred from the gel to a solid support, fixed on it, and hybridized with a labeled DNA or RNA probe. After visualization, only the DNA fragments that are complementary to the probe will be detected. This procedure corresponds to conventional RFLP. After digestion of the DNA with a restriction enzyme, several probes can be successively tested. Similarly, different restriction enzymes can be used to digest the DNA before hybridization. In this way, numerous combinations of restriction enzymes and probes can be tested until some are found that enable discrimination among the genera, species, or isolates considered. The restriction fragments resulting from all the combinations used are compared for two organisms, and the similarity between them is deduced either from their proportion of shared restriction fragments or from their proportion of common restriction sites.

Restriction analysis can be performed either with total genomic DNA, or with one of its components: nuclear DNA, ribosomal DNA (rDNA), or mitochondrial DNA (mtDNA). In any case, the procedure requires quite pure DNA samples, because they have to be accessible to, and digested by, restriction endonucleases in the first step of the protocol. Moreover, there is generally a need for high-molecular-weight DNA.

Since both quality and quantity of DNA are important criteria, small-scale or rapid preparations of DNA described for further PCR amplification (17,18) are generally not suitable for RFLP analysis. However, numerous extraction protocols of fungal DNA usable in RFLP analysis have been described (14–16). These protocols are generally usable for most fungi. Moreover, some protocols have been published for particular circumstances, such as methods for isolating DNA from endomycorrhizal fungi (114), which cannot be grown easily.

Different methods are also available for the preparation of mtDNA. As an example, Garber and Yoder (115) described a procedure for the extraction of total DNA from filamentous fungi and its separation into nuclear, mitochondrial, ribosomal, and plasmid components by cesium chloride/bisbenzimide density gradient centrifugation. Another possibility is to purify mtDNA after isolation of mitochondria (116).

RFLP analysis without hybridization has been described for molecular studies in fungi. In this case, the template DNA used is generally mitochondrial DNA. Restriction digestion of genomic DNA generates patterns with so many bands that they are difficult to directly analyze without hybridization. However, comparison of total DNA restriction patterns was described to differentiate between fungal strains at the intraspecific (117,118) or interspecific level (119). In fact, patterns of digested DNA consist of a smear, but with some distinct bands corresponding to mtDNA or nuclear rDNA since they are repeated in the fungal genome. With restriction patterns of total fungal DNA, the analysis is only based on the comparison of these distinct DNA (117,119) and does not consider the whole genome. As another example, Bridge et al. (118) undertook restriction analysis of total DNA of *Fusarium oxysporum* strains with an enzyme that cleaves the nuclear DNA into small fragments but leaves larger bands, interpreted as mtDNA, clearly visible against the nuclear DNA background.

Mitochondrial DNA has been widely used to study biological variations in fungi. Characteristics of the fungal mitochondrial genome have been reviewed recently (120). The mtDNA corresponds to a circular molecule ranging in size from 17.6 kb to 175 kb (120). Its small size makes it suitable for restriction enzyme analysis, and mtDNA has been used to show interspecific and intraspecific variability in fungi. However, because the mtDNA evolves rapidly, it is especially useful for analyzing populations at the intraspecific level.

Direct restriction analysis of purified mtDNA has been used to reveal variations among fungi. Stammler et al. (121) used differences in mtDNA restriction profiles combined with other markers to characterize *Phytophthora fragariae* and to investigate its taxonomic position, and their results confirmed the grouping of two fungi within the species and their separation at the variety level. Förster and Coffey (122,123) also based their molecular characterization of *Phytophthora* isolates on mtDNA polymorphism analysis, and described a technique to prescreen numerous isolates. After digestion with a restriction enzyme, they identified polymorphisms in mtDNA from total DNA, since mtDNA bands were distinguishable on gels against the nuclear background. This allowed them to compare the mtDNA patterns from large numbers of isolates, and they were able to group the isolates. They confirmed the results by using purified mtDNA from one isolate of each group instead of each isolate. However, comparison among the groups was performed by hybridization analysis of mtDNA patterns with various cloned mtDNA fragments, since a direct comparison of mtDNA patterns was not possible because of the high degree of mtDNA diversity among the isolates.

Thus, variability of mtDNA is commonly revealed by RFLP analysis after hybridization of total DNA with mtDNA probes. For example, RFLPs in mtDNA were useful for studying relationships between populations of *Fusarium oxysporum* and to detect genetic exchange among populations. This fungal species includes both nonpathogenic and phytopathogenic strains, the latter being divided in formae speciales and races on the basis of their host specificity. The genetic relatedness of strains among different formae speciales causing wilt in cucurbits (124) or in crucifers (125) was investigated by RFLP analysis of mtDNA. Whereas pathotypes of crucifers had specific mtDNA RFLPs, such a correlation did not exist for pathotypes of cucurbits. All the *F. oxysporum* formae speciales pathogenic on cucurbits were found to be closely related, suggesting that genetic exchange still occurred between them. In *F. oxysporum*, variations in mtDNA have also been used to detect genetically isolated populations within a particular forma specialis (126).

Variations in rDNA have also been assessed by RFLP analysis after hybridization with an rDNA probe in various fungi (125,127,128). This strategy has been used at different taxonomic levels, but especially before the emergence of the PCR. Today, RFLPs in rDNA are generally shown after PCR amplification as described above. However, it must be noted that different levels of discrimination can be obtained, depending on whether a DNA region is analyzed by PCR/RFLP or by conventional RFLP. Indeed, PCR/RFLP analysis of a DNA fragment reveals polymorphism only within this fragment, whereas the same fragment used as hybridization probe will detect polymorphisms within the fragment and also in the adjacent DNA sequences. In this way, a rDNA gene analyzed by PCR/RFLP could reveal few polymorphisms, whereas its use as probe for conventional RFLP could be more discriminating because of sequence variations in adjacent spacers.

Another approach in RFLP analysis is the use of randomly cloned DNA fragments as hybridization probes. This strategy has been developed for numerous fungal characterization studies at different taxonomic levels, especially for closely related taxa. The genomic DNA from one fungal strain is digested, and the resulting restriction fragments are cloned. Clones from this library are then tested as hybridization probes against digested DNA from other strains. As an example, Elias et al. (129) produced a genomic library for one isolate of *F. oxysporum* f. sp. *lycopersici* to estimate the genetic diversity within this forma specialis. They tested 50 clones at random as hybridization probes with digested DNA from a large collection of isolates of the same and other formae speciales. This screening allowed them to identify 10 repetitive DNA fragments, among the 50 tested clones, suggesting that 20% of the genome of *F. oxysporum* f. sp. *lycopersici* is composed of repetitive DNA. However, this high proportion of repeated DNA is not a general rule. In *Mycosphaerella fijiensis*, Carlier et al. (130) found that 98% of the tested clones contained single-copy DNA. Single or low copy number-cloned sequences were useful to confirm genetic divergence among *Mycosphaerella* species (130) and *Phaeosphaeria* species (131), and to identify species-specific probes in the genus *Hebeloma* (132). Manicom et al. (133) described a similar approach for taxonomic studies in the genus *Fusarium*. RFLPs were evaluated among different *Fusarium* species after hybridization with random probes generated from one *F. oxysporum* isolate. Even if one interesting probe, generating a multiple-banding pattern with *F. oxysporum* strains, was useful for further investigations at the intraspecific level (134), this strategy did not allow them to find a probe that recognizes the whole species *F. oxysporum*. Such an approach generally requires the screening of many isolates with many probes before detecting a specific probe for a given taxon.

Generally, randomly cloned repetitive genomic sequences are especially useful in showing DNA polymorphism at the intraspecific level, and have been successfully applied to detect genetic differences among strains of *Fusarium oxysporum* (129,134–136), *Phytophthora megasperma* (137), and *Stagonospora nodorum* (138). Similarly, in *Magnaporthe grisea*, a family of dispersed repetitive DNA sequences, called MGR for *M. grisea* repeat (139), was identified and used as a probe to generate RFLPs in *M. grisea* populations (140).

Anonymous repetitive DNA sequences have been widely used to investigate genetic variations within fungal populations, especially at the intraspecific level, since they detect many RFLPs among related strains. Because these repeated DNAs have been isolated at random, there is generally no information about the mechanisms of variation they detect, and about their sequence and their role in the genome. In many cases, their molecular characterization which will provide more information, remains to be done.

RFLP analysis can also be performed with oligonucleotide probes. This DNA-fingerprinting method is based on the detection of hypervariable repetitive sequences such as microsatellites or minisatellites [87,141]. Oligonucleotide probes corresponding to these elements will hybridize to variable number tandem repeat (VNTR) loci in the genome. This method is used to reveal polymorphism among closely related fungal taxa. For example, DNA fingerprints with the oligonucleotide probe $(CAT)_5$ were compared among isolates of *Heterobasidion annosum* [142] and *Colletotrichum orbiculare* [143] and these probes seem to be generally applicable to fungi [87]. The advantage of an oligonucleotide probe over a cloned genomic DNA probe is that the first one has only to be synthesized and can be then used for many organisms, whereas the second one must be isolated for each new taxon considered.

VIII. CONCLUSION

Many questions about taxonomic and phylogenetic relationships among fungi have been answered rapidly since PCR and RFLP applications have been developed. RFLP analyses of Southern blots have been used extensively in the last twenty years for studying fungal systematics. Depending on the taxonomic level required, different probes have been developed for many fungi, such as ribosomal or mitochondrial DNA regions, randomly cloned genomic DNA, and more recently, oligonucleotide probes. Some of these procedures can be replaced or complemented by PCR-based methods. PCR has the advantage of requiring much less DNA and avoiding time-consuming and expensive blotting and probe preparation. Even dried fungal herbarium specimens can be included in a systematic study (68), since molecular data can be obtained easily from small amounts of material. Moreover, the development of specific primers allows the selective amplification of fungal DNA from complex medium or natural samples. As methods for sequencing are continually progressing, automated sequencing of PCR products will probably be the most direct and rapid taxonomic tool of the future.

REFERENCES

1. Werman SD, Springer MS, Britten RJ. Nucleic acids I: DNA-DNA hybridization. In: Hillis DH, Moritz C, eds. Molecular Systematics. Sunderland, Mass.: Sinauer Associates, 1990:204–249.
2. Hori H, Osawa S. Origin and evolution of organisms as deduced from 5S ribosomal RNA sequences. Mol Biol Evol 1987; 4:445–472.
3. Blanz PA, Unseld M. Ribosomal RNA as a taxonomic tool in mycology. Stud Mycol 1987; 30:247–258.
4. Southern EM. Detection of specific sequences among DNA fragments separated by gel electrophoresis. J Mol Biol 1975; 98:503–517.

5. Mullis KB, Faloona FA, Scharf SJ, Saiki RK, Horn GT, Erlich HA. Specific enzymatic amplification of DNA in vitro: the polymerase chain reaction. Cold Spring Harbor Symp Quant Biol 1986: 51:263–273.
6. Mullis KB, Faloona FA. Specific synthesis of DNA in vitro via a polymerase-catalyzed chain reaction. Methods Enzymol. 1987; 155:335–350.
7. Saiki RK, Gelfand DH, Stoffel S, et al. Primer-directed enzymatic amplification of DNA with a thermostable DNA polymerase. Science 1988; 239:487–491.
8. van Belkum A. DNA fingerprinting of medically important microorganisms by use of PCR. Clin Microbiol Rev 1994; 7:174–184.
9. Williams JGK, Kubelik AR, Livak KJ, Rafalski JA, Tingey SV. DNA polymorphisms amplified by arbitrary primers are useful as genetic markers. Nucleic Acids Res 1990; 18:6531–6535.
10. Klich MA, Mullaney EJ. Molecular methods for identification and taxonomy of filamentous fungi. In: Arora DK, Elander RP, Mukerji KG, eds. Handbook of Applied Mycology. Vol. 4, Fungal Biotechnology. New York: Marcel Dekker, 1991: 35–55.
11. Bruns TD, White TJ, Taylor JW. Fungal molecular systematics. Annu Rev Ecol Syst 1991; 22:525–564.
12. Kohn LM. Developing new characters for fungal systematics: an experimental approach for determining the rank of resolution. Mycologia 1992; 84:139–153.
13. White TJ, Bruns T, Lee S, Taylor J. Amplification and direct sequencing of fungal ribosomal RNA genes for phylogenetics. In: Innis MA, Gelfand DH, Sninsky JJ, White TJ, eds. PCR Protocols: A Guide to Methods and Applications. San Diego: Academic Press, 1990:315–322.
14. Lee SB, Milgroom MG, Taylor JW. A rapid, high yield mini-prep method for isolation of total genomic DNA from fungi. Fungal Genet Newslett 1988; 35:23–24.
15. Raeder U, Broda P. Rapid preparation of DNA from filamentous fungi. Lett Appl Microbiol 1985; 1:17–20.
16. Rogers SO, Bendich AJ. Extraction of DNA from plant tissues. Plant Mol Biol Manual 1988; 6:1–10.
17. Lee SB, Taylor JW. Isolation of DNA from fungal mycelia and single spores. In: Innis MA, Gelfand DH, Sninsky JJ, White TJ, eds. San Diego: Academic Press, 1990:282–287.
18. Cenis JL. Rapid extraction of fungal DNA for PCR amplification. Nucleic Acids Res 1992; 20:2380.
19. Porteous LA, Armstrong JL. A simple mini-method to extract DNA directly from soil for use with polymerase chain reaction amplification. Curr Microbiol 1993; 27: 115–118.
20. Scharf SJ, Horn GT, Erlich HA. Direct cloning and sequence analysis of enzymatically amplified genomic sequences. Science 1986; 233:1076–1078.
21. Tindall KR, Kunkel TA. Fidelity of DNA synthesis by the *Thermus aquaticus* DNA polymerase. Biochemistry 1988; 27:6008–6013.
22. Cha RS, Thilly WG. Specificity, efficiency, and fidelity of PCR. PCR Methods Appl 1993; 3:S18–S29.
23. Winship PR. An improved method for directly sequencing PCR amplified material using dimethyl sulphoxide. Nucleic Acids Res 1989; 17:1266.
24. Dowton M, Austin AD. Direct sequencing of double-stranded PCR products without

intermediate fragment purification; digestion with mung bean nuclease. Nucleic Acids Res 1993; 21:3599–3600.
25. Takagi S, Kimura M, Katsuki M. Direct sequencing of PCR products using unlabeled primers. BioTechniques 1993; 14:218–221.
26. Bevan IS, Rapley R, Walker MR. Sequencing of PCR-amplified DNA. PCR Methods Appl 1992; 1:222–228.
27. Rao VB. Direct sequencing of polymerase chain reaction-amplified DNA. Anal Biochem 1994; 216:1–14.
28. Nei M. Molecular Evolutionary Genetics. New York: Columbia University Press, 1987.
29. Sambrook J, Fritsch EF, Maniatis T. Molecular Cloning. Cold Spring Harbor, N.Y.: Cold Spring Harbor Laboratory, 1989.
30. Martin F, Vairelles D, Henrion B. Automated ribosomal DNA fingerprinting by capillary electrophoresis of PCR products. Anal Biochem 1993; 213:30–37.
31. Fisher SG, Lerman LS. DNA fragments differing by single base-pair substitutions are separated in denaturing gradient gels: correspondence with melting theory. Proc Natl Acad Sci USA 1983; 80:1579–1583.
32. Sheffield VC, Cox DR, Lerman LS, Myers RM. Attachment of a 40-base-pair G+C-rich sequence (GC-clamp) to genomic DNA fragments by the polymerase chain reaction results in improved detection of single-base changes. Proc Natl Acad Sci USA 1989; 86:232–236.
33. Orita M, Iwahana H, Kanazawa H, Hayashi K, Sekiya T. Detection of polymorphisms of human DNA by gel electrophoresis as single-strand conformation polymorphisms. Proc Natl Acad Sci USA 1989; 86:2766–2770.
34. Simon L, Lévesque RC, Lalonde M. Identification of endomycorrhizal fungi colonizing roots by fluorescent single-strand conformation polymorphism-polymerase chain reaction. Appl Environ Microbiol 1993; 59:4211–4215.
35. Sneath PHA, Sokal RR. Numerical Taxonomy. San Francisco: W.H. Freeman, 1973.
36. Nei M, Li WH. Mathematical model for studying genetic variation in terms of restriction endonucleases. Proc Natl Acad Sci USA 1979; 76:5269–5273.
37. Saitou N, Nei M. The neighbor-joining method: a new method for reconstructing phylogenetic trees. Mol Biol Evol 1987; 4:406–425.
38. Fitch WM, Margoliash E. Construction of phylogenetic trees. Science 1967; 155: 279–284.
39. Rohlf FJ. NTSYS-pc: Numerical Taxonomy and Multivariate Analysis System. Version 1.80. Stony Brook, N.Y.: State University of New York, 1993.
40. Miller JC. RESTSITE: Program for Analyzing Restriction Site or Fragment Data. Version 1.1. Madison: University of Wisconsin, 1990.
41. Felsenstein J. PHYLIP: Phylogeny Interface Package. Version 3.5C. Seattle: University of Washington, 1993.
42. Vilgalys R, Hester M. Rapid genetic identification and mapping of enzymatically amplified ribosomal DNA from several *Cryptococcus* species. J Bacteriol 1990; 172: 4238–4246.
43. Swofford DL. PAUP: Phylogenetic Analysis Using Parsimony. Version 3-1.1. Champaign: Illinois Natural History Survey, 1993.

44. Swofford DL, Olsen GJ. Phylogeny reconstruction. In: Hillis DH, Moritz C, eds. Molecular Systematics. Sunderland, Mass.: Sinauer Associates, 1990; 411–420.
45. Li J, Heath IB. The phylogenetic relationships of the anaerobic chytridiomycetous gut fungi (Neocallimasticaceae) and the Chytridiomycota. I. Cladistic analysis of rRNA sequences. Can J Bot 1992; 70:1738–1746.
46. Dore J, Stahl DA. Phylogeny of anaerobic rumen Chytridiomycetes inferred from small subunit ribosomal RNA sequence comparisons. Can J Bot 1991; 69:1964–1971.
47. Bowman BH, Taylor JW, Brownlee AG, Lee J, Lu S-D, White TJ. Molecular evolution of the fungi: relationship of the Basidiomycetes, Ascomycetes, and Chytridiomycetes. Mol Biol Evol 1992; 9:285–296.
48. Guadet J, Julien J, Lafay JF, Brygoo Y. Phylogeny of some *Fusarium* species as determined by large-subunit rRNA sequence comparison. Mol Biol Evol 1989; 6: 227–242.
49. Cubeta MA, Echandi E, Abernethy T, Vilgalys R. Characterization of anastomosis groups of binucleate *Rhizoctonia* species using restriction analysis of an amplified ribosomal RNA gene. Phytopathology 1991; 81:1395–1400.
50. Hibbett DS, Vilgalys R. Phylogenetic relationships of *Lentinus* (Basidiomycotina) inferred from molecular and morphological characters. System Bot 1993; 18:409–433.
51. Sherriff C, Whelan MJ, Arnold GM, Lafay JF, Brygoo Y, Bailey JA. Ribosomal DNA sequence analysis reveals new species groupings in the genus *Colletotrichum*. Exp Mycol 1994; 18:121–138.
52. Rehner SA, Samuels GJ. Taxonomy and phylogeny of *Gliocladium* analysed from nuclear large subunit ribosomal DNA sequences. Mycol Res 1994; 98:625–634.
53. Zambino PJ, Szabo LJ. Phylogenetic relationships of selected cereal and grass rusts based on rRNA sequence analysis. Mycologia 1993; 85:401–414.
54. Lee SB, Taylor JW. Phylogeny of five fungus-like protoctistan *Phytophthora* species, inferred from the internal transcribed spacers of ribosomal DNA. Mol Biol Evol 1992; 9:636–653.
55. Sreenivasaprasad S, Brown AE, Mills PR. DNA sequence variation and interrelationships among *Colletotrichum* species causing strawberry anthracnose. Physiol Mol Plant Pathol 1992; 41:265–281.
56. Carbone I, Kohn LM. Ribosomal DNA sequence divergence within internal transcribed spacer 1 of the Sclerotiniaceae. Mycologia 1993; 85:415–427.
57. Lobuglio KF, Pitt JI, Taylor JW. Phylogenetic analysis of two ribosomal DNA regions indicates multiple independent losses of a sexual *Talaromyces* state among asexual *Penicillium* species in subgenus *Biverticillium*. Mycologia 1993; 85: 592–604.
58. Bernier L, Hamelin RC, Ouellette GB. Comparison of ribosomal DNA length and restriction site polymorphisms in *Gremmeniella* and *Ascocalyx* isolates. Appl Environ Microbiol 1994; 60:1279–1286.
59. Henrion B, Chevalier G, Martin F. Typing truffle species by PCR amplification of the ribosomal DNA spacers. Mycol Res 1994; 98:37–43.
60. Ward E, Akrofi AY. Identification of fungi in the *Gaeumannomyces-Phialophora* complex by RFLPs of PCR-amplified ribosomal DNAs. Mycol Res 1994; 98: 219–224.

61. Chen W. Restriction fragment length polymorphisms in enzymatically amplified ribosomal DNAs of three heterothallic *Pythium* species. Phytopathology 1992; 82: 1467–1472.
62. Chen W, Hoy JW. Molecular and morphological comparison of *Pythium arrhenomanes* and *P. graminicola*. Mycol Res 1993; 97:1371–1378.
63. O'Donnell K. Ribosomal DNA internal transcribed spacers are highly divergent in the phytopthogenic ascomycete *Fusarium sambucinum* (*Gibberella pulicaris*). Curr Genet 1992; 22:213–220.
64. Henrion B, LeTacon F, Martin F. Rapid identification of genetic variation of ectomycorrhizal fungi by amplification of ribosomal RNA genes. New Phytol 1992; 122: 289–298.
65. Edel V, Steinberg C, Avelange I, Laguerre G, Alabouvette C. Comparison of three molecular methods for the characterization of *Fusarium oxysporum* strains. Phytopathology 1995; 85:579–585.
66. Anderson JB, Stasovski E. Molecular phylogeny of northern hemisphere species of *Armillaria*. Mycologia 1992; 84:505–516.
67. Gardes M, Bruns TD. ITS primers with enhanced specificity for basidiomycetes—application to the identification of mycorrhizae and rusts. Mol Ecol 1993; 2:113–118.
68. Bruns TD, Fogel R, Taylor JW. Amplification and sequencing of DNA from herbarium specimens. Mycologia 1990; 82:175–184.
69. Glass NL, Donaldson GC. Development of primer sets designed for use with PCR to amplify conserved genes from filamentous ascomycetes. Appl Environ Microbiol 1995; 61:1323–1330.
70. Donaldson GC, Ball LA, Axelrood PE, Glass NL. Primer sets developed to amplify conserved genes from filamentous ascomycetes are useful in differentiating *Fusarium* species associated with conifers. Appl Environ Microbiol 1995; 61:1331–1340.
71. Mehmann B, Brunner I, Braus GH. Nucleotide sequence variation of chitin synthase genes among ectomycorrhizal fungi and its potential use in taxonomy. Appl Environ Microbiol 1994; 60:3105–3111.
72. Jones MJ, Dunkle LD. Analysis of *Cochliobolus carbonum* races by PCR amplification with arbitrary and gene-specific primers. Phytopathology 1993; 83:366–370.
73. Welsh J, McClelland M. Fingerprinting genomes using PCR with arbitrary primers. Nucleic Acids Res 1990; 18:7213–7218.
74. Caetano-Anollés G, Bassam BJ, Gresshoff PM. DNA amplification fingerprinting using very short arbitrary oligonucleotide primers. Biotechnology 1991; 9:553–557.
75. Hadrys H, Balick M, Schierwater B. Applications of random amplified polymorphic DNA (RAPD) in molecular ecology. Mol Ecol 1992; 1:55–63.
76. Lehmann PF, Lin D, Lasker BA. Genotypic identification and characterization of species and strains within the genus *Candida* by using random amplified polymorphic DNA. J Clin Microbiol 1992; 30:3249–3254.
77. Bayman P, Cotty PJ. Genetic diversity in *Aspergillus flavus*: association with aflatoxin production and morphology. Can J Bot 1993; 71:23–31.
78. Guthrie PAI, Magill CW, Frederiksen RA, Odvody GN. Random amplified polymorphic DNA markers: a system for identifying and differentiating isolates of *Colletotrichum graminicola*. Phytopathology 1992; 82:832–835.

79. Assigbetse KB, Fernandez D, Dubois MP, Geiger JP. Differentiation of *Fusarium oxysporum* f. sp. *vasinfectum* races on cotton by random amplified polymorphic DNA (RAPD) analysis. Phytopathology 1994; 84:622–626.
80. Crowhurst RN, Hawthorne BT, Rikkerink EHA, Templeton MD. Differentiation of *Fusarium solani* f. sp. *cucurbitae* races 1 and 2 by random amplification of polymorphic DNA. Curr Genet 1991; 20:391–396.
81. Duncan S, Barton JE, O'Brien PA. Analysis of variation in isolates of *Rhizoctonia solani* by random amplified polymorphic DNA assay. Mycol Res 1993; 97:1075–1082.
82. Wyss P, Bonfante P. Amplification of genomic DNA of arbuscular-mycorrhizal (AM) fungi by PCR using short arbitrary primers. Mycol Res 1993; 97:1351–1357.
83. Grajal-Martin MJ, Simon CJ, Muehlbauer FJ. Use of random amplified polymorphic DNA (RAPD) to characterize race 2 of *Fusarium oxysporum* f. sp. *pisi*. Phytopathology 1993; 83:612–614.
84. Manulis S, Kogan N, Reuven M, Ben-Yephet Y. Use of the RAPD technique for identification of *Fusarium oxysporum* f. sp. *dianthi* from carnation. Phytopathology 1994; 84:98–101.
85. Chen X, Line RF, Leung H. Relationship between virulence variation and DNA polymorphism in *Puccinia striiformis*. Phytopathology 1993; 83:1489–1497.
86. Nicholson P, Rezanoor HN. the use of random amplified polymorphic DNA to identify pathotype and detect variation in *Pseudocercosporella herpotrichoides*. Mycol Res 1994; 98:13–21.
87. Meyer W, Lieckfeldt E, Kuhls K, Freedman EZ, Börner T, Mitchell TG. DNA- and PCR-fingerprinting in fungi. In: Pena SDJ, Chakraborty R, Epplen JT, Jeffreys AJ, eds. DNA Fingerprinting: State of the Science. 1993:311–320.
88. Meyer W, Mitchell TG, Freedman EZ, Vilgalys R. Hybridization probes for conventional DNA fingerprinting used as single primers in the polymerase chain reaction to distinguish strains of *Cryptococcus neoformans*. J Clin Microbiol 1993; 31:2274–2280.
89. Stenlid J, Karlsson JO, Högberg N. Intraspecific genetic variation in *Heterobasidion annosum* revealed by amplification of minisatellite DNA. Mycol Res 1994; 98:57–63.
90. Van Belkum A, Quint WGV, De Pauw BE, Melchers WJG, Meis JF. Typing of *Aspergillus* species and *Aspergillus fumigatus* isolates by Interrepeat polymerase chain reaction. J Clin Microbiol 1993; 31:2502–2505.
91. Welsh J, McClelland M. Genomic fingerprints produced by PCR with consensus tRNA gene primers. Nucleic Acids Res 1991; 19:861–866.
92. Meunier JR, Grimont PAD. Factors affecting reproducibility of random amplified polymorphic DNA fingerprinting. Res Microbiol 1993; 144:373–379.
93. MacPherson JM, Eckstein PE, Scoles GJ, Gajadhar AA. Variability of the randon amplified polymorphic DNA assay among thermal cyclers, and effects of primer and DNA concentration. Mol Cell Probes 1993; 7:293–299.
94. Gardes M, White TJ, Fortin JA, Bruns TD, Taylor JW. Identification of indigenous and introduced symbiotic fungi in ectomycorrhizae by amplification of nuclear and mitochondrial ribosomal DNA. Can J Bot 1991; 69:180–190.
95. Simon L, Levesque RC, Lalonde M. Rapid quantitation by PCR of endomycorrhizal fungi colonizing roots. PCR Methods Appl 1992; 2:76–80.

96. Di Bonito R, Elliott ML, Des Jardins EA. Detection of an arbuscular mycorrhizal fungus in roots of different plant species with the PCR. Appl Environ Microbiol 1995; 61:2809–2810.
97. Tebbe CC, Vahjen W. Interference of humic acids and DNA extracted directly from soil in detection and transformation of recombinant DNA from bacteria and a yeast. Appl Environ Microbiol 1993; 59:2657–2665.
98. Porteous LA, Armstrong JL, Seidler RJ, Watrud LS. An effective method to extract DNA from environmental samples for polymerase chain reaction amplification and DNA fingerprint analysis. Curr Microbiol 1994; 29:301–307.
99. Smalla K, Cresswell N, Mendonca-Hagler LC, Wolters A, van Elsas JD. Rapid DNA extraction protocol from soil for polymerase chain reaction-mediated amplification. J Appl Bacteriol 1993; 74:78–85.
100. Nazar RN, Hu X, Schmidt J, Culham D, Robb J. Potential use of PCR-amplified ribosomal intergenic sequences in the detection and differentiation of *Verticillium* wilt pathogens. Physiol Mol Plant Pathol 1991; 39:1–11.
101. Moukhamedov R, Hu X, Nazar RN, Robb J. Use of polymerase chain reaction-amplified ribosomal intergenic sequences for the diagnosis of *Verticillium tricorpus*. Phytopathology 1994; 84:256–259.
102. Ersek T, Schoelz JE, English JT. PCR amplification of species-specific DNA sequences can distinguish among *Phytophthora* species. Appl Environ Microbiol 1994; 60:2616–2621.
103. Beck JJ, Ligon JM. Polymerase chain reaction assays for the detection of *Stagonospora nodorum* and *Septoria tritici* in wheat. Phytopathology 1995; 85:319–324.
104. Tisserat NA, Hulbert SH, Sauer KM. Selective amplification of rDNA internal transcribed spacer regions to detect *Ophiosphaerella korrae* and *O. herpotricha*. Phytopathology 1994; 84:478–482.
105. Holmes AR, Cannon RD, Shepherd MG, Jenkinson HF. Detection of *Candida albicans* and other yeasts in blood by PCR. J Clin Microbiol 1994; 32:228–231.
106. Spreadbury C, Holden D, Aufauvre-Brown A, Bainbridge B, Cohen J. Detection of *Aspergillus fumigatus* by polymerase chain reaction. J Clin Microbiol 1993; 31: 615–621.
107. Bryan GT, Daniels MI, Osbourn AE. Comparison of fungi within the *Gaeumannomyces-Phialophora* complex by analysis of ribosomal DNA sequences. Appl Environ Microbiol 1995; 61:681–689.
108. Xue B, Goodwin PH, Annis SL. Pathotype identification of *Leptosphaeria maculans* with PCR and oligonucleotide primers from ribosomal internal transcribed spacer sequences. Physiol Mol Plant Pathol 1992; 41:179–188.
109. Johnston CG, Aust SD. Detection of *Phanerochaete chrysosporium* in soil by PCR and restriction enzyme analysis. Appl Environ Microbiol 1994; 60:2350–2354.
110. Lamar RT, Schoenike B, Vanden Wymelenberg A, Stewart P, Dietrich DM, Cullen D. Quantitation of fungal mRNAs in complex substrates by reverse transcription PCR and its application to *Phanerochaete chrysosporium*-colonized soil. Appl Environ Microbiol 1995; 61:2122–2126.
111. Henson JM, French R. The polymerase chain reaction and plant disease diagnosis. Annu Rev Phytopathol 1993; 31:81–109.
112. Riedy MC, Timm EA, Stewart CC. Quantitative RT-PCR for measuring gene expression. Biotechniques 1995; 18:70–76.

113. Gilliland G, Perrin S, Bunn F. Competitive PCR for quantitation of mRNA. In: Innis MA, Gelfand DH, Sninsky JJ, White TJ, eds. PCR Protocols: A Guide to Methods and Application. San Diego: Academic Press, 1990:60–69.
114. Cummings B, Wood T. A simple and efficient method for isolating genomic DNA from endomycorrhizal spores. Gene Anal Techn 1989; 6:89–92.
115. Garber RC, Yoder OC. Isolation of DNA from filamentous fungi and separation into nuclear, mitochondrial, ribosomal, and plasmid components. Anal Biochem 1983; 135:416–422.
116. Kistler HC, Leong SA. Linear plasmidlike DNA in the plant pathogenic fungus *Fusarium oxysporum* f. sp. *conglutinans*. J Bacteriol 1986; 167:587–593.
117. Coddington A, Matthews PM, Cullis C, Smith KH. Restriction digest patterns of total DNA from different races of *Fusarium oxysporum* f. sp. *pisi*—an improved method for race classification. J Phytopathol 1987; 118:9–20.
118. Bridge PD, Ismail MA, Rutherford MA. An assessment of aesculin hydrolysis, vegetative compatibility and DNA polymorphism as criteria for characterizing pathogenic races within *Fusarium oxysporum* f. sp. *vasinfectum*. Plant Pathol 1993; 42:264–269.
119. Klich MA, Mullaney EJ, Daly CB. Analysis of intraspecific and interspecific variability of three common species in *Aspergillus* section *versicolores* using DNA restriction fragment length polymorphisms. Mycologia 1993; 85:852–855.
120. Hudspeth MES. The fungal mitochondrial genome—a broader perspective. In: Arora DK, Elander RP, Mukerji KG, eds. Handbook of Applied Mycology, Vol. 4, Fungal Biotechnology. New York: Marcel Dekker, 1991:213–241.
121. Stammler G, Seemüller E, Duncan JM. Analysis of RFLPs in nuclear and mitochondrial DNA and the taxonomy of *Phytophthora fragariae*. Mycol Res 1993; 97: 150–156.
122. Förster H, Coffey MD. Molecular characterization of *Phytophthora* isolates with non-papillate sporangia causing root rot of rasberry using mtDNA restriction fragment length polymorphisms. Mycol Res 1992; 96:571–577.
123. Förster H, Coffey MD. Molecular taxonomy of *Phytophthora megasperma* based on mitochondrial and nuclear DNA polymorphisms. Mycol Res 1993; 97:1101–1112.
124. Kim DH, Martyn RD, Magill CW. Mitochondrial DNA (mtDNA)—relatedness among formae speciales of *Fusarium oxysporum* in the Cucurbitaceae. Phytopathology 1993; 83:91–97.
125. Kistler HC, Bosland PW, Benny U, Leong S, Williams PH. Relatedness of strains of *Fusarium oxysporum* from crucifers measured by examination of mitochondrial and ribosomal DNA. Phytopathology 1987; 77:1289–1293.
126. Jacobson DJ, Gordon TR. Variability of mitochondrial DNA as an indicator of relationships between populations of *Fusarium oxysporum* f. sp. *melonis*. Mycol Res 1990; 94:734–744.
127. Adachi Y, Watanabe H, Tanabe K, Doke N, Nishimura S, Tsuge T. Nuclear ribosomal DNA as a probe for genetic variability in the Japanese pear pathotype of *Alternaria alternata*. Appl Environ Microbiol 1993; 59:3197–3205.
128. Nicholson P, Rezanoor HN, Hollins TW. Classification of a world-wide collection of isolates of *Pseudocercosporella herpotrichoides* by RFLP analysis of mitochondrial and ribosomal DNA and host range. Plant Pathol 1993; 42:58–66.
129. Elias KS, Zamir D, Lichtman-Pleban T; Katan T. Population structure of *Fusarium*

oxysporum f. sp. *lycopersici*: restriction fragment length polymorphisms provide genetic evidence that vegetative compatibility group is an indicator of evolutionary origin. Mol Plant-Microbe Interact 1993; 6:565–572.

130. Carlier J, Mourichon X, Gonzales-de-Léon D, Zapater MF, Lebrun MH. DNA restriction fragment length polymorphisms in *Mycosphaerella* species that cause banana leaf spot diseases. Phytopathology 1994; 84:751–756.
131. Ueng PP, Chen W. Genetic differentiation between *Phaeosphaeria nodorum* and *P. avenaria* using restriction fragment length polymorphisms. Phytopathology 1994; 84:800–806.
132. Marmeisse R, Debaud JC, Casselton LA. DNA probes for species and strain identification in the ectomycorrhizal fungus *Hebeloma*. Mycol Res 1992; 96:161–165.
133. Manicom BQ, Bar-Joseph M, Rosner A, Vigodsky-Haas H, Kotze JM. Potential applications of random DNA probes and restriction fragment length polymorphisms in the taxonomy of the *Fusaria*. Phytopathology 1987; 77:669–672.
134. Manicom BQ, Baayen RP. Restriction fragment length polymorphisms in *Fusarium oxysporum* f. sp. *dianthi* and other fusaria from *Dianthus* species. Plant Pathol. 1993; 42:851–857.
135. Kistler HC, Momol EA, Benny U. Repetitive genomic sequences for determining relatedness among strains of *Fusarium oxysporum*. Phytopathology 1991; 81:331–336.
136. Namiki F, Shiomi T, Kayamura T, Tsuge T. Characterization of the formae speciales of *Fusarium oxysporum* causing wilts of cucurbits by DNA fingerprinting with nuclear repetitive DNA sequences. Appl Environ Microbiol 1994; 60:2684–2691.
137. Whisson SC, Maclean DJ, Manners JM, Irwin JAG. Genetic relationships among Australian and North American isolates of *Phytophthora megasperma* f. sp. *glycinea* assessed by multicopy DNA probes. Phytopathology 1992; 82:863–868.
138. McDonald BA, Miles J, Nelson LR, Pettway RE. Genetic variability in nuclear DNA in field populations of *Stagonospora nodorum*. Phytopathology 1994; 84:250–255.
139. Hamer JE, Farrall L, Orbach MJ, Valent B, Chumley FG. Host species-specific conservation of a family of repeated DNA sequences in the genome of a fungal plant pathogen. Proc Natl Acad Sci USA 1989; 86:9981–9985.
140. Xia JQ, Correll JC, Lee FN, Marchetti MA, Rhoads DD. DNA fingerprinting to examine microgeographic variation in the *Magnaporthe grisea* (*Pyricularia grisea*) population in two rice fields in Arkansas. Phytopathology 1993; 83:1029–1035.
141. Jeffreys AJ. Highly variable minisatellite and DNA fingerprints. Biochem Soc Trans 1987; 15:309–317.
142. DeScenzo RA, Harrington TC. Use of (CAT)s as a DNA fingerprinting probe for fungi. Phytopathology 1994; 84:534–540.
143. Correll JC, Rhoads DD, Guerber JC. Examination of mitochondrial DNA restriction fragment length polymorphisms, DNA fingerprints, and randomly amplified polymorphic DNA of *Colletotrichum orbiculare*. Phytopathology 1993; 83:1199–1204.

4

Proteins in Fungal Taxonomy

Grégoire L. Hennebert
Université Catholique de Louvain, Louvain-la-Neuve, Belgium

Marc Vancanneyt
Universiteit Gent, Ghent, Belgium

I. INTRODUCTION

Many species of fungi are difficult to identify, particularly if they lack reproductive structures. Even when structures of sexual or asexual reproduction are present, isolates may exhibit atypical, intermediate, variable, or no diagnostic morphological characteristics, which makes definite identification difficult.

Morphological simplicity, as in yeasts, or similarity, as in anamorphic fungi, may mask genetic diversity. Conversely, the occurrence of distinct forms of reproduction in pleoanamorphic fungi, or dimorphism in pathogenic fungi, may mask genetic similarity. This observation is not a criticism of the morphological approach to taxonomy, but an indication of the need to consider additional taxonomic criteria.

The principal reason for selecting protein patterns for additional diagnostic characters of fungi was that their diversity is directly related to the diversity of the coding genes and may express specific differences or similarities among organisms (1). Cellular protein diversity is usually investigated through electrophoretic techniques and the mobility and similarity of separated proteins from various organisms was first demonstrated by the Tiselius electrophoretic separation. Gel electrophoresis brought macromolecular in place of micromolecular approaches in taxonomy and has been used in bacteria and fungi since 1962.

In the fungi, protein electrophoresis has aided in the resolution of taxonomic problems, such as the segregation of closely related taxa, the identification of isolates, the recognition of mutants, mating types, *formae speciales* or geographical races, the establishment of host-pathogen specificity, and the recognition of species heterogeneity.

In the present chapter, we review the taxonomic use of electrophoresis of whole-cell proteins of fungi, excluding reports on enzyme and isozyme electrophoresis and immunoelectrophoresis. Protein patterns obtained through electrophoresis are not necessarily final and comparable and they are largely dependent on the analytical techniques used. In the last three decades techniques were constantly evolving and in order to determine the value of protein patterns in fungal taxonomy it is necessary to refer to the technique applied.

II. METHODS OF PROTEIN ANALYSIS BY ELECTROPHORESIS

Fungal proteins commonly analyzed by electrophoresis for taxonomic purposes are water- or buffer-soluble native proteins (NP) or detergent-solubilized and denatured proteins in sodium dodecylsulphate (SDS) which have been extracted from fungal biomass produced in liquid culture medium. The type of buffer, gel, and stain used will differentiate whole-cell protein patterns from those of specific groups of enzymes or antigens. Gel electrophoresis was first introduced by Smithies (2) with a gel made of hydrolyzed starch that could be used in horizontal or vertical slabs (Table 1).

In 1959, Ornstein (3) and Davis (4) designed a new electrophoretic procedure (PAGE) using polyacrylamide gels proposed by Raymond and Weintraub (5). The procedure used two different buffers and two different polyacrylamide gel concentrations and was called "discontinuous" or "disc" electrophoresis. It was applied in small vertical tubes, and later in vertical or horizontal slabs. The procedure was used to separate native proteins (NP-PAGE). Chang et al. (6) were the first to apply this technique to the fungi in 1962.

In 1970, Laemmli (7) introduced denaturation of the proteins by solubilization and boiling in sodium dodecylsulphate (SDS) before polyacrylamide gel electrophoresis (SDS-PAGE). The technique increased efficiency and reproducibility. It was mainly used in a discontinuous system but also in continuous buffer systems either with constant gel concentrations or with a gradient concentration.

Both NP-PAGE and SDS-PAGE separate proteins by their molecular weight. Vesterberg and Svensson (8) created a pH gradient in the polyacrylamide gel using carrier ampholytes, separating native proteins by their isoelectric points. The technique, isoelectric focusing electrophoresis (IEF-PAGE), has only been applied a few times in fungal taxonomy.

In 1975 O'Farrell (9) combined IEF- and SDS-PAGE in cross directions, to resolve protein bands obtained by single methods. This two-dimensional electro-

Table 1 Evolution of Protein Electrophoresis and Use in the Fungi

Technique			Period of use	References
Smithies, 1955				
NP-STGE	continuous	slab	(1963–1966)	10, 37
Raymond and Weintraub, 1959				
Ornstein and Davis, 1964				
NP-PAGE	discontinuous	tube	(1962–1983)	6, 11, 13, 33, 35, 36, 39, 41, 46, 49, 53, 59, 60, 62, 64, 67, 87–89, 92, 95, 98, 99, 103, 104, 110, 119, 124, 128, 129
		slab	(1968–1994)	42, 58, 63, 78, 105, 107
	continuous	tube	(1979–1987)	45, 83
		slab	(1970–1992)	84, 94, 111, 112, 117
Vesterberg and Svenson, 1966				
IEF-PAGE		slab	(1983–1989)	56, 65, 66, 85, 90, 91, 118
Laemmli, 1970				
SDS-PAGE	discontinuous:			
	define PA conc.	tube	(1988–1994)	72, 75, 77, 93, 100, 115, 118, 121
		slab	(1985–1995)	26, 27, 29, 30, 51, 57, 63, 96, 97, 105, 116, 127, 130
	gradient conc.	slab	(1982–1989)	44, 53, 73, 114
O'Farrell, 1975				
2D-PAGE	(IEF + SDS)	slab	(1984–1985)	122, 125

phoresis (2D-PAGE), occasionally used in fungi, has been shown to resolve up to 320 proteins.

A. Extraction of Proteins

Fungal strains must be grown on identical optimal liquid medium simultaneously under the same conditions for accurate comparison.

The extraction of whole-cell proteins from fungal biomass follows some or all of the following steps: cell washing in water or in buffer, freezing or freeze-drying, disruption of fungal biomass by grinding or sonication, elimination of lipids in acetone or petroleum ether, extraction in sodium bicarbonate or various buffers, precipitation in chloroform-methanol or ethanol, centrifugation to collect the precipitate or the supernatant, dialysis in water, various buffers or polyethyleneglycol, denaturation with buffered SDS. Addition of reducing agents, protease inhibitors, deoxyribonuclease, and antifoam may also be included. A number of sequences used in the extraction of fungal proteins is given in Table 2.

B. Gel Reading and Numerical Analysis

In early applications, the analysis of protein patterns was either visual, diagrammatic, or photographic. The position of the bands was expressed as a percentage (*Rf*) of the distance from the top to the moving dye front of the profile (4). After the first use of a densitometer for gel scanning in fungi by Clare and Zentmyer (10) and Shechter et al. (11), the equalization of a scanned profile to obtain perfect matching of replicates was developed by Rouatt et al. (12). The first application of numerical analysis of fungal protein patterns was by Shechter (13), who used similarity values and UPGMA clustering (14). A more accurate gel scan analysis method was developed for bacterial systematics by Kersters and De Ley (15) that consisted of three steps:

1. Normalization. After the protein gel was scanned by a densitometer and a profile produced, the distance between two distant reference peaks is converted into a sequence of 90 equal positions, position 0 being the top of the gel. Consequently, each of the 90 positions of the profile was given the normalized optical density in mm height of the tracing.
2. Compensation. To compensate for slight positional variation of homologous bands in each profile, the bands were given a new calculated average position. This was done by stretching or narrowing the valley floor of the tracing.
3. Computer-assisted numerical analysis of the normalized and compensated scan. An *r* matrix was calculated for each scan using the Pearson product-moment correlation coefficient (*r*) (14) between any pair of scans and this was converted into a *z* matrix using the Fischer transformation. The *z* values were clustered using the UPGMA method (14,16). The Pearson correlation coefficient (*r*) was considered more suitable than similarity or dissimilarity coefficients by Kersters and De Ley (15), because it is insensitive to differences due to the amount of protein in the sample. Comparing only band positions or peak heights in profiles by similarity coefficients has generally proved unsatisfactory. It is necessary to take into account all the available information on peak and valley shape and their relative positions. This information is obtained by slicing the profile into a large number of columns and then listing heights and positions of each profile point. The success of the resulting classification was largely determined by how well the normalization-compensation technique reduced the variability of the band positions and increased the reproducibility of the electrophoretic runs (15,17,18). Jackman et al. (19) and Jackman (20–22) produced a series of computer programs to record, normalize, and compensate the profile trace and to calculate the correlation coefficient *r* matrix for analysis by UPGMA clustering.

In 1986, Plykaytis et al. (23) used mathematical normalization of SDS-PAGE protein profiles to obtain standardized absolute migration distances based on distances of internal references. Further development of an objective mathematical normalization and the compensation of profiles was achieved by Kersters and Pot (24) and Pot et al. (25) based on repeated profiles of a reference strain. The average peak positions calculated by a computer program are used to correct the raw digitized traces in different gels and within each gel. After correction, the position and height of the 400 points of the digitized trace were determined. The data were used for calculation of the correlation matrix and for clustering by UPGMA. This procedure was used in an analysis of yeast protein patterns by Vancanneyt et al. (26,27).

Vauterin and Vauterin (28) further improved the Kersters and Pot's method with a new mathematical approach in 1992. This was integrated in a package for the objective computer-aided comparison of electrophoretic patterns, with capacity for data storage, trace retrieval, diverse cluster analyses, and expert identification. The package, GELCOMPAR (Applied Maths; Kortrijk, Belgium), has been in use in bacterial identification and taxonomy by Kersters and collaborators since 1992 and applied to yeast taxonomy by Vancanneyt et al. (29,30). It is important to say that it is particularly adapted to SDS-PAGE protein patterns obtained by "extremely standardized and reproducible experimental conditions" as recommended by Pot et al. (31). Discontinuous SDS-PAGE technique is preferred against other techniques, because it provides higher resolution, a more discriminant banding pattern based on a unique characteristic (molecular weight), and higher reproducibilty (32).

III. FACTORS OF VARIABILITY IN WHOLE-CELL PROTEIN PATTERN ANALYSIS

A. Protein Pattern Resolution

Chang et al. (6) detected only six to eight bands and 18 bands, respectively, in paper and starch-gel electrophoresis from strains of *Neurospora* species. Using polyacrylamide gel disc electrophoresis, they obtained 25 protein bands from the same strains. Up to 60 protein bands have now been detected in a single SDS-PAGE profile from a fungus. Glynn and Reid (33) found lower resolution of proteins after extraction by disruption of cells in a Braun homogenizer than after grinding in a prechilled mortar and pestle at 3 to 4°C. Dialysis of the soluble protein fraction also improved the sharpness of the profile (15).

B. Protein Pattern Reproducibility

1. Cultural Conditions

The Ornstein and Davis technique was found readily reproducible when applied to replicate samplings of the same protein extract, or to different protein extracts of mycelium by Chang et al. (6) in *Neurospora* strains and by Shechter et al. (11) in dermatophytes. Good reproducibility was also observed by Durbin (34) when

Table 2 Protein Extraction Methods Used in Fungi

A. Extraction methods for NP-PAGE used in fungi								
Successive steps in protein extraction								References
W	FR	GR	LE(Pe)	C↓	EX(PB)	DL(W)	C↑	6
W	FR	GR	LE(A)	C↓	EX(PB)		C↑	39, 50
W	FR	GR	LE(A)		EX(SBC)		C↑	98, 99, 104
W	LE(A)	GR	LE(Pe)		GR.EX(PB)		C↑	11, 13, 36, 89
W	FR	GR.EX(PB)					C↑	43, 45, 46, 58, 62, 67, 107, 110
W	FR	GR.EX(PB)		C↑			C↑	59, 60
W(PB)	FR	GR.EX(PB)					C↑	63, 78
W		GR.EX(THcl)					C↑	42, 45, 103, 124
W(PB)		GR.EX(THcl)		C↑			C↑	83, 124
W		GR.EX(Tcyst)					C↑	33, 34
W		GR.EX(Tcit)					C↑	10, 37, 128
W		GR.EX(PB)		C↑		DL(PB)	C↑	95, 106
W	FR	GR.EX(PB)		C↑		DL(AmS)	C↓	129
W	(LE(A)	GR.EX(PB)		C↑		DL(W)		118
W(T20)	C↓	GR.EX(THcl)				DL(THcl)	C↑	41
W	FD	GR.EX(Tglyc)		C↑		DL(W)		94
W	FR	GR.EX(Tglyc)		C↑		DL(Tglyc)		87, 88
W		GR.EX(PB)			SO.EX(PB)		C↑	35
W		SO.EX(Tcit)					C↑	64

B. Extraction Methods for SDS-PAGE Used in Fungi

Successive steps in protein extraction					References
W	GR.EX(Phenol, SDS)	C↑	DL(W)		114
W	GR.EX(Tglyc)	C↑		DN(STB.SDS)	116
W	GR.EX(PB)	C↑		DN(STB.SDS)	77
W(PB) FR	GR.EX(PB)	C↑		DN(STB.SDS)	63
W	GR.EX(SBC)	C↑		DN(STB.SDS)	97
W	GR.EX(THcl)	C↑		DN(STB.SDS)	57
W	GR.EX(THcl) C↑	C↑ EX(CH.MT) C↓		DN(STB.SDS)	44, 115
W	GR.EX(THcl) C↑	EX(CH.MT) PR(MT) C↓		DN(THcl.SDS)	53
W	GR.EX(THcl)	C↑-EX(CH.MT)-C↓-EX(THcl)-PR(MT)-C↓–DN(SDS)			127
W PR(ET)	GR.EX(SBC)	C↑	DL(PEG)	DN(STB.SDS)	96
W	GR.EX(STB.SDS)	C↑	LE(A)	DN(STB.SDS)	105
W(PBS) FR	GR.EX(PBS)	C↑		DN(STB.SDS)	51
W(PBS)	SO.EX(STB)	C↑		DN(STB.SDS)	26, 27, 29, 30
W	GR.EX(THcl)			DN(LSD)	73

Key: W = washing (in water, if no medium indicated in brackets); FR = freezing possibly followed by freeze-drying; GR = grinding; LE = lipid elimination; C↑ = centrifugation to collect the supernatant; C↓ = centrifugation to collect the precipitate; EX = extraction; GR.EX = simultaneous grinding and extraction; SO.EX = simultaneous sonication and extraction; DL = dialysis; PR = precipitation; DN = denaturation. ( ) = in brackets, chemical components used in extraction procedures; A = acetone; Pe = petroleum ether; CH.MT = chloroform-methanol; ET = ethanol; W = water; T20 = Tween 20 solut.; SBC = sodium bicarbonate in water; ABC = ammonium bicarbonate in water; AmS = ammonium sulfate; PB = phosphate buffer; PBS = phosphate buffer saline; THcl = Tris-HCl buffer; Tcit = Tris-citrate buffer; Tcyst = Tris-cysteine buffer; Tglyc = Tris-glycine buffer; PEG = polyethyleneglycol; Phenol = phenol; SDS = sodium dodecylsulfate; LDS = lithium dodecylsulfate; STB = sample treatment buffer according to Laemmli, 1970.

cultures of *Septoria* isolates were harvested from different media at the same age. Conversely, Meyer and Renard (35) could not obtain reproducible protein patterns from extracts of replicate cultures of the same *Fusarium* strain on the same culture medium. Glynn and Reid (33) demonstrated that such variations with *Fusarium* replicate protein patterns depended on experimental conditions. Shechter et al. (11,36) also found differences in the protein patterns from cultures of the same dermatophyte strain, depending on the culture media.

In recent polyphasic taxonomic investigations of yeasts, Vancanneyt et al. (26,27,29,30) recommended standardized conditions of 2-day-old inoculum in buffered GYPP medium (glucose, yeast extract, peptone, phosphate buffer; pH 5.6), waterbath controlled temperature and the harvesting of cells at exponential growth phase to obtain high reproducibility.

2. Organic Developmental Stage and Physiological Age

Identical protein patterns were obtained from *Pythium* strains harvested at three different ages by Clare (37), *Phytophthora* strains harvested at two different ages by Clare and Zentmeyer (10), and *Septoria* isolates of different ages by Durbin (34). However, Newstedt et al. (38) pointed out that the number of protein bands from cultures of the boreal species *Typhula idahoensis* and *T. incarnata* and of a *Coprinus* strain varied with time. Differences in protein band density were observed between young conidiated mycelium and older mycelium of 11 isolates of *Glomerella* and *Colletotrichum* by McCombs and Winstead (39). The same difference was found in *Fusarium oxysporum* f. sp. *cubense* strain harvested at four different ages, from 10 to 28 days by Glynn and Reid (33). These authors also found that cultivation of the same strain either at five different temperatures or on different culture media or from a single conidium or mass conidia inoculum produced protein patterns differing by the number of bands (33). Native proteins of dimorphic *Candida albicans* examined by Dabrowa et al. (40) showed differences between the budding form and the mycelial form, with a characteristic band for the mycelial form. In *Drechslera*, Shipton and McDonald (41) found one or two bands differed between conidial and mycelial proteins. Pelletier and Hall (42) demonstrated distinct protein patterns in *Verticillium* from young mycelium with conidia, older mycelium with dark cells, or conidia alone. Russo et al. (43) and Antibus (44) showed the existence of specific proteins associated with the development of sclerotia, in Sclerotiniaceae and in *Hygrophoropsis aurantiaca*. Paranjpe and Chen (45) found different protein patterns in mycelium, fruitbody primordia, and caps and stalks of *Agaricus bisporus*, although some protein fractions were common to all stages of development.

Little is known about how protein profiles are influenced by morphogenesis or by developmental stages and cultural conditions. A partial explanation may possibly be found in the following observations of Bent (46). When mycelia of three *Penicillium* species were investigated from exponential and stationary growth phases in shaken cultures, in a medium where nitrogen supply was decreasing until exhaustion, protein band intensity decreased and some bands disappeared. In static cultures similar changes occurred at a relatively early stage, well before the nitrogen supply was used up. This may be related to the type of growth. In shaken culture, the fungus is dispersed as hyphae or hyphal pellets, and is constantly in direct contact with the nutrients, while in static conditions, most of the mycelium is remote from the nutrient supply. Similar changes in the protein patterns of *Penicillium* were observed when the culture was sporulating. Sporulation can indeed be induced by nitrogen starvation. Protein patterns of sporulating mycelia differed from those of young vegetative mycelia in the same way as those of nitrogen-starved mycelia (46). It is therefore imperative to operate from fungal material in the same organic differentiation stage and physiological age, produced under a given set of rigidly controlled conditions in order to ensure reproducibility of the protein patterns. Snider (47) suggested the use of fungal spores only, as spores would be expected to be at the same physiological state. However, one should be certain that the proportion of young and old spores does not exceed the number of mature spores. With yeasts and dimorphic fungi Dooms et al. (48) showed that fast-shaken liquid cultures at the exponential phase, allow the harvesting of a majority of actively budding and presumably biochemically similar cells. Snider (47) and Whitney et al. (49) reexamined the use of freshly isolated strains, or cultures from freeze-dried fresh isolates. Cultures, either maintained on agar slants or immersed for a longer term in mineral oil or frozen through inadequate procedures, are subject to mutations, as demonstrated by Berny and Hennebert (50), and such mutations may alter protein composition (6).

3. *Variations During Protein Extraction and Electrophoresis*

Extraction and fractionation of proteins in native conditions is critical due to subunit aggregation, disulfite bridging, carbohydrate binding, phosphorylation, acetylation, proteolysis, etc. (51). The prevention of proteolysis during cell disruption and extraction by diverse protease inhibitors (p-chloromercuribenzoate, phenyl-methylsulfonyl fluoride, pepstatin, ethylene-diamine-tetraacetate, o-phenanthroline, thiourea, polyvinylpyrrolidone) has been stressed by Benhamou et al. (52) and Insell et al. (53) Similar alterations do not occur after SDS-heat denaturation. Bent (46) tested diverse SDS-PAGE techniques on *Penicillium* species, without inducing any alteration of the protein patterns. The current good reproducibility of SDS-PAGE patterns seems to demonstrate the stability of denatured proteins. However, Petäisto et al. (51) found differences in SDS-PAGE patterns due to unprotected disruption of mycelium.

Protein degradation by endogenous protease activity may also occur during the growth of the fungus. Protease activity was found by McCombs and Winstead (39) in extracts of *Glomerella cingulata*, *G. magna*, and *Colletotrichum orbiculare* isolates. Nasuno (54) demonstrated the presence of alkaline protease in extracts of *Aspergillus oryzae* and *A. sojae* strains. Petäistö et al. (51) demonstrated that unprotected mycelial extracts of *Gremmeniella abietina* showed an increasing protease activity depending on the culture age. Petäisto et al. (51) also established that acetone or strong acid precipitation of proteins did not influence the SDS-PAGE protein pattern. Ionic impurities in the protein extract and in the gels may cause artifacts. Reducing agents like 2-mercapto-ethanol or dithiothreitol are commonly used and dialysis is another means for the elimination of ionic impurities.

IV. TAXONOMIC SIGNIFICANCE OF WHOLE-CELL PROTEIN PATTERNS

The results of electrophoresis of whole-cell fungal proteins should be seen both as a contribution to the testing of available techniques and as an approach to the resolution of specific diagnostic problems. Reliability of the results depends on the entire procedure, from the growth of the fungus to the final interpretation of results, and on the reproducibility obtained by the investigators. Genera in which protein electrophoresis has been applied in taxonomic studies are discussed and ordered below as in the most recent classification proposed in the Dictionary of the Fungi (55). Anamorphic genera are assigned to the teleomorphic family or order to which they are definitely or presumably ("cfr") linked.

A. Oomycetes

1. *Pythiales*

a. Pythium. Species of *Pythium* are not easily characterized by conventional morphological criteria. Clare (37), Adaskaveg et al. (56), and Chen et al. (57) found distinct protein patterns between isolates of 14 species of *Pythium* with one or more common generic bands and only minor differences between strains from different geographical areas. However, no differences were observed either between *P. ultimum* var. *ultimum* and *P. ultimum* var. *sporangiferum* (56) or between isolates of *P. deliense* and *P. aphanidermatum*, and between those of *P. graminicola* and *P. arrhenomanes* (57). Unreliability of species identification by classical criteria appeared evident.

b. Phytophthora. Almost 900 strains of 17 *Phytophthora* species have been investigated by different workers. Clare and Zentmyer (10), Hall et al. (58,59), Gill and Powell (59,60), Gill and Zentmyer (62), and Bielenin et al. (63) found distinct patterns with up to three common bands in a number of species investigated, and only minor differences in band density among isolates of each species. No relation to host or geographical origin nor differences between physiological races were observed. On the basis of morphology, cultural characteristics and disc electrophoretic protein patterns, Zentmyer et al. (64) assigned five *Phytophthora* isolates from avocado roots to *P. citricola*, although an additional band was detected in each of these isolates. Erlesius and De Vallavieille (65,66) used IEF-PAGE to compare isolates of six *Phytophthora* species and found their patterns differed by five to eight bands. *Phytophthora citricola* did not form a tight group and differed from *P. citrophthora* with the exception of one strain. *P. citrophthora* itself was composed of two groups of isolates regardless of origin, host, and plant age. Hamm and Hansen (67) described the new species *P. pseudotsugae* based on morphology, host specificity, pathogenicity to Douglas fir (*Pseudotsuga menziensii*), and a strongly distinct SDS-PAGE protein pattern.

The taxonomy of *P. megasperma*, an important phytopathogenic species, has for some time been controversial. The species was divided into two varieties—var. *megasperma* with larger oogonia and pathogenic to lucerne, and var. *sojae*, with smaller oogonia and pathogenic to both soybean and lucerne (68,69). Kuan and Erwin (70) rejected the varietal distinction because of overlapping oogonium size, and differentiated isolates of the species on the basis of host preference as *formae speciales P. megasperma*, f. sp. *medicaginis* from lucerne, and f. sp. *glycinea* from soybean. Hamm and Hansen (71) found oogonial sizes characteristic of both morphological varieties in more than 40% of the progeny of 280 single-zoospore strains from 35 parental isolates of *P. megasperma* originating from 14 hosts. Irwin and Dale (72) analyzed the proteins of morphologically indistinguishable *P. megasperma* isolates from soybean, lucerne, and chickpea in Australia. The protein profiles of the isolates from chickpea or lucerne were identical, but differed significantly from those of the isolates from soybean. These results supported the distinction of the established *formae speciales* in *P. megasperma*. Faris et al. (73) differentiated two groups of isolates of f. sp. *medicaginis* from lucerne characterized by their protein patterns and which were also distinct by morphology, sexual behavior, ecology, and pathogenicity. One group was lucerne-specific and severely pathogenic; the other was pathogenic to both lucerne and soybean. On the basis of protein patterns and pathogenicity, Hansen and Hamm (74) and Hansen et al. (75) recognized six groups of isolates of the *P. megasperma*, one, the "broad host range" group, being divided into four subgroups with distinct protein patterns and karyotypes, and they proposed a phylogeny of the *P. megasperma* population. Bielenin et al. (63) confirmed that grouping of isolates, characterized by host specificity or range. Each group of isolates could be recognized by both NP and SDS protein patterns and also by colony morphology, cardinal growth temperature, and oogonium size. The polyphagous *P. megasperma* is definitely a complex species.

Confusion regarding the *P. cryptogea–P. drechsleri–P. gonapodyides* complex is another example supporting the contention that a morphological approach is no longer satisfactory for the genus *Phytophthora* (76). Bielenin et al. (63) investigated both NP and SDS-denatured protein patterns and demonstrated the heterogeneity of *P. cryptogea* and *P. drechsleri*. *Phytophthora cryptogea* was divided into three groups by native protein patterns. One of these corresponded to one SDS protein pattern. The two other groups had a similar SDS pattern. *Phytophthora drechsleri* isolates were divided into two protein electrophoretic groups. The main group, including an authentic Tucker's strain, had characteristic native and denatured protein patterns. The other group consisted of two isolates that had an SDS pattern typical of *P. cryptogea*. They concluded that the two species could be merged. The occurrence of hybridization between isolates of the two taxa is an additional argument in favor of this option.

Brasier et al. (77) compared isolates that were attributed to *P. drechsleri* from trees and aquatic habitats in North America and isolates from the same habitats in Britain that were attributed to *P. gonapodyides*. The isolates from North America and Britain showed the same morphological, physiological, and sexual characteristics, and had protein patterns typical of *P. gonapodyides* that were significantly distinct from *P. drechsleri*. Furthermore, isolates of *P. gonapodyides* did not mate with *P. drechsleri* but induced oogonium formation in the latter, remaining sterile themselves. *Phytophthora gonapodyides* was divided into two groups, one of which was chlamydospore forming and exhibited some variation in characteristic protein band. Supported by genomic analysis, the chlamydospore forming group in *P. gonapodyides* was suggested to be a hybrid of *P. gonapodyides* and *P. cryptogea*, and, as presently circumscribed, *P. gonapodyides* might be polyphyletic (77). Three unusual *Phytophthora* isolates from tree roots in Britain showed a protein pattern differing broadly from those of *P. gonapodyides* and *P. cryptogea*, but had the characteristic protein band of *P. gonapodyides* and the two bands characteristic of *P. cryptogea*, together with different maximum growth temperatures and sexual behavior (77). They possibly represent a hybrid species, with *P. gonapodyides* and *P. cryptogea* as parents (77).

In another study, Zhang et al. (78) demonstrated the characteristic protein patterns of the A1 and A2 mating types of *P. colocasiae* in China. They also segregated a nonmating neutral type of isolate, designated as A0 mating type. The mating types gave protein patterns which were identical in the number and position of the major bands but which differed in their density.

B. Ascomycetes

1. *Endomycetes: Saccharomycetales*

a. Saccharomyces, Zygosaccharomyces. Since the original publication of the name *Saccharomyces cerevisiae*, more than 40 species have been described that were reduced to synonymy with *Saccharomyces cerevisiae* by Yarrow (79). This taxonomic simplification stimulated a search for other chemotaxonomic approaches. Vaughan-Martini and Kurtzman (80) and Vaughan-Martini and Martini (81,82) demonstrated four distinct species in the *S. cerevisiae* complex by nDNA/nDNA reassociation among type strains: *S. cerevisiae*, *S. bayanus*, *S. pastorianus*, and *S. paradoxus*. Van Vuuren and Van der Meer (83) investigated native proteins of isolates (no types) identified as *S. cerevisiae*, *S. bayanus*, *S. uvarum*, and *S. carlsbergensis* (*S. pastorianus*), and showed a correlation coefficient ranging from .81 to .98 among these isolates, thereby concluding their conspecificity. Van der Westhuizen and Pretorius (84) applied the same technique to South African industrial strains of *S. cerevisiae* and obtained a unique protein profile for each. They observed that protein profiles of artificial hybrids were a combination of the parent profiles. This finding is of interest for the phylogeny and the protection of industrial strains. Drawert and Bednár (85) also found similar IEF protein patterns in isolates of *S. bayanus*, *S. uvarum*, and *S. aceti* and the neotype strain of *S. cerevisiae*, and these differed significantly from the type strain of *Zygosaccharomyces rouxii*. Similarly, Cugnon and Hennebert (86) analyzed IEF proteins patterns of strains of *S. cerevisiae sensu lato*, including 16 type strains of synonyms, but could not confirm Martini's data. A clear segregation of *S. dairensis*, *S. exiguus*, *S. kluyveri*, and *S. servazzi* from each other and from a large cluster ($Ssm > 0.86$) of *S. cerevisiae* was observed, however.

b. Candida and the Teleomorphs Pichia, Kluyveromyces, Issatchenkia, Clavispora, Torulaspora, Saccharomyces, Yarrowia. The genus *Candida* is currently restricted to anamorphs of ascomycete yeasts as characterized by urease activity and the DBB reaction. However, the genus remains heterogeneous for many characters, such as coenzyme Q, which varies from Q6 to Q9 among species of the genus. Shechter et al. (14) analyzed NP-PAGE of strains assigned to seven *Candida* species and their clustering of similarity values showed clear interspecific differences. Strains of *C. stellatoidea* and *C. albicans* were grouped together, in agreement with their accepted synonymy. Clusters of *C. albicans* strains and *C. pseudotropicalis* strains were closely linked, while *C. tropicalis* was quite remote. Vancanneyt et al. (26) undertook a survey of nine anamorphic *Candida* species and 11 teleomorphic ascosporogenous species with *Candida* anamorphs in *Clavispora*, *Issatchenkia*, *Kluyveromyces*, *Pichia*, *Saccharomyces*, *Torulaspora*, and *Yarrowia*. Five basidiomycete yeasts were also included, among them *Leucosporidium scottii* with anamorph named "*Candida*" *scottii*. The 112 strains included the type strains of both anamorph and/or teleomorph species, and were compared on the basis of the numerical analysis of their SDS whole-cell protein profiles using the Pearson correlation coefficient and UPGMA clustering. A clear distinction between the ascomycetous yeasts including the anamorphic *Candida* species and the basidiomycetous yeasts indicated that the name *Candida scottii* should be transferred to *Cryptococcus*. The tree topology confirmed the polyphyletic nature of the genus *Candida*, as indicated by the coenzyme Q values. Noteworthy was the regular clustering of the anamorphic strains together with the fertile teleomorphic strains in each species. This confirmed the established anamorph-teleomorph connections and demonstrated possible links between still unrelated anamorphic strains and teleomorphic taxa. The results further confirmed the synonymy of *Candida ravautii* and *C. catenulata*, *C. vulgaris* and *C. tropicalis*, and *Saccharomyces fermentati* and *Torulaspora delbruekii*. These synonymies were supported by DNA-DNA reassociation. The clustering also led to the reidentification of four *Candida* strains.

2. Taphrinomycetes: Taphrinales

a. Taphrina. A cluster analysis of the NP patterns of isolates of 31 species of *Taphrina* carried out by Snider and Kramer (87,88) showed large intra- and interspecies variations, with no bands common to all species. Only 49% to 63% similarity was seen among species.

3. Hymenoascomycetes

a. Lecanorales: Arthrodermataceae

i. Microsporum, Nannizia, Trichophyton, Arthroderma, Epidermophyton, Coccidioides. Native protein patterns from both mycelium and culture filtrate of six dermatophytic species of *Microsporum*, *Trichophyton*, and *Epidermophyton* were investigated by Shechter et al. (11,36). Electrophoretic patterns of culture filtrates showed proteins different from those of the corresponding mycelial extracts, but led to similar results. Three common bands were exclusive to the *Microsporum* species and one to the *Trichophyton* species. *Trichophyton mentagrophytes* and *T. tonsurans* were very close, with five bands in common, and quite distinct from *T. rubrum*. Five protein bands were specific for *Epidermophyton floccosum*. The presence of common bands in each genus and the occurrence of typical bands in each of the species may indicate, respectively, generic and specific proteins. Some protein bands were common to all species studied and may be representative of the dermatophytes. In one of these studies (36) one isolate of *T. rubrum* showed high similarity with *E. floccosum*, suggesting the need for reidentification. Also, one isolate of each of *T. tonsurans* and *T. mentagrophytes* had an aberrant profile. Shechter (89) applied the simple matching similarity coefficient in the comparison of protein patterns of 21 species of dermatophytes. The results confirmed previous studies with the eight species of *Microsporum* clustered together. The genus *Trichophyton* divided into two distinct clusters, one consisting of *T. tonsurans*, *T. mentagrophytes*, *T. terrestre*, *T. ajelloi*, and *T. floccosum*, and the other of *T. rubrum, T. violaceum, T. gallinae, T. verrucosum, T. schoenleinii, T. soudanense*, and *T. yaoudei*. *Epidermophyton floccosum* remained fairly distinct from the other genera.

Water-soluble proteins of culture filtrates of isolates of *Trichophyton* and *Microsporum* species and of *Nannizzia otae* were fractionated by IEF-PAGE by Jeffries et al. (90). In each genus, the patterns of the species were distinct. Patterns for the two *Microsporum equinum* isolates were identical, but those of *M. distor-*

tum were different. Isolates of *M. canis* showed a high similarity to the (–) mating type of *Nannizzia otae*, its established teleomorph.

Symoens et al. (91) used IEF-PAGE to investigate isolates of the complex species *T. mentagrophytes* and its teleomorph *Arthroderma benhamiae*, *T. interdigitale* (once considered to be a variety of *T. mentagrophytes*) and its teleomorph *Arthroderma vanbreuseghemii*. Laser densitometric scanning, scan normalization, and numerical analysis using the Pearson correlation coefficient and UPGMA showed two clusters. The first included *Arthroderma vanbreuseghemii*, all strains of *T. interdigitale*, and also four strains identified as *T. mentagrophytes*. The second cluster grouped *A. benhamiae* and two strains of *T. mentagrophytes*. The results supported the established anamorph-teleomorph connections, and the authors concluded that *T. mentagrophytes sensu lato* is connected to two teleomorph species. As *T. interdigitale* should be maintained as a distinct anamorphic species, the results indicate the need for a more precise classical diagnosis of both of the species *T. mentagrohytes sensu stricto* and *T. interdigitale*.

Shechter and Newcomer (92) used NP-PAGE to demonstrate the specific character of *Coccidioides immitis* protein patterns.

b. Lecanorales: Parmeliaceae

i. Pseudevernia. Strobl et al. (93) analyzed lichenized specimens of *Pseudevernia furfuracea* var. *ceratea* from different habitats and altitude and only found variation in band density with altitude.

c. Eurotiales: Trichocomaceae

i. Neosartorya, Aspergillus. Isolates of species of the *Aspergillus flavus* group were compared to isolates of *A. fumigatus* and *A. ochraceus* by Kulik and Brooks (94). All species had two common protein bands, which might be of generic significance. The five species of the *flavus* group had two further bands in common, which were not found in *A. fumigatus* or *A. ochraceus*. *A. oryzae* and *A. parasiticus* appeared more closely related to *A. flavus* than were *A. leporis* and *A. tamarii*. Sorenson et al. (95) compared isolates of seven *Aspergillus* species and varieties of the *flavus, terreus*, and *wentii* groups. Similarities of 85% were found among the three isolates of *A. terreus* var. *terreus*. Similarities of 65% to 75% were found among taxa in the groups *terreus, flavus*, and *wentii*. The distinction of species and varieties in the *A. fumigatus* group, including *Neosartorya* species, was investigated by Hearn et al. (96) and Girardin and Latgé (97) through NP-PAGE and SDS-PAGE. Native protein patterns were too complex to be interpreted, but SDS protein patterns showed three major protein bands common to all except one isolate of the *A. fumigatus* varieties, and to two isolates identified as *A. unilateralis* and *A. duricaulis*. Other isolates of the latter two species were quite different from *A. fumigatus*. Similarity was also found between isolates of *A. fumigatus* var. *ellipticus* and *N. fischeri* var. *glabra*, possibly providing evidence for an anamorph-teleomorph relationship. Patterns of *N. fennelliae* and *N. aurata* differed from those of *N. fischeri* and *A. fumigatus*.

ii. Penicillium. Bent (46) found distinct protein patterns for *Penicillium griseofulvum*, *P. chrysogenum*, and *P. frequentans* and observed similar pattern alterations by varying culture conditions.

d. Microascales: Microascaceae

i. Ceratocystis. Native protein extracts of several isolates of *Ceratocystis coerulescens*, *C. fagacearum*, *C. fimbriata*, and *C. ulmi* were analyzed by Stipes (98,99) and gave these distinct, species-specific profiles, although significant intraspecific dissimilarities were noted. Five protein bands were found common to all four species.

e. Sordariales: Sordariaceae

i. Neurospora. Chang et al. (6) demonstrated differences in native proteins between a mutant and a wild-type strain of *Neurospora crassa*, which were also clearly distinct from those of *N. sitophila* and *N. intermedia*. Only minor differences in protein patterns were found between two interfertile strains of *N. crassa*, from distant geographic origin.

f. Hypocreales: cfr Hypocreaceae

i. Fusarium. Meyer and Renard (35), Glynn and Reid (33), Whitney et al. (49), and Mandeel et al. (100) observed as many variations in protein patterns among the isolates of *formae speciales* of *F. oxysporum* as between the latter and other *Fusarium* species. Protein profiles were unable to separate biotypes of different host or origin. However, there are enough recognizable differences at species level to demonstrate the large species diversity in the genus, in disagreement with the reductive taxonomy of Snyder and Hansen (101).

ii. Verticillium. Microsclerotial and dark mycelial forms of *Verticillium dahliae* are taxonomically controversial. Dark mycelial forms have been segregated as *V. albo-atrum*. Hall (102) found significantly different protein patterns for *V. dahliae* and *V. albo-atrum* together with identical protein patterns of isolates within each species, irrespective of their host or geographical origin. The results confirmed the morphological distinction of the species.

Pelletier and Hall (42) and Milton et al. (103) also demonstrated that *V. albo-atrum* and *V. dahliae* are more closely related to each other than to *V. nigrescens*, *V. nubilium*, and *V. tricorpus*, all species patterns being significantly distinct.

g. Phyllachorales: Phyllachoraceae

i. Glomerella, Colletotrichum. McCombs and Winstead (39) demonstrated distinct NP patterns in *Glomerella* species. Stipes and McCombs (104) analyzed monoascoporic strains morphologically identified as *Glomerella cingulata* from six different host plants. They recovered species-specific bands of native proteins but found no other similarities in relation to the different hosts. Similarly, Dale et al. (105) compared morphology, native and denatured protein patterns, and dsRNA patterns of Australian isolates of pathogenic types A and B of *C. gloeosporioides* (*G. cingulata*) from *Stylosanthes guyanensis*. Neither NP nor SDS protein patterns could be used to differentiate type A from type B, which was possible from dsRNA analysis.

h. Diaporthales: Valsaceae

i. Leucostoma, Cytospora. Protein patterns of six *Cytospora* isolates from peach canker were compared to those of teleomorphs *Leucostoma cincta* and *L. persoonii* by Gairola and Powell (106). The *Cytospora* protein patterns were different from each other, and none were identical to any of the reference species. The data could not resolve the taxonomic position of the *Cytospora* isolates.

i. Diaporthales: cfr Melanconiaceae

i. Melanconium. Shamoun and Sieber (107) found distinct native protein patterns for *Melanconium apiocarpum* and *M. marginale*. Separate grouping of pathogenic and endophytic isolates of *M. apiocarpum* obtained by their β-esterase profile could not be confirmed.

j. Leotiales: Leotiaceae.

i. Gremmeniella, Brunchorstia, Ascocalyx. *Gremmeniella abietina* was segregated from *Ascocalyx abietis* by Groves (108) and Morelet (109) on the basis of morphology, but Dorworth (110) showed a very high similarity of the protein patterns between the two genera. Petrini et al. (111) analyzed NP patterns of isolates of *G. abietina*, *A. laricina*, and *A. abietis*. The isolates of *G. abietina* from *Pinus* and *Abies* had the same pattern with two major common bands. *Ascocalyx abietina* and *A. laricina* showed only one of the two common bands of the *G. abietina* isolates. This corroborated the creation of the genus *Gremmeniella* by Morelet (109). However, the authors did transfer *A. laricina* to *Gremmeniella* with no reason offered in favor.

Gremmeniella abietina var. *abietina* (anamorph *Brunchorstia pinea* var. *pinea*), occurs as different pathogenic races on conifers over three geographical areas—North America, Europe, and Asia. Petrini et al. (112) showed distinct NP protein patterns among European, North American, and Asian races of *G. abietina*. var. *abietina* and demonstrated the absence of the North American race of the fungus in Europe. Petaïsto et al. (51) confirmed this from characteristic differences in minor SDS-PAGE protein bands between the North American and the North European races. Petrini et al. (112) also pointed out the high similarity of anamorph *B. pinea* var. *cembra* with *G. abietina* var. *abietina*, differing by only one protein band being of higher density. *Brunchorstia pinea* var. *cembra* is known to have the same type of conidia as *G. abietina*, differing only by the number of septa and the host range (113).

k. Leotiales: Sclerotiniaceae

i. Sclerotinia, Myriosclerotinia, Lambertella, Botryotinia, Botrytis, Sclerotium. Russo et al. (43) demonstrated the existence of particular "developmental proteins" in sclerotial and stromatal structures of Sclerotiniaceae by gradient SDS-PAGE. These were absent in young mycelium and disappeared at the conidial or apothecial production stage. Insell et al. (53) estimated their molecular weight at 32 to 37 kDa and reported their absence in the sclerotial basidomycete *Athelia rolfsii (Sclerotium rolfsii)*. Interspecies differences in two to five bands of sclerotial proteins were detected among *Sclerotinia sclerotiorum*, *S. trifoliorum*, *S. minor*, *S. homeocarpa*, *Myriosclerotinia borealis*, *Botryotinia fuckeliana*, *Lambertella subrenispora*, and *Sclerotium cepivorum* by Petersen et al. (114), Insell et al. (53), and Novak and Kohn (115). *Sclerotinia trifoliorum* was found to be so close to *S. sclerotiorum* that Tariq et al. (116) proposed their conspecificity. *Sclerotinia minor* differed significantly.

NP-PAGE of sclerotia was carried out on isolates of *Botrytis cinerea*, *B. aclada*, *B. fabae*, *B. gladiolorum*, *B. tulipae*, and *B. viciae* by Backhouse et al. (117). Patterns were composed of 61 different protein bands. Numerical analysis grouped all isolates in clusters matching their species identification based on sclerotial morphology. All isolates of *Botrytis cinerea* clustered together, with a similarity from 78% to 97%. The other species formed three separate clusters.

4. Loculoascomycetes

a. Dothideales: cfr Herpotrichiellaceae

i. Fonsecaea, Rhinocladiella. In a numerical analysis of SDS- and IEF-PAGE protein patterns by Ibrahim-Granet et al. (118), isolates of *Fonsecaea pedrosoi* (*Rhinocladiella pedrosoi*) clustered together with no strict relation to their Asian, African, and Central American origin.

b. Dothideales: cfr Mycosphaerellaceae

i. Septoria. The analysis of isolates of *Septoria avenae*, *S. nodorum*, and *S. tritici* by Durbin (34) showed high similarity between most isolates and divergence of others in each species. The author suggested a possible continuous mutation within a heterokaryotic system.

ii. Cercospora. Native protein analysis of *Cercospora* isolates from pasture legumes *Trifolium*, *Lotus*, *Melilotus*, and *Medicago* by Peterson and Latch (119) gave identical patterns in isolates from the same host and minor variation between isolates from different legume species. The data could indicate that the isolates are conspecific although they could possibly be identified to three distinct *Cercospora* species according to classical Chupp (120) taxonomy.

c. Dothideales: cfr Pleosporaceae

i. Drechslera. Protein patterns from conidial and mycelial extracts of *Drechslera erythrospilum* and *D. teres* were compared by Shipton and MacDonald (41). In both extracts, the intraspecies differences were greater than those between species. The case appears similar to that of the plant pathogenic *Fusarium* species.

d. cfr Dothideales

i. Macrophomina. Isolates of *Macrophomina phaseolina* causing legume stem blight from Puerto Rico and the Dominican Republic are morphologically similar, but vary in the size of microsclerotia and in their pathogenic behavior. Protein patterns obtained by Campo-Arana et al. (121) corroborated the distinction between the two populations and suggested that they could be considered as races.

C. Basidiomycota

1. *Ustomycetes*

a. Sporidiales: Sporidiaceae

i. Rhodosporidium, Rhodotorula, Leucosporidium, Kondoa. Vancanneyt et al. (29) analyzed the SDS protein profile, DNA base composition, and coenzyme Q of 105 type and other strains of six teleomorphic *Rhodosporidium* species and 23 anamorphic *Rhodotorula* species. Gel scans were normalized, compensated, and analyzed numerically with the GELCOMPAR (28) program. The strains of the six *Rhodosporidium* species grouped in distinct clusters. The type strains, either singly or with other identified strains, of 21 *Rhodotorula* species showed distinct unique protein profiles. Few aberrant strains which had possibly been incorrectly assigned to *Rhodotorula acheniorum*, *R. araucariae*, *R. aurantiaca*, *R. foliorum*, and *R. minuta* were clustered at some distance from their respective type cluster.

The general distribution of the anamorphic *Rhodotorula* species clusters among the teleomorphic *Rhodosporidium* clusters, supported a close mutual relationship. In one case, a high anamorph-teleomorph correlation at species level was observed. Both type and other strains of *Rhodotorula glutinis* and *R. graminis* were clustered together with the type and other strains of *Rhodosporidium diobovatum*. This supports the synonymy of the two anamorphic names and confirms their controversial connection with the named teleomorph. Other strains identified as *Rhodotorula graminis* or *R. glutinis* were found clustered with *Rhodosporidium paludigenum*, *R. kratochvilovae*, *R. sphaerocarpum*, or *R. toruloides* and need reconsideration. These results were supported by coenzyme Q analysis (29).

In a larger investigation of basidiomycetous yeasts, including Filobasidiales, by Vancanneyt et al. (27), *Rhodosporidium toruloides*, *R. paludigenum*, and *R. sphaerocarpum* on the one hand and *R. diobovatum* with *Leucosporidium antarticum* and *L. scottii* on the other hand grouped in two clusters distinct from the Filobasidiales. *R. dacryoidum* was well separated, clustering with *Kondoa malvinella*, a species first described as *Rhodosporidium malvinellum*.

b. Sporidiales: Ustilaginaceae

i. Ustilago. Kim et al. (122) examined the relationships among six species of *Ustilago*, causing smut on cereals (*Triticum, Aegilops, Hordeum, Avena*) by 2D-PAGE. More than 280 bands were detected and extracts from spores collected on the hosts and of mycelium in culture gave identical patterns. Two different patterns were found in *U. tritici* depending on the host. *U. tritici* and *U. nuda* differed significantly, by 47 bands. Variation in 13 bands was observed within the species *U. avenae*, *U. nigra*, *U. hordei*, and *U. kolleri*. These four species are sexually intercompatible. Mutual crossing of the smooth-spored species *U. kolleri* and *U. hordei* gave rise to offspring with echinulate spores, a character of *U. avenae*. The authors therefore strongly supported the Lindeberg and Nannfeldt (123) synonymy of the four species under the name *U. segetum*.

2. Teliomycetes

a. Uredinales: Pucciniaceae

i. Puccinia. Protein extracts of uredospores collected on host plants were used by Shipton and Fleischmann (124) to examine physiological races and *formae speciales* of *Puccinia graminis*, *P. recondita*, and *P. hordei*. No correlation was found between morphology and protein pattern. Most profiles differed depending on the physiological races, the *forma specialis*, or the species.

ii. Uromyces. Kim et al. (125) studied polypeptide patterns of the bean rust *Uromyces phaseoli* var. *phaseoli*, the cowpea rust *U. phaseoli* var. *vignae*, and the fababean rust *U. viciae-fabae* by 2D-PAGE. Of 335 bands found in *U. phaseoli* var. *phaseoli* and *U. phaseoli* var. *vignae* 183 were common, and 149 and 146 of the 277 bands detected in *U. viciae-fabae* were common with the former varieties respectively. Within each taxon differences ranged from zero to 19 bands only. Consequently the authors suggested raising the rank of the two varieties of *U. phaseoli* to that of species.

3. (Eu)Basidiomycetes

a. Tremellales: Filobasidiaceae

i. Filobasidium, Filobasidiella, Cystofilobasidium, Mrakia. Vancanneyt et al. (27) used SDS-PAGE to examine type and other strains of 12 basidiomycete (Filobasidiales) yeast species and varieties in *Filobasidium*, *Filobasidiella*, *Cystofilobasidium*, and *Mrakia*, compared to ustomycete (Sporidiales) species. Computation, normalization, and compensation of the profiles followed by numerical analysis based on a Pearson coefficient matrix showed a high pattern reproducibility for each strain and an excellent matching of the dendrogram with species definitions based on other criteria. Within each species the correlation between the type and other strains was generally over .87, and over .91 between mating types. The correlation coefficient between species was less than 0.87. The four species of *Mrakia* were an exception. Their type and other strains were similar, forming one possibly conspecific cluster. The five strains of *Filobasidiella neoformans* varieties were recovered as two clusters, irrespective of their varietal name. It appeared that *Cystofilobasidium capitatum* and *C. infirmominiatum* were related, while *C. bisporidii* fell in a larger cluster with *Filobasidium capsuligenum* and *F. neoformans*. It is also notable that clusters of the Filobasidiales were well distinguished from those of the Sporidiales *Rhodosporidium*, *Leucoporidium*, and *Kondoa*. This in no way contradicts the clear discrimination between the ustomycete yeasts and basidiomycete yeasts demonstrated from 18s rDNA sequences by Wilmotte et al. (126).

ii. Filobasidium, Cryptococcus. Another taxonomic investigation by Vancanneyt et al. (30) investigated the relationship between 113 type and other strains representing 22 anamorphic species of *Cryptococcus* and the presumably related teleomorphic species *Filobasidium floriforme*. Noteworthy was the clustering of the type and other strains of both *F. floriforme* and *C. ater* together. This suggests an anamorph-teleomorph connection. The type strain and some other strains of *C. elinovii* and *C. terreus* and of *C. kuetzingii* and *C. albidus* var. *albidus* were found in one cluster, indicating possible synonymies. *C. albidus* var. *aerius* was recovered quite far from *C. albidus* var. *albidus*. Strains identified as *C. albidus* var. *albidus*, *C. laurentii*, *C. luteolus*, and *C. humicola* were recovered elsewhere and need revision. Other *Cryptococcus* species were recovered in single homogeneous clusters or branches.

b. Tremellales: Tremellaceae

i. Tremella, Cryptococcus. Ten species of *Tremella* producing a yeast phase in culture were investigated together with *Cryptococcus* species (30). The yeastlike *Tremella* species were divided into two groups among the *Cryptococcus* clusters, one group of five species was recovered as three clusters, *T. encephala* and *T. subanomala* ($r > .92$), *T. fulciformis* and *T. samoensis* ($r > .92$) and *T. aurantia*, and the other group of three species was recovered in two clusters *T. coalescens* and *T. mesenterica* ($r > .92$) and *T. brasiliensis*. The synonymy of the species paired by a high correlation coefficient is supported by a close or identical DNA base composition. The remaining species *T. globospora* and *T. foliacea* formed remote branches of the tree.

c. Cantharellales: Typhulaceae

i. Typhula. Newsted and Huner (127) found characteristic protein bands in the slerotial extracts of the boreal fungi *Typhula idahoensis* and *T. incarnata*. These were of lower molecular weight than those detected in the sclerotial Ascomycetes and ranged from 16 to 22 kDa in *T. idahoensis* and from 15 to 18 kDa in *T. incarnata*.

d. Stereales: Atheliaceae

i. Athelia, Sclerotium. Sclerotial extract of *Sclerotium rolfsii*, the anamorph of *Athelia rolfsii*, was examined by Insell et al. (53) and showed three protein bands—one major one at 16 kDa, and two minor ones at 14 and 15 kDa. The characteristic developmental proteins found in sclerotia of Basidiomycetes are of significantly lower molecular weight than those from the sclerotial Ascomycetes.

e. Ceratobasidiales: Ceratobasidiaceae

i. Rhizoctonia. *Rhizoctonia solani* was found to be a complex species by Clare et al. (128) on the basis of native protein patterns, with strains showing a broad diversity of patterns.

f. Agaricales, Russulales, Boletales, Phallales

i. Hygrophoropsis, Coprinus. Antibus (44) characterized specific major bands of developmental proteins in the sclerotial phase of *Hygrophoropsis aurantiaca* at 14.1, 17.7 and 22.9 kDa molecular weight. Mycelial extracts showed similar major bands but of lower density. Sclerotial extract of *Coprinus psychromorbidus* analyzed by Newsted and Huner (127) gave three characteristic bands of 12.9, 13.8, and 14.5 kDa. Smythe and Anderson (129) found clear differences in the protein patterns of single-spore isolates from the same wild fruitbody of *Coprinus lagopus*. Two bands differed depending on the mating types, and the other 11 bands were homologous. Other profile similarities and differences could be found among other compatible and uncompatible strains from other fruitbodies of the same species. Backcrossed strains showed less variation than the wild-type strains.

g. Sclerodermatales: Sclerodermataceae

i. Pisolithus. The widespread mycorhizal fungus *Pisolithus tinctorius* is variable. One hundred isolates from Australian and non-Australian origins and from diverse symbiont species were analyzed by Burgess et al. (130) on the basis of their morphological characters and SDS-PAGE protein patterns. Data were treated by different numerical analyses, and separate and combined dendrograms were constructed. Five main groups of protein patterns were distinguished with a similarity as low as 60%, and subgroups showed a significant correlation with geographical distribution, host association, and basidiospore type. When protein and morphological data were clustered together, lower similarities were found between groups and subgroups.

V. CONCLUSION

Protein patterns of fungi are an objective chemotaxonomic approach if prepared in a reproducible standardized way. Electrophoresis is a highly sensitive technique, and the slightest modification in experimental conditions may induce variations in the resulting pattern.

Investigators have been aware of potential difficulties. Garber (131) indicated the need for an evaluation of the different cultural and extraction procedures for fungi, and this has not really been fulfilled. Hall (132) pointed out the difficulty of an objective comparative measurement of relatedness of protein profiles and expressed the need for simplification, standardization, and automation of the technique. This has occurred, particularly through bacteriology and standardized reproducible electrophoretic procedures linked to objective automated computer-aided gel scan and analysis. Protein profiles may vary significantly at the species and strain levels. In that respect the necessity of including the type strains in all studies and testing a sufficient number of isolates is recommended before taxonomic conclusions can reliably be drawn. Whole-cell protein electrophoresis is a good aid to taxonomy, identification, and population delineation. It allows determination of conspecificity of strains or definition of species heterogeneity, possibly leading to reidentification of strains or revision of classical criteria, to confirm or suggest anamorph-teleomorph connections, to confirm filiation of strains within species (mutant wild-type), to explore the structure of specific populations (*formae speciales* ...), to suggest synonymies of taxa or, conversely, segregation of varieties, to demonstrate polyphyletic nature of some genera and relationship between their respective members and, to a certain extent, to characterize higher groups of fungi. Automated methods have been applied to a significant number of fungi, thereby fulfilling conditions of reproducibility and analysis of type. The results, which cover large numbers of strains including type strains, are of real value for fungal taxonomy.

ACKNOWLEDGMENTS

The authors acknowledge the support by the Federal Office for Scientific, Technical and Cultural Affairs (OSTC) of the Belgian State, of the research project Polyphasic Taxonomic Approach of the Yeasts *Candida* and Others as a part of the Belgian program Pulse to Fundamental Research in Life Sciences, 1989–1992. They also are grateful to Professor Karel Kersters, Head of the Laboratorium voor Microbiologie, Universiteit Gent, Ghent, for critical reading of the manuscript.

REFERENCES

1. Gottlieb LD. Gel electrophoresis: new approach to the study of evolution. BioScience 1971; 21:939–944.
2. Smithies O. Zone electrophoresis in starch gels: group variations in the serum proteins of normal human adults. Biochem J 1955; 61:629–641.
3. Ornstein L. Disc electrophoresis I. Background and theory. Ann NY Acad Sci 1964; 121:321–349.
4. Davis BJ. Disc electrophoresis II. Methods and application to human serum proteins. Ann NY Acad Sci 1964; 121:404–427.

5. Raymond S, Weintraub L. Acrylamide gel as a supporting medium for zone electrophoresis. Science 1959; 130:711.
6. Chang LO, Srb AM, Steward FC. Electrophoretic patterns of the soluble proteins of *Neurospora*. Nature 1962; 193:756–759.
7. Laemmli UK. Cleavage of structural proteins during the assembly of the head of bacteriophage T4. Nature 1970; 227:680–685.
8. Vesterberg H, Svensson H. Isoelectric fractionation, analysis, and characterization of ampholytes in natural pH gradients. IV. Further studies on the resolving power in connection with separation of myoglobin. Acta Chem Scand 1966; 20:820–832.
9. O'Farrell PH. High resolution two-dimensional electrophoresis of proteins. J Biol Chem 1975; 250:4007–4021.
10. Clare BG, Zentmyer GA. Starch gel electrophoresis of proteins from species of *Phytophthora*. Phytopathology 1966; 56:1334–1335.
11. Shechter Y, Landau JW. Dabrowa N, Newcomer VD. Comparative disc electrophoretic studies of proteins from dermatophytes. Sabouraudia 1966; 5:144–149.
12. Rouatt JW. Skyring GW, Purkayastha V, Quadling C. Soil bacteria: numerical analysis of electronic protein patterns developed in acrylamide gels. Can J Microbiol 1970; 16:202–205.
13. Shechter Y, Landau JW, Dabrowa N. Comparative electrophoresis and numerical taxonomy of some *Candida* species. Mycologia 1972; 64:841–853.
14. Sokal RR, Sneath PA. Principles of Numerical Taxonomy. London: Freeman, 1963.
15. Kersters K, De Ley J. Identification and grouping of bacteria by numerical analysis of their electrophoretic protein patterns. J Gen Microbiol 1975; 87:333–342.
16. Sneath PA, Sokal RR. Numerical Taxonomy. The Principles and Practice of Numerical Taxonomy. San Francisco: Freeman, 1973.
17. Kersters K, De Ley J. Classification and identification of bacteria by electrophoresis of their proteins. In: Goodfellow M, Board RG, eds. Microbiological Classification and Identification. London: Academic Press, 1980:273–297.
18. Kersters K. Numerical methods in the classification of bacteria by protein electrophoresis. In: Goodfellow M, Jones D, Priest FG, eds. Computer-Assisted Bacterial Systematics. London: Academic Press, 1985:337–369.
19. Jackman PJH, Feltham RKA, Sneath PHA. A program in Basic for numerical taxonomy on microorganisms based on electrophoretic protein patterns. Microbios Lett 1983; 23:87–98.
20. Jackman PJH. Bacterial taxonomy based on electrophoretic whole-cell protein patterns. In: Goodfellow M, Minnilin DE, eds. Chemical Methods in Bacterial Systematics. London: Academic Press, 1985:115–128.
21. Jackman PJH. Microbial systematics based on electrophoretic whole-cell protein patterns. Methods Microbiol 1987; 19:209–225.
22. Jackman PJH. A program in Basic for numerical taxonomy on microorganisms based on electrophoretic band positions. Microbios Lett 1983; 23:119–124.
23. Plikaytis BD, Carlone GM, Plikaytis BB. Numerical analysis of normalized whole-cell protein profiles after sodium dodecylsulphate polyacrylamide gel electrophoresis. J Gen Microbiol 1986; 132:2653–2660
24. Kersters K, Pot B. Electrophoresis of proteins: data capture, analysis and construc-

tion of databanks. In: Magnien E, ed. Biotechnology Action Programme (B.P.) Progress Report. Brussels: EEC, 1988; 2:9–17.

25. Pot B, Gillis M, Hoste B. et al. Intra- and intergeneric relationships of the genus *Oceanospirillum*. Int J System Bact 1989; 39:23–24.
26. Vancanneyt M, Pot B, Hennebert GL, Kersters K. Differentiation of yeast species based on electrophoretic whole-cell protein. System Appl Microbiol 1991; 14: 23–32.
27. Vancanneyt M, Van Leberge E, Berny JF, Hennebert GL, Kersters K. The application of whole-cell protein electrophoresis for the classification and identification of basidiomycetes yeast species. Antonie Leeuwenhoek 1992; 61:69–78.
28. Vauterin L, Vauterin P. Computer-aided objective comparison of electrophoresis patterns for grouping and identification of microorganisms. Eur Microbiol 1992; 7:1–6.
29. Vancanneyt M, Coopman R, Tytgat R, Berny JF, Hennebert GL, Kersters K. Taxonomic studies of basidiomycetous yeast genera *Rhodosporidium* Banno and *Rhodotorula* Harrison based on whole-cell protein patterns, DNA base compositions and coenzyme Q types. J Gen Appl Microbiol 1992; 38:363–377.
30. Vancanneyt M, Coopman R, Tytgat R, Hennebert GL, Kersters K. Whole-cell protein patterns, DNA base compositions and coenzyme Q types in the yeast genus *Cryptococcus* Kützing and related taxa. System Appl Microbiol 1994; 17:65–75.
31. Pot B, Vandamme P, Kersters K. Analysis of electrophoretic whole-organism protein fingerprints. In: Goodfellow M, O'Donnell AG, eds. Chemical Methods in Prokaryotic Systematics. Chichester: John Wiley, 1994:493–521.
32. Vauterin L, Swings J, Kersters K. Protein electrophoresis and classification. In: Goodfellow M, O'Donnell AG, eds. Handbook of New Bacterial Systematics. London: Academic Press, 1993:251–280.
33. Glynn AN, Reid J. Electrophoretic patterns of soluble fungal proteins and their possible use as taxonomic criteria in the genus *Fusarium*. Can J Bot 1969; 47:1823–1831.
34. Durbin RD. Comparative gel-electrophoretic investigation of the protein patterns of *Septoria* species. Nature 1966; 210:1186–1187.
35. Meyer JA, Renard JL. Protein aned esterase patterns of two formae speciales of *Fusarium oxysporum*. Phytopathology 1969; 59:1409–1411.
36. Shechter Y, Landau JW, Dabrowa N, Newcomer VD. Disc electrophoretic studies of intraspecific variability of proteins from dermatophytes. Sabouraudia 1968; 6: 133–137.
37. Clare BG. Starch gel electrophoresis of proteins as an aid in identifying fungi. Nature 1963; 200:803–804.
38. Newsted WJ, Huner NPA, Insell JP, Griffith M, van Huystee RB. The effects of temperature on the growth and polypeptide composition of several snow mold species. Can J Bot 1985; 63:2311–2318.
39. McCombs CL, Winstead NN. Mycelial protein comparisons of isolates of curcubit anthracnose fungi. Phytopathology 1963; 53:882. Abstract.
40. Dabrowa N, Howard DH, Landau JW, Shechter Y. Synthesis of nucleic acid and proteins in the dimorphic forms of *Candida albicans*. Sabouraudia 1970; 8:163–169.

41. Shipton WA, McDonald WC. The electrophoretic patterns of proteins extracted from spores and mycelium of two *Drechslera* species. Can J Bot 1970; 48:1000–1002.
42. Pelletier G, Hall R. Relationships among species of *Verticillium*: protein composition of spores and mycelium. Can J Bot 1971; 49:1293–1297.
43. Russo GM, Dahlberg KR, Van Etten J. Identification of a developmental specific protein in sclerotia of *Sclerotinia sclerotiorum*. Exp Mycol 1982; 6:259–267.
44. Antibus RK. Formation and structure of sclerotia and sclerotium-specific proteins in *Hygrophoropsis aurantiaca*. Mycologia 1989; 81:905–913.
45. Paranjpe M, Chen KM, Jong SC. Morphogenesis of *Agaricus bisporus*: changes in protein and enzyme activity. Mycologia 1979; 71:469–478.
46. Bent KJ. Electrophoresis of proteins of 3 *Penicillium* species on acrylamide gels. J Gen Microbiol 1967; 49:195–200.
47. Snider RD. Electrophoresis and the taxonomy of phytopathogenic fungi. Bull Torrey Bot Club 1973; 100:272–276.
48. Dooms L, Hennebert GL, Verachtert H. Polyol synthesis and taxonomic characters in the genus *Moniliella*. Antonie Leeuwenhoek 1971; 37:107–118.
49. Whitney PJ, Vaughan JG, Heale JB. A disc electrophoretic study of the proteins of *Verticillium albo-atrum, Verticillium dahliae* and *Fusarium oxysporum* with reference to their taxonomy. J Exp Bot 1968; 19:415–426.
50. Berny JF, Hennebert GL. Variants of *Penicillium expansum*: an analysis of cultural and microscopic character as taxonomic criteria. In: Samson RA, Pitt JI, eds. Modern Concepts in *Penicillium* and *Aspergillus* classification. NATO AIS series A, Life Sciences 185. London: Plenum Press, 1990:49–65.
51. Petäistö RL, Rissanen TE, Harvima RJ, Kajander EO. Analysis of the protein of *Gremmeniella abietina* with special reference to protease activity. Mycologia 1994; 86:242–249.
52. Benhamou N, Ouelette GB, Asselin A, Maicas E. The use of polyacrylamide gel electrophoresis for rapid differentiation of *Gremmeniella* isolates. In: Manion PD. ed. Proceedings of the International Symposium on *Sclerroderris* Canker of Conifers, Syracuse, NY, June 21–24, 1983. Den Hague: Junk, 1984:68–76.
53. Insell JP, Huner NPA, Newsted WJ, Griffith M, van Huystee RB. Light microscopy and polypeptide analyses of sclerotia from mesophilic and psychrophilic pathogenic fungi. Can J Bot 1985; 63:2305–2310.
54. Nasuno S. Differentiation of *Aspergillus sojae* from *Aspergillus oryzae* by polyacrylamide gel disc electrophoresis. J Gen Microbiol 1972; 71:29–33.
55. Hawksworth DL, Kirk PM, Sutton BC, Pegler DN. Ainsworth and Bisby's Dictionary of the Fungi. 8th ed. Wallingford: CAB International, 1995.
56. Adaskaveg JE, Strangellini ME, Gilbertson RL, Egan N. Comparative protein studies of several *Pythium* species using isoelectric focusing. Mycologia 1988; 80: 665–672.
57. Chen W, Hoy JW, Schneider RW. Comparisons of soluble proteins and isozymes for seven *Pythium* species and applications of the biochemical data to *Pythium* systematics. Mycol Res 1991; 95:548–555.
58. Hall R, Zentmyer GA, Erwin DC. Acrylamide gel electrophoresis of proteins and taxonomy of *Phytophthora*. Phytopathology 1968; 58:1052. Abstract.

59. Gill HS, Powell D. Differentiation of three species of *Phytophthora* by polyacrylamide gel electrophoresis. Phytopathology 1967; 57:812. Abstract.
60. Gill HS, Powell D. The use of polyacrylamide gel disc electrophoresis in delimiting three species of *Phytophthora*. Phytopathol Z 1968; 63:23–29.
61. Hall R, Zentmyer GA, Erwin DC. Approach to taxonomy of *Phytophthora* through acrylamide gel electrophoresis of proteins. Phytopathology 1969; 59:770–774.
62. Gill HS, Zentmyer GA. Identification of *Phytophthora* species by disc electrophoresis. Phytopathology 1978; 68:163–167.
63. Bielenin A, Jeffers SN, Wilcox WF, Jones AL. Separation by protein electrophoresis of six species of *Phytophthora* associated with deciduous fruit. In: 5th International Congress of Plant Pathology and 1st International Pythium Workshop. Tokyo: Sankyo, 1988:61–63.
64. Zentmyer GA, Jefferson L, Hickman CJ, Chang-Ho Y. Studies of *Phytophthora citricola*, isolated from *Persea americana*. Mycologia 1974; 66:830–845.
65. Erselius LJ, De Vallavieille C. Variation in protein profiles of *Phytophthora*: comparison of six species. Trans Br Mycol Soc 1984; 83:653–657.
66. De Vallavieille C, Erselius LJ. Variation in protein profiles of *Phytophthora*: survey of a composite population of three species on *Citrus*. Trans Br Mycol Soc 1984; 83: 473–479.
67. Hamm PB, Hansen EM. *Phytophthora pseudotsugae*, new species causing root-rot of Douglas-fir (*Pseudostuga menziezii*). Can J Bot 1983; 61:2626–2631.
68. Waterhouse GM. The genus *Phytophthora* de Bary. Diagnosis or descriptions and figures from the original papers. Mycological Papers. Kew: CAB, 1970; 122:1–59.
69. Newhook FJ, Waterhouse GM, Stamps DJ. Tabular key to the species of *Phytophthora* de Bary. Mycological Papers. Kew: CAB, 1978; 143:1–20.
70. Kuan TL, Erwin DC. Formae speciales differentiation of *Phytophthora megasperma* isolates from soybean and alfalfa. Phytopathology 1980; 70:333–338.
71. Hamm PB, Hansen EM. Single-spore isolate variation: the effect of varietal designation in *Phytophthora megasperma*. Can J Bot 1983; 60:2931–2938.
72. Irwin JAG, Dale JL. Relationships between *Phytophthora megasperma* isolates from chickpea, lucerne and soybean. Aust J Bot 1982; 30:199–210.
73. Faris MA, Sabo FE, Cloutier Y. Intraspecific variation in gel electrophoresis patterns of soluble mycelial proteins of *Phytophthora megasperma* isolated from alfalfa. Can J Bot 1986; 64:262–265.
74. Hansen EM, Hamm PB. Morphological differetiation of host-specialized groups of *Phytophthora megasperma*. Phytopathology 1983; 73:129–134.
75. Hansen EM, Brasier CM, Shaw DS, Ham PB. Taxonomy structure of *Phytophthora megasperma*: evidence for emerging biological species groups. Trans Br Mycol Soc 1986; 87:557–573.
76. Brasier CM. Problems and prospects in *Phytophthora* research. In: Erwin DC, Barnicki-Garcia S, Tsao PH, eds. *Phytophthora*: Its Biology, Taxonomy, Ecology and Pathology. St Paul, MN: American Phytopathological Society, 1983:351–364.
77. Brasier CM, Hamm PB, Hansen EM. Cultural characters, protein patterns and unusual mating behaviour of *Phytophthora gonapodyides* isolates from Britain and North America. Mycol Res 1993; 97:1287–1298.

78. Zhang KM, Zheng FC, Li YD, Ann PJ, Ko WH. Isolates of *Phytophthora colocasiae* from Hainan Island in China: evidence suggesting an Asian origin of this species. Mycologia 1994; 86:108–112.
79. Yarrow D. *Saccharomyces* Meyen ex Rees. In: Kreger–van Rij NJW, ed. The Yeasts: A Taxonomic Study. Amsterdam: Elsevier, 1984:379–395.
80. Vaughan-Martini A, Kurtzman CP. Deoxyribonucleic acid relatedness among species of the genus *Saccharomyces sensu stricto*. Int J Syst Bacteriol 1985; 53: 508–511.
81. Vaughan-Martini A, Martini A. Three newly delimited species in *Saccharomyces sensu stricto*. Antonie Leeuwenhoek 1987; 53:77–84.
82. Vaughan-Martini A, Martini A. A proposal for correct nomenclature of the domesticated species of the genus *Saccharomyces*. In: Cantarelli C, Lanzarini G, eds. Biotechnology Applications in Beverage Production. London: Elsevier, 1989: 1–16.
83. Van Vuuren HJJ, van der Meer L. Fingerprinting of yeasts by protein electrophoresis. Am J Enol Vitic 1987; 38:49–53.
84. van der Westhuizen TJ, Pretorius IS. The value of electrophoretic fingerprinting and karyotyping in wine yeast breeding programmes. Antonie Leeuwenhoek 1992; 61: 249–257.
85. Drawert F, Bednár J. On the electrophoretical differentiation and classification of proteins. 12. Comparative investigation of yeast proteins of different *Saccharomyces* species and various strains belonging to the species *Saccharomyces cerevisiae* Hansen, by means of isoelectric focusing in polyacrylamide gels. J Agric Food Chem 1983; 31:848–851.
86. Cugnon L, Hennebert GL. Investigation in the *Saccharomyces cerevisiae* complex by isoelectric focusing electrophoresis of whole-cell proteins. (Unpublished.)
87. Snider RD, Kramer CL. Polyacrylamide gel electrophoresis and numerical taxonomic of *Taphrina coerulescens* and *Taphrina deformans*. Mycologia 1974; 66: 743–753.
88. Snider RD, Kramer CL. An electrophoretic protein analysis and numerical taxonomic study of the genus *Taphrina*. Mycologia 1974; 66:754–772.
89. Shechter Y. Electrophoresis and taxonomy of medically important fungi. Bull Torrey Bot Club 1973; 100:277–287.
90. Jeffries CD, Reiss E, Ajello L. Analytical isoelectric focusing of secreted dermatophyte proteins applied to taxonomic differentiation of *Microsporum* and *Trichophyton* species (preliminary studies). Sabouraudia 1984; 22:369–380.
91. Symoens F, Willenz P, Rouma V, Planard C, Nolard N. Isoelectric focusing applied to taxonomic differentiation of the *Trichophyton mentagrophytes* complex and the related *Trichophyton interdigitale*. Mycoses 1989; 32:652–663.
92. Shechter Y, Newcomer VD. Disc electrophoretic studies of proteins from *Coccidioides immitis*. J Invest Dermatol 1967; 48:119–123.
93. Strobl A, Turk R, Thalhamer J. Investigations on the protein composition of the lichen *Pseudevernia furfuracea* (L.) Zopf var. *ceratea* (Ach.) Hawksw. from different altitudes. Phyton 1994; 34:67–83.
94. Kulik MM, Brooks AG. Electrophoretic studies of soluble proteins from *Aspergillus* spp. Mycologia 1970; 62:365–376.

95. Sorenson WG, Larsh HW. Hamp S. Acrylamide gel electrophoresis of proteins from *Aspergillus* species. Am J Bot 1971; 58:588–593.
96. Hearn VM, Moutaouakil M, Latgé JP. Analysis of components of *Aspergillus* and *Neosartorya* mycelial preparations by gel electrophoresis and Western blotting procedures. In: Samson RA, Pitt JI, eds. Modern Concepts in *Penicillium* and *Aspergillus* Classification, New York: Plenum Press, 1990:235–245.
97. Girardin H, Latgé JP. Comparison of *Neosartorya fischeri* varieties based on protein profiles and immunoactivities. In: Samson RA, ed. Modern Methods in Food Mycology. Development in Food Science 31. Amsterdam: Elsevier, 1992:177–182.
98. Stipes RJ. Disc electrophoresis of mycelial proteins from *Ceratocystis* species. Phytopathology 1967; 57:833. Abstract.
99. Stipes RJ. Comparative mycelial protein and enzyme patterns in four species of *Ceratocystis*. Mycologia 1970; 62:987–995.
100. Mandeel QA, Gamal El-Din AY, Mohammed SA. Analysis of SDS-dissociated proteins of pathogenic and non pathogenic *Fusarium* species. Mycopathologia 1994; 127:159–166.
101. Snyder WC, Hansen HN. The species concept in *Fusarium*. Am J Bot 1940; 27: 64–67.
102. Hall R. *Verticillium albo-atrum* and *V. dahliae* distinguished by acrylamide gel electrophoresis of proteins. Can J Bot 1969; 47:2110–2111.
103. Milton JM, Rogers WG, Isaac I. Application of acrylamide gel electrophoresis of soluble fungal proteins to taxonomy of *Verticillium* species. Trans Br Mycol Soc 1971; 56:61–65.
104. Stipes RJ, McCombs CL. Comparative protein and enzyme profile on polyacrylamide gels of perithecial isolates of *Glomerella cingulata*. Phytopathology 1965; 55:1078. Abstract.
105. Dale JL, Manners JM, Irwin JAG. *Colletotrichum gloeosporioides* isolates causing different anthracnose diseases on *Stylosanthes* in Australia carry distinct double-stranded RNAs. Trans Br Mycol Soc 1988; 91:671–676.
106. Gairola G, Powell D. Eletrophoretic protein patterns of *Cytospora* fungi. Phytopathol Z 1971; 71:135–140.
107. Shamoun SJ, Sieber TN. Isoenzyme and protein patterns of endophytic and disease syndrome associated isolates of *Melanconium apiocarpum* and *Melanconium marginale* collected from alder. Mycotaxon 1993; 49:151–166.
108. Groves JW. Two new species of *Ascocalyx*. Can J Bot 1968; 46:1273–1278.
109. Morelet M. Un Discomycète inoperculé nouveau. Bull Soc Sci Nat Archeol Toulon 1969; 183:9.
110. Dorworth CE, Comparison if soluble proteins of *Ascocalyx abietis* and *Gremmeniella abietina* by serology and electrophoresis. Can J Bot 1974; 52:919–922.
111. Petrini O, Petrini LE, Laflamme G, Ouellette GB. Taxonomic position of *Gremmeniella abietina* and related species: a reappraisal. Can J Bot 1989; 68:2629–2635.
112. Petrini O, Toti L Petrini LE. *Gremmeniella abietina* and *G. laricina* in Europe: characterization and identification of isolates and laboratory strains by soluble protein electrophoresis. Can J Bot 1990; 68:2629–2635.
113. Morelet M. La maladie à *Brunchorstia* I. Position systématique et nomenclature du pathogène. Eur J For Pathol 1980; 10:268–277.

114. Petersen GR, Russo GM, van Etten J. Identification of a developmental specific protein in sclerotia of *Sclerotinia minor* and *Sclerotinia trifoliorum*. Exp Mycol 1982; 6:268–273.
115. Novak LA, Kohn LM. Electrophoresis of major proteins in stromata of members of the Sclerotiniaceae. Trans Br Mycol Soc 1988; 91:639–647.
116. Tariq VN, Gutterridge CS, Jeffries P. Comparative studies of cultural and biochemical characteristics used for distinguishing species within *Sclerotinia*. Trans Br Mycol Soc 1985; 84:381–397.
117. Backhouse D, Willets HJ, Adam P. Electrophoretic studies of *Botrytis* species. Trans Br Mycol Soc 1984; 82:625–630.
118. Ibrahim-Granet O, De Bièvre C, Romain F, Letoffe S. Comparative electrophoresis, isoelectric focusing and numerical taxonomy of some isolates of *Fonsecaea pedrosoi* and allied fungi. Sabouraudia J Med Vet Mycol 1985; 23:253–263.
119. Peterson PJ, Latch CCM. Polyacrylamide gel electrophoresis of cellular proteins of *Cercospora* isolates from some pasture legumes. NZ J Sci 1969; 12:3–12.
120. Chupp C. A monograph of the fungus genus *Cercospora*. Ithaca, New York: Cornell Univ, 1953:1–667.
121. Campo-Arana R, Echavez-Badel R, Schroeder EC, Sanchez-Paniagua A. Characteristics of *Macrophomina phaseolina* isolates from Puerto-Rico and the Dominican Republic. Phytopathology 1994; 84:866–867. Abstract.
122. Kim WK, Rohringer R, Nielsen J. Comparison of polypeptides in *Ustilago* spp. pathogenic on wheat, barley and oats: a chemotaxonomic study. Can J Bot 1984; 62: 1431–1437.
123. Lindeberg B, Nannfeldt JA. Ustilaginales in Sweden. Symb Ups 1959; 16(2):1–175.
124. Shipton WA. Fleischmann G. Taxonomic significance of protein patterns of rust species and formae speciales obtained by disc electrophoresis. Can J Bot 1969; 47: 1351–1358.
125. Kim WK, Heath MC, Rohringer R. Comparative analysis if proteins of *Uromyces phaseoli* var. *typica, U. phaseoli* var. *vignae*, and *U. viciae-fabae*: polypeptide mapping by two dimensional electrophoresis. Can J Bot 1985; 63:2144–2149.
126. Wilmotte A, Van de Peer Y, Goris A, et al. Evolutionary relationships among higher fungi inferred from small ribosomal subunit RNA sequence analysis. Syst Appl Microbiol 1993; 16:436–444.
127. Newsted WJ, Huner NPA. Major polypeptides associated with differentiation in psychrophilic fungi. Can J Bot 1987; 65:233–241.
128. Clare BG, Flentje NT, Atkinson MR. Electrophoretic patterns or oxidoreductases and other proteins as criteria in fungal taxonomy. Aust J Biol Sci 1968; 21:275–295.
129. Smythe R, Anderson GE. Electrophoretic protein spectra of wild-type and isogenic monokaryons of *Coprinus lagopus*. J Gen Microbiol 1971; 66:251–253.
130. Burgess T, Malajczuk N, Dell B. Variation in *Pisolithus* based on basidiome and basidiospore morphology, culture characteristics and analysis of polypeptides using 1D SDS-PAGE. Mycol Res 1995; 99:1–13.
131. Garber ED. Electrophoresis as a taxonomic and genetic tool for fungi. Bull Torrey Bot Club 1973; 100:308–312.
132. Hall R. Molecular approaches to taxonomy of fungi. Bot Rev 1971; 35:285–304.

5

Use of Isozymes in Fungal Taxonomy and Population Studies

Søren Rosendahl and Søren Banke
Botanical Institute, University of Copenhagen, Copenhagen, Denmark

I. INTRODUCTION

Isozyme electrophoresis is a frequently used tool in systematics, genetics, and population biology. The technique has been widely applied in plant biology to study gene flow and genetic diversity, but the limited knowledge of fungal population genetics has made the application of isozymes in mycology controversial.

Several handbooks and reviews on isozyme techniques are available (1–4). These reviews explain the background for the techniques and some guidelines as to how the results can be interpreted. Protocols developed for studies of plants and animals are not always directly applicable in mycology. One major difference between fungi and other groups of organisms is the dominant mitotic reproduction in fungi, which complicates the interpretation of locus and allele variation. The reviews on the application of isozymes in mycology (5,6) mainly discuss the use of isozymes for identification and detection of fungi. Isozyme bands are genetic markers useful to detect and identify specific fungi, and specific enzyme bands represent the activity of metabolic pathways in the fungus. This information can be combined to quantify the activity of mycorrhizal (7–9) or pathogenic fungi (10) directly in plant material. This application is important and in many respects superior to the far more expensive DNA techniques.

The application of isozymes in fungal taxonomy and phylogeny may still be controversial, but several studies have used differences in isozyme patterns of fungi to clarify taxonomic problems at the species level. The PCR-based molecular methods are widely used for developing new taxonomic characters and in studies of phylogenetic relationships among fungi (11,12). PCR-based techniques can be used to determine sequences of nucleic acids to give precise measures of the number of substitutions in specific genes. The techniques are time-consuming, and often only a few genes coding for one or two proteins are sequenced. In comparison, the loci that can be detected by isozyme analysis will cover several evolutionary events, which should give more information about the organism.

II. ISOZYME CHARACTERS

A. Isozymes in Fungal Taxonomy

Isozyme characters are most often used as phenetic (unweighted) characters. This is done by simply listing the resolved bands as independent characters. In traditional fungal taxonomy characteristics from morphology, life cycles and physiology are used as taxonomic characters. The lack of sexual reproductive structures in certain groups of fungi has limited the number of available morphological characters. This has led to a controversial taxonomy, with several polyphyletic taxonomic groups: Deuteromycotina, dematiaceous hyphomycetes, *Rhizoctonia* spp, etc. A major advantage of isozymes as phenetic characters is the access to more than 100 independent characters. This is of great value where sufficient traditional diagnostic characters are not available, as in *Phytophthora* taxonomy (13).

There is not always a good correlation between isozyme data and morphologically defined species. Isozyme data revealed clearly defined clusters in a study of 10 *Pythium* species (14), but these clusters did not all reflect the morphological species studied. Such results may suggest that the morphological key characters used in the genus are questionable. In studies of *Phytophthora cryptogea* and *P. drechsleri*, morphological species were not clearly defined by isozymes. This result was supported by data obtained from restriction length fragment polymorphisms (RFLP) in the mitochondrial DNA (15).

Isozyme data have been used in several studies of plant endophytes. In a study of *Xylaria* and xylariaceous endophytes, the taxonomy based on morphology was confirmed, but most of the endophytic isolates could not be linked to these species (16). In this study no correlation was found between production of various secondary metabolites and the classification based on morphology and isozymes. This is in contrast to a study of xerophilic penicillia where isozyme data correlated with both the morphological species and the ability to produce various secondary metabolites (17). In several other studies, isozyme data have been correlated to other parameters, such as somatic incompatibility in *Suillus* (18) and intersterility groups of *Heterobasidion* (19–21). In plant pathology, genetic groups defined by isozyme analysis have been correlated to virulence of the pathogen (22–25). Isozyme variation has further been compared with formae speciales of *Puccinia* (26) and to physiological races of *Ustilago hordei* (27) and *Cochliobolus carbonium* (28).

The taxonomy of the fungi in the order Glomales forming arbuscular mycorrhizas is highly controversial. Isozymes have been used with these nonculturable fungi and have revealed a considerable variation among the spore-cluster forming *Glomus* spp. (29). The study probably included several morphological species. Another study that reported a high intraspecific variation within *Glomus mosseae* (30) did not mention the morphology of the isolates included in the study. Recent studies based on the presence of putative isozyme loci show that morphological species in *Glomus* are remarkably similar throughout the world (31). A study of the genetic similarity in *G. mosseae* and *G. coronatum* revealed a very low divergence between isolates from different continents (32). Similar results have been obtained with *G. fistulosum* and *G. claroideum* (31). These findings agree with the suggested long asexual evolution line of arbuscular mycorrhizal fungi (11).

The fruitbody of the controversial basidiomycetous species *Rhodocybe stangliana* is characterized by its bulbous stem base. The bulbous base and the cap of the basidiocarp show different enzyme loci, and it has been suggested that the species is composed of two different basidiomycete taxa, one parasitizing the other (33). Further studies of other Basidiomycetes may show that this biotrophic relationship may occur among other species.

Isozyme data can be useful in species descriptions. In the redescription of *Phytopthtora citrophthora*, isozyme data are included to support the morphological characters that can sometimes be difficult to assess (34). This application of isozyme data will be particularly useful in the future in the species descriptions of fungi where the key morphological characters are questionable.

A large number of enzymes have been used in studies of fungal taxonomy These enzymes are mostly those involved in the primary metabolism, but some studies have used banding patterns of extracellular enzymes for the identification of fungi (35). In contrast to general metabolic enzymes present in hyphae and spores, the extracellular enzymes are found in the culture broth. This limits the application of extracellular enzymes to culturable fungi. A genetic interpretation of extracellular enzyme bands can be difficult, as the appearance and the intensity of the bands may vary among electrophoretic runs (36).

Patterns of extracellular enzymes, zymograms, have been used in *Penicillium* taxonomy, and the technique was found useful for classification of *Penicillium* species with questionable morphological characters (37). A statistical analysis of the data was not performed, nor was the variation within the species presented. In *Aspergillus* taxonomy, the extracellular enzyme patterns were used to group four species in the Flavi section into four well-defined clusters (38). The authors stated that their results justified the species *Aspergillus parasiticus*, *A. flavus*, and *A. oryzae*, but the study did not mention the variation within the species. This could be important as from the banding patterns on the reproduced gels it is difficult to distinguish between *A. oryzae* and *A. flavus*. The selection pressure on genes coding for extra and intracellular enzymes may be different as the extracellular enzymes are secreted to the environment. A comparison of genetic variation between the two types of enzymes has not been made.

B. Isozymes in Phylogenetic Studies

Isozymes are valuable as unweighted phenetic characters, but an evolutionary interpretation of isozyme data can be obtained by using them in phylogenetic reconstructions (39). Allele frequencies can be used to estimate the number of changes (substitutions) that have occurred since the two lines diverged, and a genetic distance between taxa can then be calculated from the occurrence of putative loci and alleles (40) (Eq. 3). The formulas developed for the calculation of genetic distances are intended for sexually reproducing organisms where alleles are exchanged between individuals in a population. In mitotic organisms, the allele frequencies cannot readily be used as estimates of genetic distances, as the reproductive barriers restrain exchange of genetic material. This will be discussed further in the last section of this chapter.

Stasz et al. (41) used isozymes to evaluate relationships between morphological species of *Trichoderma* and *Gliocladium*. They used the isozyme data in a cladistic analysis and did not find a good correlation between the monophyletic groups and the morphologically defined species. The study revealed a considerable number of loci and alleles. They suggested that the high number of alleles may be the result of genetic exchange between the strains. However, the high number of alleles could also be found in a clonal population where the individual mycelia maintain their genetic integrity through the lack of sexual recombination. The taxonomic and phylogenetic interpretation of allele and locus variation in mitotic fungi is difficult, and should be addressed in further studies. The taxonomy and evolutionary biology of 10 *Pythium* spp. was studied based on isozymes. The results showed that whereas some morphological species, such as *P. ultimum* and *P. irregulare*, form well-defined clusters, other species could not be identified by their isozyme pattern (42). Low levels of genetic diversity was found within *P. irregulare* and *P. spinosum* and were correlated to the facultative parasitic life history of these species. Alleles and loci were identified from the species, but unfortunately the information was not used in the discussion of the evolutionary biology of the species. Isozymes have been used at the family level in the lichens *Umbilicaria* and *Lasallia*. The enzyme pattern similarities did not support the current taxonomy based on morphological characters and suggested that the two genera should not be separated (43).

C. Isozymes in Population Studies

Population studies must be based on interpretation of putative loci and alleles. Isozymes have been used in this way in population studies of plants and animals, but there are only a few studies on fungi. This is partly due to the difficulties in defining alleles and loci, which can be done by back-crosses for plants, but is difficult in fungi and impossible in strictly mitotic fungi. A well-documented population study with fungi involved the sexual and asexual populations of *Phytophthora infestans* (44). This study confirmed that *Phytophthora* has a diploid vegetative stage, and suggested random mating and recombination in the sexual population. No recombination could be detected in the asexual population.

The dematiaceous hyphomycete *Leptographium wageneri* is known to occur as three distinct, host-specialized variants. Isozyme analysis and ordination from genetic distances have been used to confirm these groupings (45). The study demonstrated low genetic diversity within the three variants and a considerable genetic distance between the variants. This suggests rare or absent genetic recombination in this fungus. Conversely, sexual reproduction has been confirmed in *Morchella esculenta* and *M. deliciosa* (46), where the results showed that the two species constitute separate gene pools and could be regarded as distinct species. A later study, including more collections, by the same authors showed that the two taxa could not be separated (47). Isozyme analysis of single ascospore isolates of the canker pathogen *Crumenulopsis sororia* in Scotland showed that the genetic variation was high among isolates but that the genetic distance between the populations was small. This suggests a random biparental outcrossing of the fungus (48).

Isozyme analysis can also be used in studies of environmental effects on genetic diversification in populations. Principal-components analysis and cluster analysis of isozyme data obtained from the ectomycorrhizal fungus *Suillus tormentosus* (49) showed that habitat isolation and host selection could be responsible for the genetic variation among forest regions. The population structure of the grass endophyte *Atkinsoniella hypoxylon* (50) suggested that the host plant species and genus are major factors responsible for genetic variation within the fungus. Similar results were obtained with endophytic *Acremonium* isolates, where the isozyme data were used to estimate genetic identities between strains isolated from different host plants (51). In that study the author stressed the importance of a correct genetic interpretation of isozyme banding patterns. Multiband phenotypes were found consistently and were interpreted as the isolates possessing multiple copies of a gene because of heterocaryosis or aneuploidy (unstable number of chromosomes). The multiband phenotypes in that study were regarded as heterozygous, although another plausible explanation could be that the isolation method has led to a sectioning of a heterocaryotic mycelium. In planta detection of the multiband phenotypes is necessary to confirm the existence of diploid mycelia.

Before isozymes can be used in population studies, polymorphic loci must be identified. This can be difficult, as it is not possible to make back-crosses. The polymorphic loci can by identified only by studying several isolates of the same species. Such intraspecific variation was documented in *Tuber melanosporum* as allelic variation in two out of four enzyme loci (52). In other studies it may be difficult to interpret the variation as intraspecific allelic variation (29) possibly due to incorrect identification of some of the studied isolates.

III. GEL ELECTROPHORESIS—METHODS

Several electrophoretic systems for isozyme analysis have been described (1,6). Most protocols use horizontal starch gel electrophoresis, but the advantages of polyacrylamide as the gel matrix make it a better choice. The genetic interpretation of the bands resolved in the gels can be controversial, and a good resolution is important. This can generally be obtained by vertical zone electrophoresis in a discontinuous buffer system (53). In some cases, as in experiments with the arbuscular mycorrhizal fungi, the fungi cannot be cultured and the amount of material available can be limited. Studies on such fungi must be done on thin (0.5 to 0.7 mm) polyacrylamide gels where small volumes can be loaded. The hazards of the neurotoxic acrylamide can be minimized by using stock solutions of acrylamide and bis-acrylamide and by casting the gels directly on the electrophoretic unit. The protocols for casting acrylamide gels can be found in the above references (53).

A. Extraction of Proteins from Fungal Material

Proteins must be extracted from the fungus, and the most suitable extraction procedure depends on the fungal material available. Fruitbodies of Ascomycetes or Basidiomycetes can be used directly by grinding in an ice-cold mortar or a glass homogenizer with extraction buffer and PVPP (polyvinylpolypyrrolidone) (8). When field-collected fruitbodies are used, possible parasites should be avoided. The extraction procedure can also be performed with fungal mycelium grown in liquid culture or agar cultures. If liquid nitrogen is available, frozen samples can be ground in a mortar with liquid nitrogen and the powder transferred in ice. The material is then centrifuged for 20 min at 20,000g at 4°C and the supernatant transferred to tubes kept in ice. The extraction protocol is quick, but as the material can only be quantified as fresh weight, it is difficult to obtain the same protein concentration in all extracts. Differences in protein concentration of samples can in some cases explain differences between faint bands in some isolates. To overcome this problem it is necessary to adjust the protein concentration in the samples. This is done colorimetrically using bovine serum albumin (BSA) as standard (54). The method quantifies all proteins in the sample including the structural proteins, and cannot be used directly to quantify specific enzyme activities in the extracts.

The fungal material can be freeze-dried before grinding in a mortar. The procedure will provide a dry mycelium powder that can be weighted. The method can be used with both fruit bodies and mycelium from liquid cultures. Mycelium from liquid cultures should be washed several times in double distilled water after harvesting, before it is freeze-dried. This will avoid high salt concentrations from the growth medium, which can interfere with the electrophoresis. The freeze-dried mycelium is ground with 10 mg of PVPP. The optimal proportion of fungal mycelium to extraction buffer may differ between fungi and the enzymes studied. As a standard, 50 mg dry mycelium powder can be transferred to a 1.5-ml Eppendorf tube containing 500 μl cold (4°C) extraction buffer. The tubes are centrifuged at 20,000g for 20 min at 4°C. The supernatant is recovered and transferred to new Eppendorf tubes. It is recommended to make several aliquots of extracts, e.g., 10 tubes with 10 or 25 μl extract each, and store these at −80°C. We have tested several extraction buffers, but the buffer that yields the highest protein concentration is a Tris-HCl buffer (8). Several extraction buffers contain the protease inhibitors (PMSF) phenylmethylsulfonylfluoride. This compound is toxic and possibly carcinogenic, and generally we are not able to detect any effect of this compound on the resolution of enzyme bands.

B. Staining the Gels

The gels are stained after the electrophoretic separation. Several enzymes can be detected on the gels; the choice of stain depends on the fungi studied. Recipes for enzyme stains can be found in several books (1–3,55). The selection of useful enzymes will include a screening of the various protocols available. All these stains will rarely be applicable, and of 20 to 25 stains tested, only 10 to 12 may give an appropriate resolution. Several enzymes may not show on the gels, and even enzymes in the central part of the primary metabolism, e.g., malate dehydrogenase, cannot always be detected.

The genetic information that can be obtained from the enzymes may differ. Some enzymes as malate dehydrogenase and glucose-6-phosphate dehydrogenase are often monomorphic, whereas esterase and diaphorase can be polymorphic. Other stains can be difficult to interpret, as several different enzymes may show activity with the substrate. The stain for phosphoglucomutase may also detect the enzyme gluco-6-phosphate dehydrogenase, and these loci may then appear on the gel. To make a correct interpretation of the stained gel, the two enzymes should be detected individually and the results compared.

The time of harvesting the mycelium can be important for expression of the enzymes, when cultures grown in liquid media are used. Several Oomycetes only express the enzyme glucose-6-phosphate dehydrogenase during the first days of growth, and the enzyme cannot be detected after 6 or 7 days (unpublished results). Different expression of isozymes have also been reported from different parts of the fungus (57).

IV. DATA ANALYSIS

Before gel data are analyzed the different types of isozymes should be identified. First, the secondary bands should be identified, as these should not be included in the analysis. Second, putative loci and alleles should be identified. This interpretation is crucial in the use of isozyme data for estimating genetic identities and genetic distances.

A. Calculation of Similarity

The simplest method to analyze isozyme data is to score each band as a character that can be either present or absent, with the value 1 or 0, respectively. Several coefficients have been suggested for calculation of similarity between operational taxonomic units; the most frequently used is the Jaccard coefficient. The Jaccard coefficient gives the same weight to the observed characters, but does not include negative matches. Other coefficients may give more weight to the shared characters; this can be done by multiplying the number of shared bands by 2 (Sørensen or Dice coefficients).

In the Kulczynski similarity coefficient, S_K (56) (Eq. 1):

$$S_K = \frac{1}{2}\left[\left(\frac{A}{A+B}\right) + \left(\frac{A}{A+C}\right)\right] \tag{1}$$

A is the number of shared characters, B the characters unique to one of the isolates, and C the characters unique to the other isolate. The Kulczynski coefficient gives more weight to shared characters, but maintains the total number of characters. This is in contrast to the Soerensen coefficient that gives more weight to shared characters by increasing the number of shared bands.

A difference between the similarity coefficients can be found when isolates with an uneven number of bands are compared. The Kulczynski coefficient will then give the highest percentage of similarity, as this coefficient gives more weight to the characters shared by the two organisms. This is relevant in several cases, as some bands may not show on the gel due to insufficient loading on the gel, or because the genes responsible for the protein are not expressed.

B. Cluster Analysis

Cluster analysis of isozyme data can be performed by using a suitable computer program or package. The NTSYS program (58) contains several clustering methods. The UPGMA (unweighted pair group method arithmetic mean) is used most often, but other methods can be information. Single or complete-linking method can be used as an alternative, but it is always a good idea to compare more than one clustering method. The neighbor-joining method can be applied to isozyme data in phylogenetic reconstructions (59). This method is based on parsimony, and the resulting tree may not have equal branch lengths, as the ultrametric trees constructed by the UPGMA method will.

C. Genetic Distance

Genetic distance among groups, morphological species, or geographical populations can be estimated from the general enzyme patterns of each group. The groups can be defined on basis of the result of a cluster analysis. Several estimates of genetic distances have been published. In an isozyme study the phylogenetic relationship between 11 *Pleurotus* species was estimated using the genetic distance estimator D_z (60):

$$D_Z = 1 - \frac{1}{n}\sum_{i=1}^{L} S_i \qquad (2)$$

where L is the total number of loci examined; n is the number of loci possessing nonsilent alleles for the two electrophoretic types compared. S_i is 1 if the two groups share all alleles in the examined locus, and 0 if no alleles are shared. In any other case the value is 0.5.

The calculation of genetic distances had previously been described only for sexually reproducing organisms (39). Enzyme patterns from populations of mitotic fungi may differ from the patterns obtained from populations with meiotic states, as asexual organisms do not exchange alleles. Previously published methods for estimating genetic distances between organisms are based on information on allele frequencies at each locus (39,40,61), and this implies that the organisms studied have a sexual reproduction system. In studies of mitotic fungi, allele frequencies are most often 1·0 or 0·0 (50). This is either due to the lack of sexual reproduction, or because the mitotic phase is dominating in the fungal population.

In strictly mitotic fungi it is still possible to detect different alleles as well as different loci. This interpretation may provide additional evolutionary information that is not accounted for by using traditional formulas for calculation of genetic distances. If the Rogers genetic distance *Dr* (40,61) (Eq. 3) is applied to data obtained from a group of mitotic fungi, the result will be similar to a phenetic analysis, as the allele frequencies will have binary characters.

$$D_R = \left(\frac{1}{L}\right) \Sigma_L \frac{\sqrt{\Sigma (X_i - Y_i)^2}}{2} \tag{3}$$

where X_i and Y_i = the frequencies of the *i*th allele at a particular locus in taxa *X* and *Y*, respectively, and *L* total number of loci.

The lack of evolutionary information on fungi when using Rogers genetic distance or other distances based on genetics has been solved by other workers by regarding each band in the enzyme patterns as a character (14). By doing this, all loci are treated as monomorphic. This assumption may give unreliable results at the species level and above, although it may provide information at population level.

We have suggested that the genetic distance between mitotic fungi is calculated from a formula including information on allele frequencies and on differences in loci (17) (Eq. 4). This genetic distance, D_s, will give twice as much weight to differences in loci as differences in alleles at the same locus. The maximum distance 1 can thus be obtained only if all loci are different.

$$D_s = \left(\frac{1}{L}\right)\left(\left(\Sigma X_l + \Sigma Y_l\right) + \left(\frac{\Sigma X_{lfA} - \Sigma Y_{lfA}}{4}\right)\right) \tag{4}$$

Where L = total number of loci. ΣX_l : is the sum of allele frequencies in loci only recognized in taxa *X*. ΣY_i : sum of allele frequencies in loci only recognized in taxa Y. ΣX_{lfA} : sum of allele frequencies in taxa *X*, in loci recognized in both *X* and *Y*. ΣY_{lfA} : sum of allele frequencies in taxa *Y*, in loci recognized in both *X* and *Y*.

The principles of the Rogers genetic distance are used in the calculation of the new genetic distance Ds. The maximum distance per locus is *1/L*, the total number of loci is maintained, and the genetic distances range from 0 to 1. As most frequencies are either 1 or 0, the square sum and square root of the differences in allele frequencies are omitted.

V. CONCLUSION

Isozymes are still important characters in molecular taxonomy of fungi, although techniques based on comparisons of nucleotide sequences have developed rapidly in recent years. Detection of isozymes allows a genetic interpretation of variations in alleles and loci, but this can only be done if it is possible to differentiate between the different events leading to isozyme variation. As in morphological taxonomy, the number of available characters is important. This is particularly relevant for the use of isozymes as taxonomic characters, as different enzyme loci exhibit variations at various taxonomic levels. These differences may range from the individual isolates to the genus.

The characters generated by enzyme electrophoresis must be subjected to a data analysis to describe the variation and similarity among the tested isolates. This can be done by phenetic analysis, where the technique may give access to more independent characters than a morphological analysis. The results of isozyme analysis will in most cases confirm the morphological taxonomy, but may also be contradictory.

The use of isozyme in phylogenetic studies of fungi is still controversial, but a proper interpretation of bands may provide information on fungal microevolution. Isozyme bands represent both allelic and locus variation; by using mathematic algorithms where the bands are given different weights, a possible phylogeny of the fungi can be generated. It is important that the algorithm take into account the lack of sexual reproduction in many fungi.

The application of isozyme bands as genetic markers in population studies is an area with a great potential. Isozymes are easy and inexpensive to generate, and a large number of isolates can be screened. The detection of the genetic isozyme markers can be combined with measurements of activity of the specific enzymes. The techniques can then be used both to qualify and to quantify fungal mycelium.

ACKNOWLEDGMENTS

We would like to thank Lotte Dahl, Rasmus Kjøller, Ann-Berith Petersen, Kerstin Skovgaard, Dorte Sørensen, and Ida Thingstrup for their encouraging work on isozymes that has contributed to this work. We also thank Kim B. Petersen for reviewing the manuscript.

REFERENCES

1. Harris H, Hopkinson DA. Handbook of Enzyme Electrophoresis in Human Genetics (with Supplements). New York: Oxford American Publishing Co., 1976.
2. Murphy RW, Sites JW, Buth DG, Haufler CH. Isozyme electrophoresis. In: Hillis DM, Morit C, eds. Molecular Systematics. Sinauer Associates Inc. Publishers, Sunderland, Massachusetts, 1990:45–126.
3. Soltis DE, Soltis PS. Isozymes in Plant Biology. London: Chapman & Hall, 1989.
4. Tanksley SD, Orton TJ. Isozymes in Plant Genetics and Breeding (Part A). Amsterdam: Elsevier, 1983.
5. Bonde MR, Micales JA, Peterson GL. The use of isozyme analysis for identification of plant-pathogenic fungi. Plant Dis 1993; 77:961–968.
6. Micales JA, Bonde MR, Peterson GL. The use of isozyme analysis in fungal taxonomy and genetics. Mycotax 1986; 27:405–449.

7. Rosendahl S. Influence of three vesicular-arbuscular mycorrhizal fungi (Glomales) on the activity of metabolic enzymes of the host plant. Plant Soil 1992; 144:219–226.
8. Rosendahl S, Sen R. Isozyme analysis of mycorrhizal fungi and their mycorrhizas. In: Varma AK, Read DJ, Norris JR, eds. Methods in Microbiology. Vol. 24. Experiments with Mycorrhizae. New York: Academic Press, 1992:169–194.
9. Thingstrup I, Rosendahl S. Quantification of fungal activity in arbuscular mycorrhizal symbiosis by polyacrylamide gel electrophoresis and densitometry of malate dehydrogenase. Soil Biol Biochem 1994; 26:1483–1489.
10. Kjøller R, Rosendahl S. The presence of the arbuscular mycorrhizal fungus *Glomus intraradices* influences enzymatic activities of the root pathogen *Aphanomyces euteiches* in pea roots. Mycorrhiza 1997; 6:487.
11. Berbee ML, Taylor JW. Dating the evolutionary radiation's of the true fungi. Can J Bot 1993; 71:1114–1127.
12. Gargas A, DePriest PT, Grube M, Tehler A. Multiple origins of lichen symbiosis in fungi suggested bu SSU rDNA phylogeny. Science 1995; 268:1492–1495.
13. Oudemans P, Coffey MD. Isozyme comparison within and among worldwide sources of three morphologically distinct species of *Phytophthora*. Mycol Res 1991; 95: 19–30.
14. Chen W, Hoy JW, Schneider RW. Comparisons of soluble proteins and isozymes for seven *Pythium* species and applications of the biochemical data to *Pythium* systematics. Mycol Res 1991; 95:548–555.
15. Mills SD, Forster H, Coffey MD. Taxonomic structure of *Phytophthora cryptogea* and *P. drechsleri* based on isozyme and mitochondrial DNA analyses. Mycol Res 1991; 95:31–48.
16. Brunner F, Petrini O. Taxonomy of some *Xylaria* species and xylariaceous endophytes by isozyme electrophoresis. Mycol Res 1992; 96:723–733.
17. Banke S, Frisvad JC, Rosendahl S. Taxonomy of *Penicillium chrysogenum* and related xerophilic species based on isozyme analysis. Mycol Res 1997; 101:617–624.
18. Sen R. Intraspecific variation in two species of *Suillus* from Scots pine (*Pinus sylvestris L.*) forests based on somatic incompatibility and isozyme analyses. New Phytol 1990; 114:607–616.
19. Karlson JO, Stenlid J. Pectic isozyme profile of intersterility groups in *Heterobasidion annosum*. Mycol Res 1991; 95:531–536.
20. Otrosina WJ, Chase TE, Cobb FW. Allozyme differentiation of intersterility groups of *Heterobasidion annosum* isolated from conifers in the western United States. Phytopathology 1992; 82:540–545.
21. Stenlid J. Population structure of *Heterobasidion annosum* as determined by somatic incompatibility, sexual incompatibility and isoenzyme patterns. Can J Bot 1985; 63: 2268–2273.
22. Ferreira JF, Bosland PW, Williams PH. The variability of *Pyrenochaeta terrestris* isolates based on isozyme polymorphism cultural characteristics and virulence on differential onion breeding lines. J Phytopathol 1991; 133:289–296.
23. Gall C, Balesdent MH, Desthieux I, Robin P, Rouxel T. Polymorphism of Toxo *Leptosphaeria maculans* isolates as revealed by soluble protein and isozyme electrophoresis. Mycol Res 1995; 99:221–229.
24. Koch G, Kohler W. Isoenzyme variation and genetic distances of *Erysiphe graminis* DC. Formae speciales. J Phytopathol 1990; 129:89–101.

25. Julian AM, Lucas JA. Isoenzyme polymorphism in pathotypes of *Pseudocercosporella herpoetrichoides* and related species from cereals. Plant Pathol 1990; 39: 178–190.
26. Burdon JJ, Marshall DR. Isozyme variation between species and formae speciales of the genus *Puccinia*. Can J Bot 1981; 59:2628–2634.
27. Hellmann R, Christ BJ. Isozyme variation of physiologic races of *Ustilago hordei*. Phytopathology 1991; 81:1536–1540.
28. Simcox KD, Nickrent D, Pedersen WL. Comparison of isozyme polymorphism in races of *Cochliobolus carbonum*. Phytopathology 1992; 82:621–624.
29. Rosendahl S. Comparisons of spore-cluster forming *Glomus* species (Endogonaceae) based on morphological characteristics and isoenzyme banding patterns. Opera Bot 1989; 100:215–223.
30. Hepper CM, Sen R, Azcon-Aguilar C, Grace C. Variation in certain isozymes amongst different geographical isolates of the vesicular-arbuscular mycorrhizal fungi *Glomus clarum*, *Glomus monosporum* and *Glomus mosseae*. Soil Biol Biochem 1988; 20:51–59.
31. Rosendahl S, Dodd JC, Walker C. Taxonomy and phylogeny of the *Glomales*. In: Gianinazzi S, Schuepp H, eds. Impact of Arbuscular Mycorrhizas on Sustainable Agriculture and Natural Ecosystems. Basel: Birkhauser Verlag, 1994:1–12.
32. Dodd JC, Rosendahl S, Giovannetti M, Broome A, Lanfranco L, Walker C. Inter- and intraspecific variation within the morphologically-similar arbuscular mycorrhizal fungi *Glomus mosseae* and *Glomus coronatum*. New Phytol 1996; 133:113–122.
33. Laessøe T, Rosendahl S. *Rhodocybe stangliana*: a parasite on other agarics? Mycol Res 1994; 98:88–90.
34. Mchau GR, Coffey MD. An integrated study of morphological and isozyme patterns found within a world-wide collection of *Phytophthora citriphthora* and a redescription of the species. Mycol Res 1994; 98:1291–1299.
35. Cruickshank RH, Wade GC. Detection of pectic enzymes in pectin-acrylamide gels. Anal Biochem 1980; 107:177–181.
36. Paterson RRM, Bridge PD, Crosswaite MJ, Hawksworth DL. A reappraisal of the terverticillate penicillia using biochemical, physiological and morpological features. III. An evaluation of pectinase and amylase isoenzymes for species characterization. J Gen Microbiol 1989; 135:2979–2991.
37. Cruickshank RH, Pitt JI. The zymogram technique: isoenzyme patterns as a aid in *Penicillium* classification. Microbiol Sci 1987; 4:14–17.
38. Cruikshank RH, Pitt JI. Isoenzyme patterns in *Aspergillus flavus* and closely related species. In: Samson RA, Pitt JI, eds. Modern Concept in *Penicillium* and *Aspergillus* classification. New York: Plenum Press, 1990:259–265.
39. Swofford DL, Olsen GJ. Phylogeny reconstruction. In: Hillis DM, Moritz C, eds. Molecular Systematics. Sinauer Associates Inc. Publishers, Sunderland, Massachusetts. 1990:411–500.
40. Rogers JS. Deriving phylogenetic trees from allele frequencies: a comparison of nine genetic distances. Syst Zool 1986; 35:297–310.
41. Stasz TF, Nixon K, Harman GE, Weeden F, Kuter GA. Evaluation of phenetic species and phylogenetic relationships in the genus *Trichoderma* by cladistic analysis of isozyme polymorphism. Mycologia 1989; 81:391–403.
42. Chen W, Schneider RW, Hoy JW. Taxonomic and phylogenetic analyses of ten

Pythium species using isozyme polymorphisms. Phytopathology 1992; 82:1234–1244.
43. Hageman C, Fahselt D. Relationships within the lichen family Umbilicariaceae based on enzyme electromorph data. Lichenologist 1992; 24:91–100.
44. Tooley PW, Fry WE, Villarreal-Gonzalez MJ. Isozyme characterisation of sexual and asexual *Phytophthora infestans* populations. J Hered 1985; 76:431–435.
45. Zambino PJ, Harrington TC. Isozyme variation within and among host-specialized varieties of *Leptographium wageneri*. Mycologia 1989; 81:122–133.
46. Gessner RV, Romano MA, Schultz RW. Allelic variation and segregation in *Morchella deliciosa* and *M. esculenta*. Mycologia 1987; 79:683–687.
47. Yoon CS, Gessner RV, Romano MA. Population genetics and systematics of the *Morchella esculenta* complex. Mycologia 1990; 82:227–235.
48. Ennos RA, Swales KW. Genetic variability and population structure in the canker pathogen *Crumenulopsis sororia*. Mycologia 1991; 95:521–525.
49. Zhu H, Higginbotham KO, Dancik BP, Navratil S. Intraspecific genetic variability of isoenzymes in the ectomycorrhizal fungus *Suillus tomentosus*. Can J Bot 1988; 66: 588–594.
50. Leuchtmann A, Clay K. Isoenzyme variation in the fungus *Akinsonella hypoxylon* within and among populations of its host grasses. Can J Bot 1989; 67:2600–2607.
51. Leuchtmann A. Isozyme relationships of *Acremonium* endophytes from twelve *Festuca* species. Mycol Res 1994; 98:25–33.
52. Cameleyre I, Olivier JM. Evidence for intraspecific isozymes variations among French isolates of *Tuber melanosporum* (Vitt.) FEMS Microbiol Lett 1993; 110:159–162.
53. Hames BD, Rickwood D, eds. Gel Electrophoresis of Proteins—A Practical Approach. 2nd ed. Oxford: Oxford University Press, 1990.
54. Bradford MM. A rapid and sensitive method for the quantification of microgram quantities of protein utilizing the principle of protein-dye binding. Anal Biochem 1976; 72:248–254.
55. Shaw CR, Ballal SK, Harris JW. Starch gel electrophoresis of enzymes: a compilation of recipes. Biochem Genet 1973; 4:297–320.
56. Kulczynski S. Die pflanzenassoziationen der Pieninen. Bull Int Acad Polish Sci Lett Cl Sci Math Nat 1927; B:57–203.
57. Paranjpe MS, Chen PK, Jong SC. Morphogenesis of *Agaricus bisporus*: changes in proteins and enzyme activity. Mycologia 1979; 71:469–478.
58. Rohlf FJ. NTSYS. Numerical taxonomy and multivariate analysis system. Exeter Software ver. 1.80. 1994.
59. Saitou N, Nei M. The neighbor-joining method: a new method for reconstructing phylogenetic trees. Mol Biol Evol 1987; 4:406–425.
60. Zervakis G, Sourdis J, Balis C. Genetic variability and systematics of eleven *Pleurotus* species based on isozyme analysis. Mycol Res 1994; 98:329–341.
61. Rogers JS. Measures of genetic similarity and genetic distance. Stud Genet 1972; VII(7213):145–153.

6

Fungal Immunotaxonomy

S. H. W. Notermans
National Institute of Public Health and the Environment, Bilthoven, The Netherlands

M. A. Cousin
Purdue University, West Lafayette, Indiana

G. A. De Ruiter
Hercules European Research Center, Barneveld, The Netherlands

F. M. Rombouts
Wageningen Agricultural University, Wageningen, The Netherlands

I. INTRODUCTION

The taxonomy of fungi has become very important because they are used in various industrial, medical, food, and other biotechnological applications. They are also involved in plant diseases, human and animal medical problems, and food spoilage and toxigeneses. Hence, the correct identification of fungi has taken on added significance in recent years. The identification of filamentous fungi, which involves the isolation, subculturing, and morphological study of the colonial characteristics after a set time at a given temperature, can take up to 2 weeks or more. In addition, the microscopic characteristics of the fungus, particularly the hyphae and spores, are carefully studied. Since the identification of fungi is so time-consuming, new methods for rapid identification have been studied. Some of these methods include the use of biochemical techniques: chemical profiling of the secondary metabolites (including mycotoxins); scanning electron microscopic identification of surface textures; computer-assisted keys; immunological methods; enzyme, carbohydrate, and protein profiles; and various molecular identification techniques (G + C molar ratios, DNA-DNA hybridization, nucleic acid probes, rRNA sequences, restriction fragment length polymorphism [RFLP], polymerase chain reaction [PCR], randomly amplified polymorphic DNA [RAPD]) (1–4).

One of the earliest uses of immunological methods for fungal taxonomic purposes was the work of Hayashi et al. (5) and of the French group headed by Biguet (6). They used immunoelectrophoresis to study species of *Absidia*, *Aspergillus*, *Candida*, *Penicillium*, and various genera of dermatophytes. Other researchers have also used immunological methods to study fungi, especially those involved in disease production in humans where rapid identification and diagnosis of illness are essential. Fungal antigens can be prepared from whole cells or parts of cells, spores, metabolic products, or cultural media used to grow fungi (7,8). Most fungal antigens have been obtained from the soluble component of cells or the medium in which they have grown (8). The soluble immunogenic molecules secreted by fungi into the growth medium have been termed exoantigens (8–10). Many fungi produce antigens that are unique to a specific genus and/or species; therefore, those antigens can be used in identification (8–10). Kaufman et al. (9) used exoantigen tests to identify several pathogenic fungi with some being specific to single genera and species. The exoantigen is produced by any state of the fungi from typical to atypical growth and sporulating to nonsporulating cultures. A further advantage of using exoantigens to identify fungi, especially for medical diagnosis, is that they can be detected in both nonviable and mixed cultures (10–14). Microorganisms produce a number of substances that have immunological properties. In the past many investigators have worked out the immunogenicity of bacterial polysaccharides (15,16). These investigations have demonstrated that immunological assays can be used to identify groups of bacteria. Filamentous fungi, like bacteria, are able to excrete a variety of polysaccharide exoantigens which show immunological properties, and as a consequence they may be suited for taxonomic identification of fungi. Chaumeton et al. (17) studied the water-soluble polysaccharides produced by over 300 species of higher fungi to determine if they could be used to describe taxonomic groups. They concluded that the production of these polysaccharides could be a useful taxonomic tool because some groups produced them while others did not.

In addition to using exoantigens to identify fungi, exoantigens have been used to differentiate various fungi that are taxonomically similar (11,13). Kaufman and Standard (11) suggested that the antigenic characteristics of fungi could be used as a taxonomic tool because they are produced during fungal growth and generally do not depend on culture medium, temperature, or age, as do some morphological characteristics. Since some polyclonal antibodies show cross-reaction with antigens from related genera and species, the use of selected monoclonal antibodies may be of interest (4,12,18). Ferguson et al. (18) found that a cluster analysis technique was able to separate various genera of fungi based on the specific carbohydrate makeup of the monoclonal antibody. Most of the fungal antigens have been classified as being mainly polysaccharide and glycoprotein in nature (7,8,19–23). These fungal polysaccharides are responsible for the specificity; however, the type of extraction used for them can affect their final specificity (8). Soluble polysaccharides from bacteria were shown to produce an immune response as early as 1917, and now most immunologically active polysaccharides are associated with microorganisms (24). These immune responses are reviewed by Bishop and Jennings (24). Commercial test kits, based on polyclonal and monoclonal antibodies, are now available for the identification of both medical and plant pathogens.

Bacterial polysaccharides are generally composed of repeating units of specific oligosaccharides, resulting in a high epitope density and a high molecular weight. This usually leads to a T-cell-independent immune response in which B-cells are directly activated, resulting in mainly IgM antibodies. In contrast, fungal polysaccharides or glycoproteins are in general not composed of repeating units but are much more heterogeneous and have a much lower molecular weight, similar to plant polysaccharides. These phenomena lead to a T-cell-dependent activation of B-cells, often resulting in IgG antibodies. In many cases of raising antibodies against fungal antigens, the so-called carrier effect occurs. The protein part of a glycoprotein or the protein to which synthetic antigens are coupled (25) acts as carrier—i.e., a molecule that renders a hapten linked to it to stimulate antibody production without activating the B-cells themselves (26). As this protein carrier is usually essential to obtain antibodies against fungal carbohydrate residues, it is often extremely difficult to obtain antibodies against pure fungal polysaccharides. The immunological properties of polysaccharides and the knowledge that fungi also produce extracellular and water-soluble polysaccharides has resulted in many studies concerning the structural analysis of polysaccharides produced by fungi. Work carried out by Notermans and Soentoro (19), Tsai and Cousin (23), Kamphuis (27), de Ruiter (28), and de Ruiter et al. (29) has shown that the immunogenic properties of extracellular, water-soluble polysaccharides and glycoproteins produced by fungi can, due to their specificity, be applied for immunotaxonomical purposes.

II. PRINCIPLES OF IMMUNOTAXONOMY FOR FUNGI

A. Immunology as a Taxonomic Tool

Antigens can be useful in taxonomy because they are a normal part of fungal growth (11). Initial studies concerning the immunological activity of polysaccharides produced by fungi were carried out with glycoproteins isolated from the cell wall. The carbohydrate composition of the cell wall is useful in delineating taxa at various levels (8,12,18). Using immunology, it was possible to correlate or rearrange species into the same genus or into different genera (6,30–33). As a result of these findings, chemical structures of cell wall polysaccharides were analyzed. Preston and Gander (34) and Preston et al. (35,36) determined that the polysaccharides of *Penicillium charlesii* contained primarily mannose, galactose, and glucose. Galactose was present in the furanose configuration, and this galactofuranose was immunodominant in the polysaccharide (21,37). Polysaccharides produced by the species belonging to the order Mucorales contained mainly glucuronic acid, mannose, galactose, glucose, and fucose (23,28,38,40). De Ruiter (28) demonstrated that the 2-O-methyl D-mannose residues at the nonreducing terminal end were involved in the immunological properties of the polysaccharide.

Initial work carried out by Notermans and Soentoro (19) showed that antibodies raised against extracellular polysaccharides of *P. digitatum* were reactive with culture fluid of other species belonging to the genera *Penicillium* and *Aspergillus*. No reactions were observed with culture fluids of other fungi. An exception was the culture fluids of *Penicillium* subgenus *Biverticillium*, containing species such as *P. islandicum*, *P. funiculosum*, and *P. rubrum*, which did not react with antibodies raised against the extracellular polysaccharides of *P. digitatum*. Antibodies raised against the polysaccharides produced by *Cladosporium cladosporoides* and *C. herbarum* were specific for all species belonging to the genus *Cladosporium* (19,39). De Ruiter et al. (22,40) demonstrated that antibodies raised against the polysaccharide antigens of *Mucor racemosus* and *M. circinelloides* were reactive with fungi of the genera *Mucor*, *Rhizomucor*, *Rhizopus*, *Thamnidium*, *Absidia*, and *Syncephalastrum* and species belonging to the *Mortierella isabellina* group (29), all of which belong to the order Mucorales. Tsai and Cousin (39) showed that *Mucor circinelloides* antibodies reacted only with other *Mucor* species. These findings show that extracellular polysaccharides of fungi are antigenic and that immunotaxonomy is possible to the genus level. With further refinement of the technique it may be possible to achieve species specificity.

B. Selection of Antigens for Immunotaxonomy

Before selecting the method to produce antibodies for immunotaxonomy, one must decide whether the method is intended for genus, species, or strain level specificity. This will dictate whether the antigens will be from the culture fluid, mycelial cell wall, spore, or some metabolite produced by the fungus. For medical (6,8,11,13) and plant pathological (4,14) taxonomy, the fungus that causes the disease is chosen as the antigen. However, in the area of spoilage fungi involved in deterioration of foods and feeds, the choice of antigen will depend on several species that are routinely isolated from spoiled products. Notermans and Heuvelman (41) and Notermans and Soentoro (19) randomly selected fungal strains and used the polysaccharide fraction of the culture fluid to produce antibodies in rabbits. Polysaccharides produced by *P. digitatum* and *P. verrucosum* var. *cyclopium* gave identical results (20). Initial immunization experiments of rabbits with the polysaccharide fraction of *M. racemosus* resulted in antibodies that were also reactive with certain *Penicillium* and *Aspergillus* species. In further immunization experiments carried out by de Ruiter et al. (40), antibodies were obtained that were specific only for fungi belonging to the order Mucorales. Similar specificity has been observed for polysaccharides produced by *Fusarium* species. These findings indicate that well-selected strains and the immunization of different rabbits will be necessary for obtaining specific antibodies when using polyclonal techniques. Antigens that show genus specificity have also been prepared to fungal mycelia (23,42–44).

Specific antibodies may also be obtained by producing monoclonal antibodies or by using synthetic polysaccharide antigens. Monoclonal antibodies against exoantigens of species of *Penicillium*, *Aspergillus*, *Botrytis*, and *Mucor* have been described (45–57). Monoclonal antibodies show specificity comparable to, or greater than, that of polyclonal antibodies. The work of de Ruiter (28) showed that 2-O-methyl-D-mannose residues of the exoantigens of *M. racemosus* are immunoreactive with polyclonal antibodies raised in rabbits; however, the antigenicity of a murine monoclonal antibody raised against the exoantigens of *M. racemosus* was not based on 2-O-methyl-D-mannose residues. Nevertheless, the monoclonal antibody was still very specific for fungi belonging to the order Mucorales (57).

Since it became clear that *Penicillium* and *Aspergillus* species produce exoantigens of which $\beta(1 \rightarrow 5)$-linked D-galactofuranosides are immunodominant (19), synthetic tetramers and heptamers of $\beta(1 \rightarrow 5)$-linked D-galactofuranosides conjugated to tetanus toxoid have been applied to produce antibodies in rabbits (25). The antibodies obtained with the tetramer conjugate reacted only with the exoantigens of a few *Aspergillus* and *Penicillium* species. Antibodies obtained with the heptamer conjugate were reactive with the exoantigens of all strains of *Penicillium* and *Aspergillus* tested. Again the exoantigens of the species of *Penicillium* subgenus *Biverticillium* did not react. No reactions were observed with the exoantigens of fungi belonging to any other species tested. The selection of antigens needs to be carefully evaluated if the taxonomic identification to genus, species, and strain level is desired. Since immunoassays are being used to detect fungi rapidly in medical, agricultural, and food systems, it follows that antigens could be developed that have taxonomic potential.

C. Genus/Species Specificity of Antibodies

To use immonology in the systematics of fungi, it is necessary to show genus, species, and even strain specificity. Early work in both medical and plant pathology has shown that this can be accomplished by immunological analysis. Several researchers have shown that immunologically based tests can be specific for a given species by using antisera to specific fungal exoantigens. Sekhon et al. (58) found that antisera to *Penicillium marneffei* reacted only with strains of this species and one unidentified *Penicillium* species. There was no cross-reaction with other *Penicillium* species or with *Aspergillus* spp. Similarly, Kaufman and Standard (11) and Sekhon and Padhye (13) reviewed the specificity of immunological assays developed to the exoantigen of several important medical fungi. Polonelli and Morace (12) concluded that serology should become a rapid and accurate method to identify medical pathogens because mycoimmunology has advanced rapidly within the past two decades. Correll (4) reviewed the early use of immunology to detect and identify specific plant pathogens. It was not until the introduction of enzyme-linked immunosorbent assays (ELISA) and monoclonal antibodies that the identification of a specific genus and/or species was possible. Now there are commercial diagnostic kits to detect specific plant pathogens.

Within the past decade research has focused on the immunological identification of fungi in foods because several assays were genus-specific. Genus specificity has been shown for *Botrytis* and *Monascus* species (59), *Cladosporium*, *Geotrichum* (39,43), and for the order Mucorales (57). Other assays had cross-reactions between related genera. For example, antibodies against exoantigens of *P. digitatum* are reactive with exoantigens of all *Penicillium* and *Aspergillus* species except those in the subgenus *Biverticillium* (19,20).

This specificity is caused by the presence of β(1 → 5)-linked D-galactofuranosides in the exoantigens of these fungi (20). A trimer of β(1 → 5)-linked D-galactofuranose inhibited the reaction between antibodies raised in rabbits and exoantigens, indicating that the antibodies were specific. Kamphuis et al. (60) showed that when these antigenic β(1 → 5)-linked D-galactofuranosides were removed by acid hydrolysis, the antigenicity from IgG antinative exoantigens and acid-hydrolyzed exoantigens disappeared. Antibodies raised against the acid-hydrolyzed exoantigens revealed new antigenic determinants, which were no longer directed to the galactofuranose residues. Furthermore, immunological tests demonstrated that these antibodies were more species-specific. These initial findings demonstrate that acid hydrolysis of exoantigens of *Penicillium* will result in raising antibodies that allow a more species-specific detection.

Another way to specifically detect individual species of fungi may be to use monoclonal antibodies (MAb) that are well selected for this purpose. Careful selection of clones is necessary since the MAb against exoantigens of *M. racemosus* was reactive with all species in the order Mucorales (57). The immunodominant epitopes of exoantigens of *M. racemosus* are composed of 2-O-methyl-D-mannose residues linked to the 2 position of the next mannose residue (28). Complete inhibition of the reaction of rabbit antibodies and exoantigens was obtained after addition of the trimer of this mannose residue.

All these results support the theory that different epitopes are present on exoantigens produced by fungi. Certain epitopes, such as $\beta(1 \rightarrow 5)$-linked D-galactofuranose (*Penicillium/Aspergillus*) and 2-O-methyl-D-mannose residues (Mucorales), which are terminally situated in the exoantigen molecule, are dominant in the production of antibodies. If further refinement of the immunodominant site can be achieved, then species or strain specificity can be realized. Currently, the immunotaxonomic potential to the genus level would even be helpful for food microbiologists, who are forced to act as food mycologists when identification is needed.

III. IMMUNOLOGICAL IDENTIFICATION OF *PENICILLIUM* AND *ASPERGILLUS*

A. Production and Isolation of Antigens and Antibodies

There are no standard methods for the production of fungal antigens. Antigens can come from crude culture filtrates, ground whole cells or spores, and partially purified extracellular material or cells. These antigens have been used as either live or killed cells. Initially, polysaccharides produced by fungi were isolated by extraction with boiling aqueous potassium hydroxide (61,62), by extraction with a trichloroacetic acid solution (63) or even by total disruption of the mycelium (37,64). Galactomannans were isolated from *Hormodendrum* species and shown to be immunologically active (65). Notermans and Heuvelman (41) observed that polysaccharides are released to the surrounding environment during the growth of fungi. In contrast to the findings of Suzuki and Takeda (65), it was shown that these extracellular polysaccharides were almost genus-specific (19). Hence, they may be suitable for immunotaxonomy with groups of fungi.

1. Production of Antigens

Various methods have been used to produce exoantigens for antibody production. The medium used, temperature, time of incubation, and static or agitated conditions have depended on the fungi considered, the main concern being to produce a culture with a broad array of antigens. Longbottom and Austwick (8) recommended a dialyzable, chemically defined medium for fungal antigen production to eliminate antigens that may be present in the components of the medium. Some of the media used to produce exoantigens have included dialyzed Czapek Dox, Sabauroud's medium, Bacto-synthetic broth, malt extract broth, and media supplemented with casein hydrolysate, beef extract, or V-8 juice (8,20). Kamphuis (27) and de Ruiter et al. (22) successfully used a complete synthetic medium composed of defined quantities of minerals, amino acids, vitamins, and trace elements (yeast-nitogen base) supplemented with glucose. Mycelial antigens have been produced in brain heart infusion broth (23,39) and Czapek medium (42,56), washed, dried, and injected into rabbits or mice. Cells and spores can also be crushed, sonicated, homogenized, or otherwise broken to release antigens from the cell walls, membranes, or other parts of the cell. The resulting antigens can be frozen in liquid nitrogen, lyophilized, or frozen at −20°C. The main concern in producing either exoantigens or mycelial antigens is to have a medium that is devoid of antigens that can cross-react in assays.

2. Purification of Antigens

There are many different ways to prepare and purify the antigens. After the fungus has grown for the desired time at the appropriate temperature, then the medium can be decanted, recovered after centrifugation or filtration or similar procedure (8). The resulting filtrate can be further purified and concentrated by membrane or ultrafiltration followed by dialysis. These crude antigens can be further purified by precipitation, affinity chromatography, or similar methods. Some antigens have been deproteinated (48) or extracted with various chemicals such as urea, ammonium oxalate, ethanolamine, Triton X-100, or other chemicals selected to remove specific antigens (51). To produce antigens to spores or mycelia, the fungal material is washed, centrifuged, and dried. In some cases it is treated chemically (merthiolate, formalin, or formaldehyde) or heated to kill the cells; however, Kaufman et al. (9) found that formaldehyde inactivated some antigens. The purity that is desired will determine the extent to which these antigens are treated.

3. Production of Antibodies

Polyclonal antibodies are prepared by immunizing rabbits, and monoclonal antibodies by immunizing mice. Immunization of rabbits is performed by subcutaneous injection with the exoantigen over several different time periods. In the first injection the portions are mixed with Freund's complete adjuvant, and in the latter injections the portions are mixed with Freund's incomplete adjuvant. After the last injection, blood is collected, allowed to clot, and centrifuged to collect serum, and the antibody fraction is isolated from the serum by methods such as those of Steinbuch and Audran (66). Antibodies can be purified using protein A or G, gel filtration, affinity chromatography, anion exchange resins, hydroxyapatite, ammonium sulfate precipitation, and combinations of these methods (69). Immunization of mice has been by intravenous or intraperitoneal injections with the immunogen in adjuvant, as described for rabbits. After the mice are sacrificed, the spleen is removed and used to produce monoclonal antibodies in tissue culture. Various procedures have been used for hybridoma production (45–48, 50–57). Several other ways to produce and purify antibodies are possible. A new approach for producing polyclonal antibodies in rabbits is to implant subcutaneous chambers (sterilized plastic whiffle balls) to collect the antibodies which can easily be removed as fluid without blood by a sterile syringe (67,68). Also, various chemical and biotechnological companies have rapid test kits that can be purchased for easy purification of antibodies.

B. Characterization of Some Antigens

1. Composition of Antigens

Some of the earliest work in the immunological identification of *Aspergillus* and *Penicillium* species showed that the immunodominant portion is a galactomannan containing mainly galactose and mannose (65,70,71). Reiss and Lehmann (71) showed that the antigen from *A. fumigatus* contained galactose and mannose in a 1:1.17 ratio with a 1 → 6-linked mannan backbone and oligogalactoside side chains terminating in galactofuranose. Other researchers also identified galactofuranosyl groups in *Aspergillus* and *Penicillium* antigens (34–37,72). In later research, in addition to mannose, and galactose glucose were identified in *Penicillium* antigens from several species (73–75). Rupérez et al. (74) postulated that the extracellular fraction from *P. erythromellis* was β-glucan with a 1 → 6 linkage and two malonyl hemiesters separated by five glucose residues. This laid the groundwork for further research on the chemistry of the antigens.

The exoantigen fractions of *Penicillium* and *Aspergillus* strains contain primarily mannose, galactose, and glucose (Table 1). The molecular weight range of the exoantigens was estimated between 10 and 65 kD (41,73). A model of the exoantigen is presented in Figure 1. The mannose residues are linked primarily through α(1 → 2)- and α(1 → 6)-O-glycosidic linkages. In this model the galactose residues occur in the furanosyl configuration and are β(1 → 5)-linked. They contribute to the antigenicity of the molecule, as was demonstrated by immunological inhibition experiments carried out by Bennett et al. (37) and Notermans et al. (21). Van Bruggen-van der Lugt et al. (77) made use of a purified exo-β-D-galactofuranosidase combined with a reductive-cleavage technique to

Table 1 Monosaccharide Composition of Exoantigens Produced by *Penicillium* and *Aspergillus* Species

Strain	Polysacch. fraction (%)	Monosaccharides (mol%) Man	Gal	Glc	Protein %	Reference
P. verrucosum var. cyclopium	70	76	17	7	N.D.[a]	20
P. digitatum	64	64	60	12	N.D.	20
P. chrysogenum	55	79	20	2	8	73
A. repens	71	30	66	2	N.D.	20
A. niger	46	50	45	5	N.D.	20
A. versicolor	38[b]	62[b]	31[b]	7[b]	15[b]	23

Source: Ref. 23.
[a]N.D. not determined.
[b]Average of three peaks from a sepharose CL-4B column that could not be absorbed by a Con A column (23).

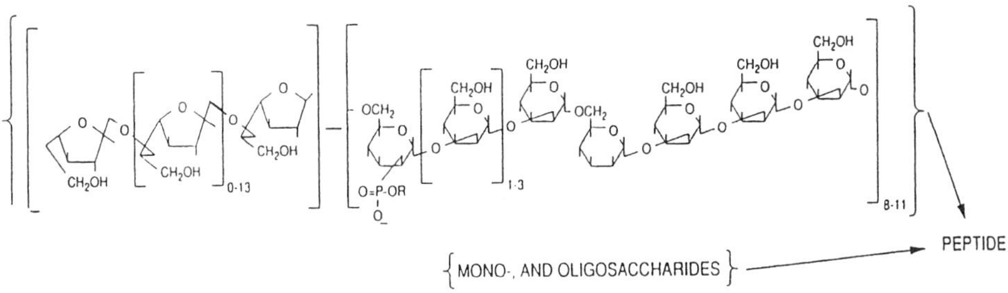

Figure 1 Model of the exoantigen from *P. charlesii* to show that the terminal β-(1,5)-linked D-galactofuranose is immunodominant. Source: Gander et al. (76) and Notermans et al. (21).

produce a new structural model for the antigenic galactofuranose residues of *Penicillium* species, as shown in Figure 2. Tsai and Cousin (23) found that antigens from *A. versicolor* produced three separate ELISA-positive peaks that did not absorb to concanavalin A (Con A)-sepharose. One of these antigenic peaks was used to develop a specific antibody for *Aspergillus* species that did not cross-react with *Penicillium* antigens (39).

Mycelial antigens of *P. chrysogenum* have more galactose (39%) and glucose (7%) and less mannose (54%) than the extracellular fractions shown in Table 1 (73). The mycelial antigens from *A. versicolor* show two active peaks, one absorbed by Con A that was 53% mannose, 39% galactose, 8% glucose, and 17% protein; and one not adsorbed by Con A that was 11% mannose, 48% galactose, 41% glucose, and 50% protein (23). Therefore, the type of antigen produced may have different absolute sugar and protein concentrations, but the immunodominant fraction may be similar (73).

The site of the antigen in the fungus has been briefly studied by Cole et al. (78). They found the antigen in the cell walls of vegetative hyphae, but not sporulating hyphae. It was found in young and old hyphae and conidiophores, but was not present in phialides and conidia. It was postulated that this antigen was an autolysin from the hyphae. Therefore, antigens produced to the spore and mycelium may not always recognize each other because they may not be structurally similar.

2. *Stability*

Many of the polysaccharide antigens produced by fungi are glycoproteins (8). The stability of these antigens to various environmental factors is criticial if they are to be used in assays to determine taxonomic relationships. They resist hydrolysis by some enzymes (8), and the exoantigens produced by *Penicillium* and *Aspergillus* are highly heat-stable (41). The activity of these antigens was only removed by

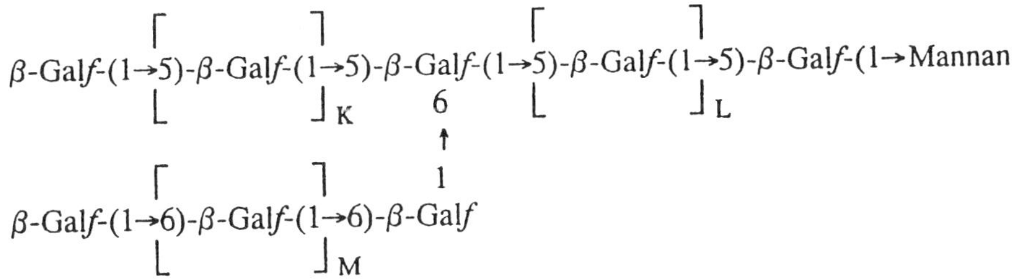

Figure 2 New structural model for the antigenic galactofuranose side chains of the extracellular polysaccharides from *Penicillium* species with the values for K, L, and M varying from 0 to 8 residues. Source: Ref. 77.

acid hydrolysis (pH 1.8 at 100°C for 1 hour). Kamphuis (25) demonstrated that acid hydrolysis caused the galactose contents of the exoantigen to decrease by 19% to 36%. Digestion with protease decreases the activity of *Aspergillus* and *Penicillium* antigens by about 50% to 70% (23,73). This strongly suggested that these antigens were glycoproteins.

Treatment of the exoantigens of *P. digitatum* and *A. fumigatus* with exo-β-D-galactofuranosidase resulted in a complete disappearance of the antigenicity of the polysaccharides when using antibodies raised against the native exoantigen (27). This again demonstrates that the terminal galactofuranose residues are the immunodominant part. In a study by Cousin et al. (79), it was demonstrated that although most *Penicillium* and *Aspergillus* species produced small quantities of β-D-galactofuranosidase, it did not interfere with the detection of the exoantigen produced by these fungi. In contrast, substantial quantities of this enzyme were produced by the biverticillate *Penicillium* spp.—namely, *P. funiculosum*, *P. islandicum*, *P. rubrum*, and *P. tardum* (79). Antigens of these *Penicillium* spp. were not detected by antibodies produced to the exoantigens of *P. digitatum*. All these results show that fungal antigens are stable enough to use in developing immunoassays for identifying fungi.

C. Sensitivity and Specificity of Antibodies

Several studies have been conducted to test the exoantigen production by *Penicillium* and *Aspergillus* fungi under different conditions (27,39,41,80,81). These conditions included the effect of water activity (80), incubation temperature (39,80,81), type of carbon and nitrogen source, and glucose concentration (27,41,81). In general it was found that if growth of *Penicillium* and *Aspergillus* occurred, then immunologically active exoantigens were produced. However, the quantity of exoantigens produced varied and depended on the species involved

Table 2 Production of Exoantigens (EPS) by Several Fungal Species, as Estimated by Applying Antibodies Raised Against the Exoantigen of *P. digitatum* (19,27,80)

Fungi	Production of EPS	Fungi	Production of EPS
Penicillium atramentosum	++	*Botrytis aclada*	–
P. aurantiogriseum	++	*B. cinerea*	–
P. camembertii	++	*B. tulipae*	–
P. canensis	++		
P. capsulatum	++	*Fusarium dimerum*	–
P. chermesinum	++	*F. graminearum*	–
P. chrysogenum	++	*F. moniliforme*	–
P. citreonigrum	++	*F. oxysporum*	–
P. citrinum	++	*F. sambucinum*	–
P. clavigerum	++	*F. verticilloides*	–
P. commune	++		
P. corylophilum	++	*Mucor flavus*	–
P. cyaneum	++	*M. circinelloides*	–
P. decumbens	++	*M. hiemalis*	–
P. dierckxii	++	*M. mucedo*	–
P. digitatum	++	*M. racemosus*	–
P. diversum	++	*M. plumbeus*	–
P. echinulatum	++		
P. expansum	++	*Cladosporium herbarum*	–
P. frequentans	++	*C. cucumerinum*	–
P. gladioli	++	*C. cladosporioides*	–
P. granulatum	++	*C. sphaerospermium*	–
P. grisiofulvum	++		
P. implicatum	++	*Rhizopus oryzae*	–
P. italicum var. *italicum*	++	*R. nigricans*	–
P. janthinellum	+	*R. oligosporus*	–
P. lavendulum	++		
P. lividum	++		
P. nalgiovense	++	*Geotrichum flei*	–
P. namyslowski	++	*G. candidum*	–
P. ochrachloron	++	*G. capitatum*	–
P. oxalicum	++	*G. fermentans*	–
P. palitans	++	*G. klebahnimorenz*	–
P. paxillii	++		
P. purpurrescens	++	*Scopulariopsis brevis*	–
P. raistrickii	++	*S. candida*	–
P. regulosum	++		
P. roquefortii	++	*Alternaria alternata*	–
P. simplissininum	++		
P. tardum	++		
P. terlikowskii	++		

Table 2 Continued

Fungi	Production of EPS	Fungi	Production of EPS
P. thomii var. *flavescens*	++		
P. trzebinskii	++		
P. turbatum	++		
P. variable	++		
P. varians	++		
P. velutinum	++		
P. veridicatum	++		
P. verrucosum var. *cyclopium*	++		
P. funiculosum	−		
P. islandicum	−		
P. rubrum	−		
P. tardum	−		
Aspergillus candidus	++		
A. clavatus	++		
A. fischeri	++		
A. flavus	++		
A. fumigatus	++		
A. nidulans	++		
A. niger	++		
A. parasiticus	++		
A. ochraceus	++		
A. ostianus	++		
A. repens	++		
A. sydow	++		
A. tamarii	++		
A. versicolor	++		

− No ELISA reaction in 1/100 dilution of culture fluids.
+ ELISA reaction in 1/100 dilution not in 1/1000 of culture fluid.
++ ELISA reaction in 1/1000 dilution of culture fluid.

(81). Kamphuis (27) observed that the carbon source influenced the monosaccharide composition of the exoantigens, but not its immunological activity.

Table 2 shows the specificity of the antibodies raised against the exoantigen of *P. digitatum*, which only reacted with culture fluid from *Penicillium* and *Aspergillus* species; however, strains of the *Biverticillium* group did not react (19,27,80). Identical results were obtained with antibodies raised against the exoantigen of *P. verrucosum* var. *cyclopium*. In general, *Penicillium* and *Aspergillus* produce high quantities of immunologically active exoantigens; in most cases, a 10^4 to 10^5 dilution reacted positively in the ELISA test applied. Tsai and Cousin (39) showed that antibodies raised to *A. versicolor* cross-reacted with antigens from species of *Aspergillus* and *Penicillium*. Similar activities were shown for antibodies raised against *P. chrysogenum* (39). Several cross-reactions between antibodies and antigens of *Penicillium* and *Aspergillus* species have been described (42,47,57,82). Fuhrmann et al. (42) noted that one polyclonal antibody produced to *Penicillium verrucosum* var. *cyclopium* recognized antigens from both *Penicillium* and *Aspergillus* species in addition to giving minor reactions with species of *Fusarium*, *Mucor*, and *Trichoderma*. A second antibody only recognized *Penicillium* antigens from species in the subgenera *Furcatum* and *Penicillium*, but not *Aspergilloides*. When monoclonal antibodies were raised to *P. glabrum*, one clone shared epitopes with antigens from both *Penicillium* and *Aspergillus* species; however, other clones were specific for certain groups or species within the two genera (82). Classifications based on morphology could be confirmed by using several antigens produced by these two genera. Stynen et al. (50) found that monoclonal antibodies to *A. fumigatus* cross-reacted with antigens from *Penicillium digitatum*, *Trichophyton rubrum*, *T. interdigitalis*, *Botrytis tulipae*, *Wallemia sebi*, and *Cladosporium cladosporioides*.

Some examples of specificity have been demonstrated for *Penicillium* and *Aspergillus* antibodies. Dewey et al. (47) produced a monoclonal antibody to *P. islandicum* that only recognized other *Penicillium* species. An antibody produced to a pathogenic *A. fumigatus* strain only recognized other *Aspergillus* species and not *P. notatum* or yeasts (83). Sekhon et al. (58) found that a polyclonal antibody raised against *Penicillium marneffei* reacted only with antigens from *P. marneffei* and an unnamed *Penicillium* species. The antibody did not react with sera from people suffering from other mycoses, such as aspergillosis, blastomycosis, and coccidioidomycosis, suggesting that the antibody was highly specific. When Polonelli et al. (84) studied *Penicillium camemberti* and other penicillia from cheese, they found that many strains were antigenically related, and they could group them into nine classes. They suggested that the profiles of exoantigens could be used to differentiate fermentation cultures from contaminants. Similarly, Sekhon et al. (85) produced antibodies to five *Aspergillus* species (*A. flavus*, *A. fumigatus*, *A. nidulans*, *A. niger*, and *A. terreus*) in an effort to group medically important *Aspergillus* species. They were able to group the antigens into categories that agreed with traditional taxonomy by absorbing out cross-reactions using heterologous antigens. A assay specific for *A. flavus* and not *A. parasiticus* was developed by producing antibodies to surface proteins that were only weakly recognized by the other species (86). Therefore, genus- or species-specific antibodies can be produced for *Aspergillus* and *Penicillium* species, and they can be used for identification in these genera.

IV. IMMUNOLOGICAL IDENTIFICATION OF MUCORALES

A. Production and Isolation of Antigens and Antibodies

1. Production and Purification of Antigens

Notermans and Heuvelman (41) and Notermans et al. (81) demonstrated that exoantigens from *Mucor* were immunologically active. Similarly, antigens prepared from *Rhizopus stolonifer* mycelium had antigenic activity (43,44). As a result, extracellular or mycelial antigens have been used to produce antibodies to Mucorales. Dialyzed malt extract broth, yeast nitrogen base supplemented with D-glucose, and brain heart broth have been used to produce exoantigens to *Mucor* species (22,39,41,81), and *Rhizopus stolonifer* mycelial antigens have been produced in tomato juice broth (44,87). The exoantigen from culture fluid of *M. racemosus* was purified in an identical way to that described for *Penicillium* species (39). De Ruiter et al. (40) used ethanol precipitation and anion-exchange chromatography to further purify the Mucorales antigens. Neither the *R. stolonifer* nor the *M. circinelloides* mycelial antigens were further purified after being freeze-dried; the ground powder was used directly to immunize rabbits (39,44).

2. Production of Antibodies

Polyclonal antibodies were produced to the exoantigens or mycelium from rabbits immunized with a certain amount of antigen in Freund's complete adjuvant followed by booster shots in incomplete adjuvant in the same way as described for *Penicillium* and *Aspergillus* species. De Ruiter et al. (57) also produced monoclonal antibodies by immunizing BALB/c mice with the exoantigen from *M. racemosus*. Clones were made by using spleen cells fused to a mouse myeloma cell line; subclones were made until a sufficient quantity was produced (57).

B. Characterization of Some Antigens

1. Composition of Antigens

Bartnicki-Garcia (30) proposed a classification of fungi according to their cell-wall polysaccharide composition. The hyphal walls of Mucorales are mainly composed of uronic acids, neutral sugars, hexosamines, and proteins. Typically, members of the Mucorales contain heteropolysaccharides composed of glucuronic acid, fucose, mannose, and some other, minor neutral sugars. Members of Mucorales also produce extracellular polymers consisting mainly of carbohydrates, such as glucuronic acid, galactose, fucose, and mannose (88). Miyazaki and Irino (89,90) showed that Mucorales also contain *N*-acetyl glucosamine and *N*-acetyl galactosamine. The exoantigen fractions of representatives of the Mucorales contain fucose in addition to mannose, galactose, and glucose (Table 3)

Table 3 Monosaccharide Composition of Exoantigens Produced by Fungal Species Belonging to the Order Mucorales

Strain	Polysacch. fraction (%)	Monosaccharides (mol%)				
		Fuc[a]	Man[b]	Gal[c]	Glc[d]	GlcA[e]
Mucor hiemalis	50	28	18	15	2	37
Rhizopus oryzae	52	22	17	4	15	42
Absidia corymbifera	44	13	10	8	52	16
Syncephalastrum racemosus	37	35	32	9	3	21
Thamnidium elegans	35	16	14	9	32	29

Source: De Ruiter et al. (191).
[a]Fucose.
[b]Mannose.
[c]Galactose.
[d]Glucose.
[e]Glucuronic acid.

(23,40,91) and <2% arabinose (40). A characteristic component of the extracellular polysaccharides of the Mucorales is a polymer composed of a β(1-4)-D-glucoran (94).

Tsai and Cousin (23) showed that the *M. circinelloides* mycelial antigens combined 53.8% mannose, 32.4% glucose, 8.6% galactose, and 5.2% fucose, compared to the 61.6% mannose, 10.8% glucose, 13.9% galactose, and 13.7% fucose found in the extracellular antigens. About 25% of these antigens are protein, suggesting that they are glycoproteins. L-fucosidase does not reduce the antigenic activity; therefore, fucose is not part of the immunodominant site (23). Research suggests that in the *Mucor* mycelial antigens α,1-4-glucose is linked to the mannosyl residues at the immunodominant site and that β-linked glucose is at the nonreducing terminal end of the *Mucor* extracellular antigens (23).

Specific degradation experiments with purified exo-α-D-mannanase and a α-mannosidase have shown the presence of 2-O-methyl-D-mannose residues in the epitopes of all exoantigens of the genera of Mucorales (49). Hapten inhibition experiments carried out later with synthetic oligosaccharides showed that a trimer based on α(1 → 2)-linked mannose residues with a 2-O-methyl-D-mannose residue at the nonreducing terminal could completely inhibit the immunological reaction with rabbit antibodies (92,93). These experiments show that these oligosaccharides are part of the immunochemical reaction. De Ruiter et al. (93) demonstrated that this immunologically dominant oligosaccharide represents less than 0.5% (w/w) of the exoantigen fraction.

2. *Stability*

The immunological activity of exoantigens from some strains (*Mucor racemosus* and *Rhizopus oligosporus*) is not lost after acid hydrolysis at pH 1.8 and 100°C for 4 hours (94) or after steaming for 30 min (39). The *R. stolonifer* antigen was stable to 100°C for 90 min (43). Exo-α-D-mannanase will completely destroy the ELISA activity of the polysaccharides produced by species of the order of Mucorales (93). About 60% of the antigenic activity is lost after reacting with microbial protease for 24 hours at 37°C (23). Reaction with protease followed by either α-amylase and α-mannosidase or cellulase will decrease the antigenicity by about 40% (23).

C. Sensitivity and Specificity of Antibodies

In general, the quantity of the immunologically active exoantigen produced by species of the order Mucorales is lower than that produced by *Penicillium* and *Aspergillus* (22). Nevertheless, the antigen is readily detectable in culture fluids diluted 1/100 to 1/1000 (22). The specificity of antibodies produced in rabbits against the exoantigen of *M. racemosus* has been examined by de Ruiter et al. (22) by screening against 34 different species of ascomycetous yeasts, five species of basidiomycetous yeasts, and 18 non-Mucorales fungal species. No cross-reactions were observed with the strains of yeasts and fungi tested, except with *Pichia membranaefaciens* (Hansen) (22). On the other hand, antigens from all strains belonging to the order of Mucorales showed clear positive reactions with the antibody raised to *M. racemosus*.

Lin and Cousin (43) raised antibodies to *R. stolonifer* mycelium and found that they did not cross-react with yeasts but that they did with *Mucor* species. *R. stolonifer* could be detected in less than 1000 ng/ml in this assay (44). Tsai and Cousin (39) produced antibodies to *Mucor circinelloides* mycelia and extracellular antigens and found that they cross-reacted with other *Mucor* species only in the presence of other fungi; however, when no other fungi were present, there was some cross-reactivity with some *Penicillium* and *Aspergillus* antigens. Subsequently, De Ruiter et al. (57) produced monoclonal antibodies to *M. racemosus* and found that they reacted with all fungi from the Mucorales plus species belonging to the *Mortierella isabellina* group. A clear example of the usefulness of the immunochemical properties of fungal exoantigens to establish a taxonomy has been described for the so-called *Mortierella isabellina* group. This group was recognized in 1977 by Gams and consists of the species *M. nana*, *M. roseonana*, *M. ramanniana*, *M. isabellina*, *M. ovata*, and *M. vinacea* (95). The chemical and antigenic features of the exoantigen preparations of the *Mortierella isabellina* group were compared with those derived from three other *Mortierella* species and with *Mucor racemosus*. Immunological characterization of the exoantigen preparations from the *Mortierella isabellina* group gave a positive ELISA reaction with anti-*Mucor* IgG only. The analysis of the exoantigens showed a chemotaxonomic difference between species of the *Mortierella isabellina* group and the species of *Mortierella* subgenus *Mortierella*. It is therefore unlikely that the *M. isabellina* group is related to the Mortierellaceae, as isolates show antigenic similarity with mucoralean species. The exoantigen from *Mortierella* subgenus *Mortierella* also differs from both other groups as the β(1–4)-D-glucuronan polymer is lacking. However, the *M. isabellina* group also deviates from Mucoraceae and related families in that the antigenic activity of the exoantigen is lower. In conclusion, the immunological properties of the exoantigens were used successfully to demonstrate that the *Mortierella isabellina* group were related to the Mucoraceae and not to the Mortierellaceae (29).

V. IMMUNOLOGICAL IDENTIFICATION OF OTHER FUNGI

A. Identification of *Alternaria* species

Early research was undertaken with *Alternaria* species as they can cause allergies in people who suffer from respiratory problems (96). *AV. tenuis* antibodies cross-reacted with fungal genera other than *Alternaria*—*Aspergillus fumigatus*, *Curvularia* sp., and *Stemphylium* sp. (96). This antigen had a large number of carbohydrate residues with vicinal −OH groups (demonstrated by oxidation with sodium metaperiodate that destroyed the antigenicity). However, the antigens were not affected much by pronase, pepsin, or neuraminidase treatment.

In the food detection area, polyclonal antibodies have been raised to *A. alternata* that also recognized antigens from *Colletotrichum*, *Epicoccum*, *Leptosphaerulina*, *Schizophyllum*, and *Trichoderma* species (44). Preliminary research showed that these antigens were high-molecular-weight polysaccharides with 1 → 3 linkages (43). Clark et al. (97) mentioned only one attempt to detect *A. alternata* immunologically in plant materials. Further research is required in this genus before immunotaxonomic tools can be proposed.

B. Identification of *Botrytis* Species

Botrytis species are plant pathogens and food spoilage microorganisms, and there is interest in their rapid identification. Savage and Sal (98) developed an antibody to unbroken *B. cinerea* thalli that reacted with species of *Botrytis* and *Sclerotinia* and could be detected to 900 ng/ml. Cousin et al. (59) produced a polyclonal antibody to the extracellular antigen of *B. tulipae* that could be detected to 1 ng/ml and did not cross-react with other fungal genera. This antigen was stable to heat and in systems with pH from 2 to 10 and had mannose, glucose, and galactose as the major sugars, with minor contributions from xylose and arabinose (59). However, Ricker et al. (99) found that their polyclonal antibody to *B. cinerea* cross-reacted with *A. niger*.

Monoclonal antibodies have also been produced to *B. cinerea* (45,47). Bossi and Dewey (45) produced monoclonal antibodies to *B. cinerea* surface washings that recognized *B. cinerea* and *B. fabae* but not *B. allii*. They speculated that the antigens were carbohydrates since they were stable to heat, periodate, and slightly protein sensitive. They also speculated that the two *Botrytis* species that cross-reacted may be more closely related to each other than to the *B. allii*. All of the research done so far suggests that specific antibodies can be produced to *Botrytis* species that may be useful in taxonomic work if further refinements for genus, species, and strain are developed.

C. Identification of *Cladosporium* species

Very little research has been undertaken with the immunological identification of *Cladosporium* species, although some work in food spoilage detection has shed light on the immunological characteristics of this genus. Notermans and Soentoro (19) produced a polyclonal antibody to *Cladosporium cladosporioides* and found that it only reacted to antigens from *Cladosporium* species. Similar results were shown by Tsai and Cousin (39) for *C. herbarum* with polyclonal antibodies. Further analysis of the antigens showed that they were polysaccharides with very different compositions (Table 4) depending on the species and culture form (20,23). Notermans et al. (20) found that the antigens from *C. cladosporioides* were not inhibited by either the methyl α- and β-D-galactofuranosides nor the methyl α- and β-D-galactopyranosides, suggesting that galactose was not immunodominant. However, Tsai and Cousin (23) found that β-galactosidase reduced the activity of their *C. herbarum* antigens from both the mycelial and extracellular fractions, suggesting that galactosyl groups were immunodominant. Further enzymatic analysis suggested that the galactosyl units were β-linked to glucose at the nonreducing ends (23). Since all of these assays for *Cladosporium* species

Table 4 Chemical Composition of Antigens Produced by *Cladosporium* Species

Species	Polysaccharide fraction (%)	Protein (%)	Monosaccharide (mol %) Man	Gal	Glc	Ref.
C. cladosporioides extracellular	92	N.D.[a]	56	8	33	20
C. herbarum						
extracellular	59	20	73	20	8	23
mycelial	62	17	73	20	8	23

[a]Not done.

were genus-specific, serology may be useful in the taxonomy for this genus. Further research will be needed to produce antibodies that may even be species-specific.

D. Identification of *Fusarium* Species

The taxonomy of *Fusarium* species is still very difficult, especially in speciation. Brayford (100) stressed that there is much work to be done before the taxonomy of *Fusarium* species is acceptable and that a wide range of characteristics must be used to classify the species and strains within this genus. Immunology as a taxonomic tool was suggested as early as the late 1920s; however, groups based on immunology were not proposed until the late 1950s through to the mid-1980s (100). In fact, Brayford (100) concluded that the immunological methods produced to this point were too variable to be used for taxonomic purposes. Clark et al. (97) also summarized several attempts at using immunological methods for taxonomy.

In the food mycology area, it is important to identify *Fusarium* species because many produce mycotoxins. Early work by Notermans and Heuvelman (41) and Notermans and Soentoro (19) showed that the antibody to *Fusarium oxysporum* extracellular antigen reacted with antisera from species of *Fusarium*, *Aspergillus*, *Penicillium*, and *Scopulariopsis*, and *Trichothecium*. Yong (101) produced antibodies to *F. poae* mycelium that cross-reacted only with antigens from other *Fusarium* species, and two *Aspergillus* species—*A. versicolor* and *A. wenti*. All *Fusarium* species tested were recognized by the assay. This cross-reactivity with the two *Aspergillus* species has subsequently been eliminated (unpublished data). Two immunoassays have been developed that can detect strains of *F. oxysporum* in plant tissues (102,103), and both these assays had some cross-reactivity with other *Fusarium* species. More research will need to be done on the use of immunology to develop assays that can be used to both detect and classify *Fusarium* species, as this is a taxonomically complex group.

E. Identification of *Geotrichum* Species

Geotrichum species are yeastlike fungi that are important in food contamination and food processing plant sanitation. All the immunoassays that have been developed to *Geotrichum* show that they are very specific to the genus (19,20,23,39,43). The chemical composition of some *Geotrichum* antigens is shown in Table 5. Further characterization by Tsai and Cousin (23) has shown that protease digestion decreased the antigenic activity by about 40%. Selective removal of the sugar residues by β-galactosidase and β-glucosidase suggested that galactosyl fractions were immunodominant and that they were β-1 → 4-linked to glucosyl residues (23). Since the immunoassays that have been done with *G. candidum* have been

Table 5 Chemical Composition of *Geotrichum candidum* Antigens

Antigen	Polysaccharide (%)	Protein (%)	Monosaccharide (mol %) Man	Glc	Gal	Ref.
G. candidum extracellular	68	N.D.[a]	56	8	33	19
G. candidum						
extracellular	47	10	62	12	26	22
mycelial	75	17	72	9	19	22

[a]Not determined.

genus-specific, immunology may be useful for taxonomy. Further research in this area would help to develop that potential.

F. Identification of Other Fungi

There are several other references that focus on either the identification or detection of fungi using various immunological methods; however, these fungal antigens have not been well defined. Some of these references have focused on taxonomy directly; however, many have been used in applied taxonomy to identify fungi involved in medical mycoses (6,9,11–14), plant pathogenicity (18,51–56,97), or food mycology (7,19–23,27–29). The immunological identification of fungi seems to be a field that will continue to grow in these three areas, and information gained from this research can be used to develop immunotaxonomy for some genera of fungi.

VI. APPLICATIONS OF IMMUNOTAXONOMY FOR FUNGI

A. Medical Applications

The need to identify medically important fungi that cause mycoses in humans and animals is increasing, as immunocompromised hosts are more susceptible to fungus infection. In addition, allergies due to fungi are very prevalant in the environment and may be increasing due to the "sick building syndrome" where new construction, air conditioning, and related concerns make exposure to fungi more likely. The use of fungal antigens can be very important in the identification of these fungal pathogens, and this identification will undoubtly overlap with fungal taxonomy (9–13). In several reviews (9–11,13) the authors concluded that the production of antigens by fungi can be genus- and species-specific and could therefore be used to resolve taxonomic problems.

There are many examples of the use of exoantigens to study fungi that are important in human and animal diseases; only a few examples will be cited here. Sekhon and Padhye (13) noted that in 1991 there were only a few test kits commercially available in the United States and Canada to identify fungi based on exoantigens. The major ones included *Histoplasma capsulatum*, *Blastomyces dermatitidis*, *Coccidioides immitis*, and some *Aspergillus* species (13). Most work on identification of fungi by exoantigen production has been confined to research laboratories. Kaufman et al. (104) used exoantigen analysis to identify over 100 isolates of *Blastomyces dermatitidis* and found that all of them produced a specific antigen, except for isolates from Africa. Hence, they were able to identify two serotypes to *B. dermatitidis* based on antigen production that could be identified by immunology and could be used for immunodiagnosis of disease (104).

The controversy over the taxonomy of *Basidiobolus* has lead some researchers to use immunology to distinguish between species and closely related genera. Yangco et al. (105) found that there were enough distinct antigenic characteristics among several isolates of *Basidiobolus* to use them in serodiagnosis and serotaxonomy. Species of *Basidiobolus* and *Conidiobolus* share a common antigen, suggesting that these two genera are taxonomically related (105). Further study has shown that the exoantigens produced by *Basidiobolus* were heterogeneous and a clear distinction between species was not possible (106). However, Espinel-Ingroff et al. (107) were able to differentiate between two dematiaceous black yeasts by microimmunodiffusion. These fungi are difficult to identify by classical methods because they grow slowly and must be studied morphologically by phase contrast microscopy (107). In summary, for medically important fungi, immunotaxonomy may be a good way to differentiate among genera that are difficult to identify. Immunotaxonomy can also be used to confirm the identity of closely related genera.

B. Plant Pathology Applications

Many fungi cause diseases in plants that are used for food, feed, fiber, or leisure. It is important to identify these fungi rapidly to be able to prevent the loss of a crop. The use of antibodies in various immunological techniques has become an important diagnostic tool for the identification of plant pathogens (4,97,108). There are many literature references to research on immunotaxonomy, immunoidentification, and immunodiagnosis of fungi in plant tissues; hence, only a few articles will be cited here to show the importance and use of immunology based on fungal antigens.

Jung et al. (109) used an ELISA to study the taxonomic problems related to the *Morchella esculenta* group. They found that the tan, gray, and large tan forms of the *M. esculenta* were not immunologically different, but that *M. esculenta* and *M. semilibera* were different. In addition to immunology, they used phylogenetic techniques and numerical taxonomy to differentiate these fungi. Belisario et al. (110) used morphology, cultural characteristics, protein profiles, and immunology to confirm a new habitat for *Phytophthora iranica* that had never been recorded outside of Iran. Hence, immunological methods can be used to help identify fungi isolated from various environments. Immunology has also been used to show that species cannot be differentiated because of antigenic relationships, such as between *Tilletia controversa* and *T. caries* teliospores (51). Other researchers have found that species and strain specificity can be demonstrated using monoclonal antibodies. Wright et al. (52) were able to separate *Glomus occultum* from other *Glomus* species. Some degree of strain difference was noted by the reactions with the monoclonal antibodies; however, more research needs to be done in this area (52).

Most of the immunological research in the field of plant pathology has been to identify plant pathogens to genus or species. Selected research for different genera will be summarized briefly. Yuen et al. (56) developed monoclonal antibodies that could identify *Pythium ultimum* at species level and differentiate it from related *Pythium* species. This assay gave 99% accuracy in identifying *P. ultimum* in sugar beet seedling roots (56). These authors concluded that their work was the first to show immunological diversity with *P. ultimum* because two subgroups could be differentiated based on these monoclonal antibodies. Hardham et al. (111) showed that monoclonal antibodies to *Phytophthora cinnamomi* cysts and zoospores were isolate-specific and could be used in taxonomy of the genus. The taxonomic relationship between *Phomopsis longicolla* and species of the *Diaporthe/Phomopsis* group was confirmed by an immunoassay developed to detect *Phomopsis* species in soybean plants (112,113).

Detection of fungi in soil has lead to several immunological assays. Thornton et al. (53) were able to detect *Rhizoctonia solani* in soil and to differentiate it from other *Rhizoctonia* species by use of a monoclonal antibody assay. Novak and Kohn (114) used stromatal proteins to differentiate between the plant pathogenic and saprophytic *Sclerotiniaceae*. They found different immunological and biochemical profiles that could suggest two distinct genera within the *Sclerotiniaceae*. The sclerotial and stomatal proteins were also antigenically distinct. The sclerotial species of *Aspergillus* and *Sclerotium* were antigenically different from those in Sclerotiniaceae (114). It was suggested that the differences in these proteins can be a powerful taxonomic tool for this family. Priestley and Dewey (54) found that high-molecular-weight glycoproteins and carbohydrates needed to be removed from the immunogen before it became species-specific for *Pseudocercosporella herpotrichoides*. Gabor et al. (55) also discussed the nature of antigenic preparations for their usefulness in producing species-specific antibodies. They suggested that antigens associated with the plasma membrane may be more species-specific than those from the cell wall. Furthermore, they suggested that the use of immunology for taxonomy could mean that extensive knowledge of fungi is not needed and that a large number of samples could be considered with little time and expense (55).

All the research on the immunological detection of fungi in plant tissues is leading to new information that either confirms or sheds new light on their taxonomic placing. More research than ever is being undertaken on the immunological aspects of fungal plant pathogens, and this is likely to continue into the foreseeable future. Regardless of whether the research has been initiated for taxonomy, detection, or diagnosis of fungal plant pathogens, the basic information that is generated can be useful as an immunotaxonomic tool.

C. Food Mycological Applications

Rapid detection and identification of fungi in foods are important because the three most common genera found in foods—*Aspergillus*, *Fusarium*, and *Penicillium*—have species that produce mycotoxins and spoil foods. Within the past decade, immunological detection of fungi in foods has resulted in several research projects that have shed new light on the use of immunoassays to detect fungi in foods and have indicated some of the important immunodominant sites. The early research looked at simply the ability of immunoassays to detect fungi in foods (19,39,41,43,44,80,81). Attention then focused on the characterization of the fungal antigens in an effort to determine immunodominant sites. Sugar and protein compositions were identified for species of *Aspergillus*, *Botrytis*, *Cladosporium*, *Fusarium*, *Geotrichum*, *Monascus*, *Mucor*, *Penicillium*, and *Rhizopus* (20,21, 23,27,28,59,60). There has been success in partially characterizing the immunodominant site in some of the fungal antigens (21,23,27,28,49,77,79,93,94). More research is needed on the fungi found in foods before information can be for immunotaxonomy. Although there is overlap with both the medical and plant pathology fields, there are some genera and species that are specific to food mycology. More research is needed with these fungi.

D. Biotechnological Applications

There are many products that result from the direct growth of fungi; therefore, biotechnological applications are important. Two examples where immunological methods have been used for screening cultures used in biotechnology are production of fermented foods (84) and biocontrol agents (115). Polonelli et al. (84) studied the antigenic characteristics of *Penicillium camemberti* in an effort to distinguish between this fermentation species and common cheese contaminants. They were able to separate *Penicillium* contaminants from *P. camemberti* by immunodiffusion. It was suggested that immunological methods could be used to analyze cultures to be used in fermentations (84). Toriello et al. (115) immunologically separated genera within the Entomophthorales that are potential biocontrol agents. They noted that these genera are difficult to identify because of their variable morphology and physiological characteristics. In the future, immunology may be used to identify and show purity of cultures used in biotechnology.

VII. FUTURE DEVELOPMENTS IN IMMUNOTAXONOMY

All the information reviewed here shows that immunological methods could be very useful in the classification and differentiation of some fungi. These immunotaxonomic tools may be particularly useful for problematic families, genera, species, or strains that are difficult to identify by traditional morphological and microscopic methods. Significant advances have been made in the understanding of fungal antigens, particularly in the fields of medicine and plant pathology. Currently, there are commercially available immunology-based, rapid diagnostic kits that can help to identify fungi in these disciplines (49). As research continues on the elucidation of fungal antigens that are specific for particular genera or species, the usefulness of immunotaxonomy will become more evident. Research is needed on several different isolates from many different regions of the world to determine if the antigenic component is the same. Also, it would be beneficial to have collaborative research by several laboratories on the immunological methods in order to determine their reproducibility and reliability. As fungal taxonomy grows and evolves, it is likely that immunology will be one method among many that will be used to identify fungi.

VIII. CONCLUSIONS

There are many reasons for the correct indentification of fungi. Traditional identification methods are generally laborious and therefore expensive. During the last two decades new methods allowing a more rapid identification have been studied, and some of these rapid methods are based on immunology. Like many microorganisms, fungi produce a number of antigens which can be used in identification. The desired specificity of the immunological identification method dictates the type of antigens to be applied for antibody production. Extracellular polysaccharides, for example, are highly suited for the identification of fungi at genus level. Species-specific antigens and monoclonal antibodies against these are suited for identification at species level. In the course of time, a number of methods have been described as to how antigens can be extracted from fungi and purified. The same applies for production of antibodies. Methods have been described for *Aspergillus*, *Penicillium*, Mucorales, *Alternaria*, *Botrytis*, *Cladosporium*, and *Geotrichum* among others. Immunotaxonomy is now applied routinely in several disciplines such as in medical science, food research, plant pathology, and biotechnology.

REFERENCES

1. Samson RA, Frisvad JC. New taxonomic approaches for identification of food-borne fungi. Inter Biodeterior Biodegrad 1993; 32:99–113.
2. Samson RA, Frisvad JC, Arora DK. Taxonomy of filamentous fungi in foods and feeds. In: Arora DK, Mukerji KG, Marth EH, eds. Handbook of Applied Mycology. Vol. 3: Foods and Feeds. New York: Marcel Dekker, 1991:1–29.
3. Williams AP. Fungi in foods—rapid detection methods. Prog Indust Microbiol 1989; 26:255–272.
4. Corel JC. Genetic, biochemical, and molecular techniques for the identification and detection of soilborne plant pathogenic fungi. In: Singleton LL, Mihail JD, Rush CM, eds. Methods for Research on Soilborne Phytopathogenic Fungi. St. Paul, MN: APS Press, 1992:7–16.
5. Hayashi O, Yadomae T, Yamada H, Myazaki T. Cross-reactivity of antiserum to *Absidia cylindrospora* among some Mucorales and other fungi. J Gen Microbiol 1978; 108:345–347.
6. Longbottom JL. Applications of immunological methods in mycology. In: Weir DM, Herzenberg LA, Blackwell C, Herzenberg LA, eds. Handbook of Experimental Immunology. Vol. 4: Applications of Immunological Methods in Biomedical Sciences. Oxford: Blackwell Scientific, 1986:121.1–121.30.
7. Cousin MA. Development of the enzyme-linked immunosorbent assay for detection of molds in foods: a review. Dev Indust Microbiol 1986; 31:157–163.
8. Longbottom JL, Austwick PKC. Fungal antigens. In: Weir DM, Herzenberg LA, Blackwell C, Herzenberg LA, eds. Handbook of Experimental Immunology. Vol. 1: Immunochemistry. Oxford: Blackwell Scientific, 1986:7.1–7.11.
9. Kaufman L, Standard P, Padhye AA. Exoantigen tests for the immuno-identification of fungal cultures. Mycopathologia 1983; 82:3–12.
10. Kaufman L, Standard P. Fungal exoantigens. In: Drouhet E, et al., eds. Fungal Antigens: Isolation, Purification, and Detection. New York: Plenum Press, 1988: 111–117.
11. Kaufman L, Standard PG. Specific and rapid identification of medically important fungi by exoantigen detection. Annu Rev Microbiol 1987; 41:209–225.
12. Polonelli L, Morace G. Serological potential for fungal identification. Bot J Linnean Soc 1989; 99:33–38.
13. Sekhon AS, Padhye AA. Exoantigens in the identification of medically important fungi. In: Arora DK, Ajello L, Mukerji KG, eds. Handbook of Applied Mycology. Vol. 2: Humans, Animals, and Insects. New York: Marcel Dekker, 1991:757–764.
14. Richardson MD, Warnock DW. Enzyme-linked immunosorbent assay and its application to the serological diagnosis of fungal infection. Sabouraudia 1983; 21:1–14.
15. Jann K, Westphal O. Microbial polysaccharides. In: Sela M, ed. The Antigens, Vol. 3. New York: Academic Press, 1975:1–127.
16. Sutherland IW. Immunochemical aspects of polysaccharide antigens. In: Glynn LE, Steward MW, eds. Immunochemistry: An Advanced Textbook. New York: John Wiley & Sons, 1977:399–443.
17. Chaumeton J-P, Chauveau C, Chavaut L. Water-soluble polysaccharides excreted by

mycelium of higher fungi: relationship with taxonomy and physiology. Biochem System Ecol 1993; 21:227–239.

18. Ferguson MW, Wycoff KL, Ayers AR. Use of cluster analysis with monoclonal antibodies for taxonomic differentiation of phytopathogenic fungi and for screening and clustering antibodies. Curr Microbiol 1988; 17:127–132.
19. Notermans S, Soentoro PSS. Immunological relationship of extracellular polysaccharide antigens produced by different mould species. Antonie van Leeuwenhoek 1986; 52:593–401.
20. Notermans S, Wieten G, Engel HWB, Rombouts FM, Hoogerhout P, van Boom JH. Purification and properties of extracellular polysaccharide (EPS) antigens produced by different mould species. J Appl Bacteriol 1987; 62:157–166.
21. Notermans S, Veeneman GH, van Zuylen CWEM, Hoogerhout P, van Boom JH. (1 → 5)-Linked β-D-galactofuranosides are immunodominant in extracellular polysaccharides of *Penicillium* and *Aspergillus* species. Mol Immunol 1988; 25:975–978.
22. De Ruiter GA, van Bruggen–van der Lugt AW, Nout MJR, et al. Formation of antigenic extracellular polysaccharides by selected strains of *Mucor* spp., *Rhizopus* spp., *Rhizomucor* spp., *Absidia corymbifera* and *Syncephalastrum racemosum*. Antonie van Leeuwenhoek 1992; 62:189–199.
23. Tsai G-J, Cousin MA. Partial purification and characterization of mold antigens commonly found in foods. Appl Environ Microbiol 1993; 59:2563–2571.
24. Bishop CT, Jennings HJ. Immunology of polysaccharides. In: Aspinall GO, ed. The Polysaccharides. New York: Academic Press, 1982:291–330.
25. Kamphuis HJ, Veeneman GH, Rombouts FM, van Boom JH, Notermans S. Antibodies against synthetic oligosaccharide antigens reactive with extracellular polysaccharides produced by moulds. Food Agric Immunol 1989; 1:235–242.
26. Feldmann M, Mole D. Cell cooperation in the immune response. In: Roitt I, Brostoff J, Male D, eds. Immunology. 2nd ed. London: Gower Medical Publishing, 1989: 8.1–8.12.
27. Kamphuis HJ. Extracellular polysaccharides as target compounds for the immunological detection of *Aspergillus* and *Penicillium* in food. Ph.D. Thesis, Wageningen Agricultural University, Wageningen, Netherlands, 1992.
28. De Ruiter GA. Immunological and biochemical characterization of extracellular polysaccharides of mucoralean moulds. Ph.D. Thesis, Wageningen Agricultural University, Wageningen, Netherlands, 1993.
29. De Ruiter GA, Van Bruggen–Van der Lugt AW, Rombouts FM, Gams W. Approaches to the classification of the *Mortierella isabellina* group: antigenic extracellular polysaccharides. Mycol Res 1993; 97:690–696.
30. Bartnicki-Garcia S. Cell wall chemistry, morphogenesis, and taxonomy of fungi. Annu Rev Microbiol 1968; 22:87–108.
31. Rosenberger RF. The cell wall. In: Smith JE, Berry DR, eds. The Filamentous Fungi. Vol. 2. Biosynthesis and Metabolism. New York: John Wiley & Sons, 1976: 328–344.
32. Bobbit TF, Nordin JH. Hyphal nigeran as a potential phylogenetic marker for *Aspergillus* and *Penicillium* species. Mycologia 1978; 70:1201–1211.
33. Leal JA, Moya A, Gomez-Miranda B, Ruperez P, Guerrero C. Differences in cell

wall polysaccharides in some species of *Penicillium*. In: Nombela C, ed. Microbial Cell Wall Synthesis and Autolysis. Amsterdam: Elsevier Science, 1984:149–155.
34. Preston JF III, Gander JE. Isolation and partial purification of the extracellular polysaccharides of *Penicillium charlesii*. I. Occurrence of galactofuranose in high molecular weight polymers. Arch Biochem Biophys 1968; 124:504–512.
35. Preston JF III, Lapis E, Gander JF. Isolation and partial purification of the extracellular polysaccharides of *Penicillium charlesii*. III. Heterogeneity in size and composition of high molecular weight extracellular polysaccharides. Arch Biochem Biophys 1969; 134:324–334.
36. Preston JF III, Lapis E, Westerhouse S, Gander JE. Isolation and partial purification of the extracellular polysaccharides of *Penicillium charlesii*. II. The occurrence of phosphate groups in high molecular weight polysaccharides. Arch Biochem Biophys 1969; 134:316–323.
37. Bennett JE, Bhattacharjee AK, Glaudemans CPJ. Galactofuranosyl groups are immunodominant in *Aspergillus fumigatus* galactomannan. Mol Immunol 1985; 22: 251–254.
38. Martin SM, Adams GA. Survey of fungal polysaccharides. Can J Microbiol 1956; 34:715–721.
39. Tsai G-J, Cousin MA. Enzyme-linked immunosorbent assay for detection of molds in cheese and yogurt. J Dairy Sci 1990; 73:3366–3378.
40. De Ruiter GA, van der Lugt AW, Voragen AGJ, Rombouts FM, Notermans SHW. High-performance size-exclusion chromatography and ELISA detection of extracellular polysaccharides from Mucorales. Carbohydr Res 1991; 215:47–57.
41. Notermans S, Heuvelman CJ. Immunological detection of enzyme-linked immunosorbent assay (ELISA); preparation of antigens. Int J Food Microbiol 1986; 2: 247–258.
42. Fuhrmann B, Roquebert MF, van Hoegaerden M, Strosberg AD. Immunological differentition of *Penicillium* species. Can J Microbiol 1989; 35:1043–1047.
43. Lin HH, Cousin MA. Evaluation of enzyme-linked immunosorbent assay for detection of molds in foods. J Food Sci 1987; 52:1089–1094, 1096.
44. Lin HH, Lister RM, Cousin MA. Enzyme-linked immunosorbent assay for detection of molds in tomato puree. J Food Sci 1986; 51:180–183, 192.
45. Bossi R, Dewey FM. Development of a monoclonal antibody-based immunodetection assay for *Botrytis cinerea*. Plant Pathol 1992; 41:472–482.
46. Dewey FM, MacDonald MM, Philips SI, Priestley RA. Development of monoclonal antibody, ELISA and dip-stick immunoassays for *Penicillium islandicum* in rice grains. J Gen Microbiol 1990; 136:753–760.
47. Dewey FM, Banham AH, Priestley RA, et al. Monoclonal antibodies for the detection of spoilage fungi. Int Biodeterior Biodegrad 1993; 32:127–136.
48. Stynen D, Meulemans L, Goris A, Braendlin N, Symons N. Characteristics of a latex agglutination test based on monoclonal antibodies for the detection of fungal antigens in foods. In: Samson RA, Hocking AD, Pitt JI, King AD, eds. Modern Methods in Food Mycology. Amsterdam: Elsevier, 1992:213–219.
49. De Ruiter GA, Notermans SHW, Rombouts FM. Review: new methods in food mycology. Trends Food Sci Technol 1993; 4:91–97.

50. Stynen D, Sarfati J, Goris A, et al. Rat monoclonal antibodies against *Aspergillus* galactomannan. Infect Immun 1992; 60:2237–2245.
51. Banowetz GM, Trione EJ, Krygier BB. Immunological comparisons of teliospores of two wheat bunt fungi. *Tilletia* species, using monoclonal antibodies and antisera. Mycologia 1984; 76:51–62.
52. Wright SF, Morton JB, Sworobuk JE. Identification of a vesicular-arbuscular mycorrhizal fungus by using monoclonal antibodies in an enzyme-linked immunosorbent assay. Appl Environ Microbiol 1987; 53:2222–2225.
53. Thornton CR, Dewey FM, Gilligan CA. Development of monoclonal antibody-based immunological assays for the detection of live propagules of *Rhizoctonia solani* in soil. Plant Pathol 1993; 42:763–773.
54. Priestley RA, Dewey FM. Development of a monoclonal antibody immunoassay for the eyespot pathogen *Pseudocercosporella herpotrichoides*. Plant Pathol 1993; 42: 403–412.
55. Gabor BK, O'Gara ET, Philip BA, Horan DP, Hardham AR. Specificities of monoclonal antibodies to *Phytophthora cinnamomi* in two rapid diagnostic assays. Plant Dis 1993; 77:1189–1197.
56. Yuen GY, Criag ML, Avila F. Detection of *Phythium ultimum* with a species specific monoclonal antibody. Plant Dis 1993; 77:692–698.
57. De Ruiter GA, Van Bruggen–van der Lugt AW, Bos W, Notermans SHW, Rombouts FM, Hofstra H. The production and partial characterization of a monoclonal IgG antibody specific for moulds belonging to the order Mucorales. J Gen Microbiol 1993; 139:1557–1564.
58. Sekhon AS, Li JSK, Garg AK. *Penicillosis marneffei*: Serological and exoantigen studies. Mycopathologia 1982; 77:51–57.
59. Cousin MA, Dufrenne J, Rombouts FM, Notermans S. Immunological detection of *Botrytis* and *Monascus* species in food. Food Microbiol 1990; 7:227–235.
60. Kamphuis HJ, De Ruiter GA, Veeneman GH, van Boom JH, Rombouts FM, Notermans SHW. Detection of *Aspergillus* and *Penicillium* extracellular polysaccharides (EPS) by ELISA: using antibodies raised against acid hydrolysed EPS. Antonie van Leeuwenhoek 1992; 61:323–332.
61. Barreto-Bergter EM, Travossos LR, Gorin PAJ. Chemical structure of the D-galacto-D-mannan component from hyphae of *Aspergillus niger* and other *Aspergillus* spp. Carbohydr Res 1980; 86:273–285.
62. Barreto-Bergter EM, Gorin PAJ, Travassos LR. Cell constituents of mycelia and conidia of *Aspergillus fumigatus*. Carbohydr Res 1981; 95:205–218.
63. Webster SF, McGinley KJ. Serological analysis of the extractable carbohydrate antigens of *Pitgrosporum ovale*. Microbes 1980; 28:41–45.
64. Hearn VM. Surface antigens of intact *Aspergillus fumigatus* mycelium: their location using radiolabelled protein A as marker. J Gen Microbiol 1984; 130:907–917.
65. Suzuki S, Takeda N. Serological cross-reactivity of the D-galacto-D-mannans isolated from several pathogenic fungi against anti-*Hermodendrum pedrosoi* serum. Carbohydr Res 1975; 40:193–197.
66. Steinbuch M, Audran R. The isolation of IgG from mammalian sera with the aid of caprylic acid. Arch Biochem Biophys 1969; 134:279–284.

67. Clemons DJ, Besch-Williford C, Steffen EK, Riley LK, Moore DH. Evaluation of a subcutaneously implanted chamber for antibody production in rabbits. Lab Anim Sci 1992; 42:307–311.
68. Reid JL, Walker-Simmons MK, Everard JD, Diani J. Production of polyclonal antibodies in rabbits is simplified using perforated plastic golf balls. Biotechnology 1992; 12:660–666.
69. Harlow E, Lane D. Antibodies. A Laboratory Manual. Cold Spring Harbor, NY: Cold Spring Harbor Laboratory, 1988.
70. Sakaguchi O, Yokota K, Suzuki M. Immunochemical and biochemical studies of fungi. XIII. On the galactomannans isolated from mycelia and culture filtrates of several filamentous fungi. Jpn J Microbiol 1969; 13:1–7.
71. Reiss E, Lehman PF. Galactomannan antigenemia in invassive aspergillosis. Infect Immun 1979; 25:357–365.
72. Mischnick P, De Ruiter GA. Application of reductive cleavage in the structural investigation of the antigenic polysaccharides of *Aspergillus fumigatus* and *Penicillium digitatum* with respect to the determination of the ring size of the galactose moieties. Carbohydr Polym 1994; 23:5–12.
73. Tsai G-J. Development and evaluation of enzyme-linked immunosorbent assay for detection of molds in dairy products. Ph.D. Thesis, Purdue University, West Lafayette, IN. 1990.
74. Rupérez P, Gómez-Miranda B, Leal JA. Extracellular β-malonoglucan from *Penicillium erythromellis*. Trans Br Mycol Soc 1983; 80:313–318.
75. Leal JA, Moya A, Gómez-Miranda B, Rupe-Rez P, Guerrero C. Differences in cell wall polysaccharides in some species of *Penicillium*. In: Nombela C, ed. Microbial Cell Wall Synthesis and Autolysis. Amsterdam: Elsevier Science, 1984:149–155.
76. Gander JE, Jentoff NH, Drewes LR, Rick PD. The 5-O-β-D-galactofuranosyl-containing exocellular glycopeptide of *Penicillium charlesii*. J Biol Chem 1974; 249:2063–2072.
77. Van Bruggen–van der Lugt AW, Kamphuis HJ, de Ruiter GA, Mischnick P, van Boom JH, Rombouts FM. New structural features of the antigenic extracellular polysaccharides of *Penicillium* and *Aspergillus* species revealed with exo-β-D-galactofuranosidase. J Bacteriol 1992; 174:6096–6102.
78. Cole L, Hawes C, Dewey FM. Immunocytochemical localization of a specific antigen in cell walls of *Penicillium islandicum*. Mycol Res 1991; 12:1369–1374.
79. Cousin MA, Notermans S, Hoogerhout P, van Boom JH. Detection of β-galactofuranosidase production by *Penicillium* and *Aspergillus* species using 4-nitrophenyl β-D-galactofuranoside. J Appl Bacteriol 1989; 66:311–317.
80. Notermans S, Heuvelman CJ, van Egmond HP, Paulsch WE, Besling JR. Detection of mold in food by enzyme-linked immunosorbent assay. J Food Prot 1986; 49: 786–791.
81. Notermans S, Heuvelman CJ, Beumer RR, Maas R. Immunological detection of moulds in foods: relation between antigen production and growth. Int J Food Microbiol 1986; 3:253–261.
82. Fuhrmann B, Roquebert MF, Lebreton V, van Hoegaerden M. Immunological differentiation between *Penicillium* and *Aspergillus* taxa. In: Samson RA, Pitt JI, eds.

Modern Concepts in *Penicillium* and *Aspergillus* Classification. New York: Plenum Press, 1990:423–432.
83. Piechura JE, Kurup VP, Fink JN, Calvanico NJ. Antigens of *Aspergillus fumigatus*. III. Comparative immunochemical analyses of clinically relevant aspergilli and related fungal taxa. Clin Exp Immunol 1985; 59:716–724.
84. Polonelli L, Morace G, Rosa R, Castagnola M, Frisvad JC. Antigenic characterization of *Penicillium camemberti* and related common cheese contaminantes. Appl Environ Microbiol 1987; 53:872–878.
85. Sekhon AS, Standard PG, Kaufman L, Garg AK, Cifuentes P. Grouping of *Aspergillus* species with exoantigens. Diagn Immunol 1986; 4:112–116.
86. Neucere JN, Ullah AHJ, Cleveland TE. Surface proteins of two aflatoxin producing isolates of *Aspergillus flavus* and *Aspergillus parasiticus* mycelia. 1. A comparative immunochemical profile. J Agric Chem 1992; 40:1610–1612.
87. Cousin MA, Zeidler CS, Nelson PE. Chemical detection of mold in processed food. J Food Sci 1984; 49:439–445.
88. Bartnicki-Garcia S, Reyes E. Polyuronides in the cell wall of *Mucor rouxii*. Biochim Biophys Acta 1968; 170:54–62.
89. Miyazaki T, Irino T. Studies on fungal polysaccharides. VIII. Extracellular heteroglycans of *Rhizopus nigricans*. Chem Pharm Bull 1971; 19:1450–1454.
90. Miyazaki T, Irino T. Studies on fungal polysaccharides. X. Extracellular heteroglycans of *Absidia cylindrospora* and *Mucor mucendo*. Chem Pharm Bull 1972; 20: 330–335.
91. De Ruiter GA, Schols HA, Voragen AGJ, Rombouts FM. Carbohydrate analysis of water-soluble uronic acid containing polysaccharides with high-performance anion-exchange chromatography using methanolysis combined with TFA hydrolysis is superior to four other methods. Anal Biochem 1992; 207:176–185.
92. De Ruiter GA, Hoopman T, van der Lugt AW, Notermans SHW, Nout MJR. Immunochemical detection of Mucorales species in foods. In: Samson RA, Hocking AD, Pitt JI, King AD, eds. Modern Methods in Food Mycology. Amsterdam: Elsevier, 1992:221–227.
93. De Ruiter GA, van Bruggen–van der Lugt AW, Mischnick P, et al. 2-O-Methyl-D-mannose residues are immunodominant in extracellular polysaccharides of *Mucor racemosus* and related molds. J Biol Chem 1994; 269:4299–4306.
94. De Ruiter GA, Josso SL, Colquhoun IJ, Voragen AGJ, Rombouts FM. Isolation and characterization of $\beta(1 \rightarrow 4)$-D-glucuronans from extracellular polysaccharide of moulds belonging to Mucorales. Carbohydr Polym 1992; 18:1–7.
95. Gams W. A key to the species *Mortierella*. Persoonia 1977; 9:381–391.
96. Schumacher MJ, McClatchy JK, Farr RS, Minden P. Primary interaction between antibody and components of *Alternaria*. I. Immunological and chemical characteristics of labeled antigens. J Allergy Clin Immunol 1975; 56:39–53.
97. Clarke JH, MacNicoll AD, Norman JA. Immunological detection of fungi in plants, including stored cereals. Int Biodeterior Suppl 1986; 22:123–130.
98. Savage SD, Sall MA. Radioimmunosorbent assay for *Botrytis cinerea*. Phytopathology 1981; 71:411–415.
99. Ricker RW, Marois JJ, Dlott JW, Bostock RM, Morrison JC. Immunodetection and

quantification of *Botrytis cinerea* on harvested wine grapes. Phytopathology 1991; 81:404–411.
100. Brayford D. Progress in the study of *Fusarium* and some related genera. J Appl Bacteriol Symp Suppl 1989:47s–60s.
101. Yong RE. Enzyme-linked immunosorbent assays for detecting molds in foods. M.S. Thesis, Purdue University, West Lafayette, IN, 1993.
102. Kitagawa T, Sakamoto Y, Furumi E, Ogura H. Novel enzyme immunoassays for specific detection of *Fusarium oxysporum* f. sp. *cucumerinum* and for general detection of various *Fusarium* species. Phytopathology 1989; 79:162–165.
103. Linfield CA. A rapid serological test for detecting *Fusarium oxysporum* f. sp. *narcissi* in *Narcissus*. Ann Appl Biol 1993; 123:685–693.
104. Kaufman L, Standard PG, Weeks RJ, Padhye AA. Detection of two *Blastomyces dermatitidis* serotypes by exoantigen analysis. J Clin Microbiol 1983; 18:110–114.
105. Yangco BG, Nettlow A, Okafor JI, Park J, TeStrake D. Comparative antigenic studies of species of *Basidiobolus* and other medically important fungi. J Clin Microbiol 1986; 23:679–682.
106. Te Strake D, Park JY, Yangco BC. Exoantigen comparisons of selected isolates of *Basidiobolus* species. Mycologia 1989; 8:284–288.
107. Espinel-Ingroff A, Shadomy S, Kerkering TM, Shadomy HJ. Exoantigen test for differentiation of *Exophiala jeanselmei* and *Wagiella dermatitidis* isolates from other dematiaceous fungi. J Clin Microbiol 1984; 20:23–27.
108. Clark MF. Immunosorbent assays in plant pathology. Annu Rev Phytopathol 1981; 19:83–106.
109. Jung SW, Gessner RV, Kendell KC, Romano MA. Systemics of *Morchella esculenta* complex using enzyme-linked immunosorbent assay. Mycopathologia 1993; 85: 677–684.
110. Belisario A, Magnano di San Lio G, Pane A, Cacciola SO. *Phytophthora iranica*, a new root pathogen of myrtle from Italy. Plant Dis 1993; 77:1050–1055.
111. Hardham AR, Suzaki E, Perkin JL. Monoclonal antibodies to isolate-, species-, and genus-specific components on the surface of zoospores and cysts of the fungus *Phytophthora cinnamomi*. Can J Bot 1986; 64:311–321.
112. Velicheti RK, Lamison C, Brill LM, Sinclair JB. Immunodetection of *Phomopsis* species in asymptomatic soybean plants. Plant Dis 1993; 7:70–73.
113. Brill LM, McClary RD, Sinclair JB. Analysis of two ELISA formats and antigen preparations using polyclonal antibodies against *Phomopsis longicolla*. Phytopathology 1994; 84:173–179.
114. Novak LA, Kohn LM. Electrophoretic and immunological comparisons of developmentally regulated proteins in members of the Sclerotiniaceae and other sclerotial fungi. Appl Environ Microbiol 1991; 57:525–534.
115. Toriello C, Zerón E, Latgé JP, Mier T. Immunological separation of Entomophthorales genera. J Invert Pathol 1989; 53:358–360.

7

Taxonomic Applications of Polysaccharides

J. Antonio Leal and Manuel Bernabé
Centro de Investigaciones Biológicas (CSIC), Madrid, Spain

I. INTRODUCTION

The fungal cell wall, with few exceptions, is composed of 80% to 90% polysaccharides. The skeletal components of the cell wall of most fungi are the microfibrils of chitin or cellulose, which are cemented by glucans, mannans, galactans, xylans, heteropolysaccharides, and proteins. The early literature on fungal cell wall chemistry was reviewed by Aronson (1), Bartnicki-Garcia (2), and Rosenberger (3). Bartnicki-García (2) stated:

> Although the data are exceedingly limited in the number of related organisms examined, diversity of groups selected, and analytical refinement it has become increasingly clear that the entire spectrum of fungi may be subdivided into various categories according to the chemical nature of the walls, and that those categories closely parallel conventional taxonomic boundaries.

He proposed a classification of fungal cell walls based on dual combinations of the polysaccharides which had been described as the principal components of the cell wall. He went further and predicted: "The correlation between wall chemistry and taxonomy may be effectively extended to the genus level, although in such close proximity the differences are apt to be minor and chiefly, if not entirely, quantitative." Cell wall knowledge is advancing slowly, however, in spite of the usefulness of cell wall polysaccharides in taxonomy and the advances in the instrumentation for the characterization of polysaccharides (4).

In order to utilize polysaccharides as chemotaxonomic markers at genus or lower taxonomic levels, a large number must be present in fungal cell walls or in the culture media. Ideally, each genus may have at least one characteristic polysaccharide. Nigeran, a hot-water-soluble, cold-water-insoluble α-(1 → 3)(1 → 4)-glucan (5) was considered a potential phylogenetic marker for *Aspergillus* and *Penicillium* species (6). Galactofuranosyl residues had been reported in the cell walls of *P. charlesii* (7) and *P. ochro-chloron* (8) and in various *Aspergillus* species (9,10). The analysis of the cell wall and polysaccharide fractions of species of *Penicillium* (11), *Eupenicillium* (12), and *Talaromyces* (13) showed two types of cell walls: type A, which has a high glucose and low galactose content; and type B, which has a high galactose content. Cell walls of type A were found in *E. crustaceum*, and type B in *T. flavus*, the type species of these genera (14). Heteropolysaccharides rich in galactofuranose were isolated and partially characterized from the cell wall of *P. erythromellis* (15) and *T. helicus* (16). Alkali- and water-soluble galactomannoglucans were isolated from the cell walls of *Gliocladium viride* (17). Alkali- and water-soluble polysaccharides have been isolated and characterized from the cell walls of fungi belonging to several genera (18–26).

The differences in chemical composition and structure of the polysaccharides extracted from the cell walls of fungi indicate that each genus may have its own characteristic polysaccharide. In this review it will be shown that polysaccharides, especially alkali- and water-soluble forms from the cell wall, can be used as chemotaxonomic characters at the genus or subgenus level, and they may help in establishing relationships among anamorphic genera and their teleomorphs.

II. ISOLATION AND CHARACTERIZATION OF POLYSACCHARIDES

A. Fungal Growth

Growth conditions and the composition of the culture medium are of paramount importance in fungal systematics. The identification of an isolate depends on the utilization of the media and on the growth conditions in which morphological, physiological, or biochemical characters have been described (27).

The effect of fungal growth conditions on cell wall composition has received little attention. It seems that the composition of the cell wall may not change during the growth phase, and in certain fungi only small differences in composition are found after a long incubation period. The glucose and glucosamine content does not change during the incubation period once they have been incorporated into the wall polymers of *Aspergillus clavatus* (28). Glucose decreases and glucosamine increases in the cell walls of *Paecilomyces persicinus* during growth (29). The amount of the polysaccharide fractions obtained from the cell walls of *Penicillium expansum* does not change during a 43-day incubation period (30). The glucose and galactose content of the cell wall of *P. allahabadense* decreases slightly during a growth period of 59 days, and the mannose content increases (31). The amount of nigeran deposited in the cell walls of species of *Aspergillus* and *Penicillium* increases in response to nitrogen depletion (6). The composition of the culture media and growth conditions affect the production of the mycelial and yeast forms of dimorphic fungi. The cell wall composition differs in each of these forms (32,33) and so growth conditions will affect the composition of the cell wall in these particular fungi. The composition of the culture medium and growth conditions do, however, considerably affect the production of exocellular polysaccharides (34–39). To avoid variations in cell wall composition which could be due to the culture medium or growth conditions, microorganisms must always be grown in the same medium and environmental conditions except for thermophiles, which are grown at their required temperature (30).

B. Cell Wall Preparation

Cell walls are obtained by fragmentation of the mycelium which is repeatedly washed to eliminate the cytoplasmic debris. Mycelium is disrupted mechanically with a sonicator, a French press, or with homogenizers with or without glass beads (e.g., Manton-Gaulin, the Mickle disintegrator, the Braun MSK, or the Sorvall Omnimixer). Other procedures can involve the breakage of cells in frozen blocks or by disruption in ball mills (40). One method for obtaining large amounts of cell walls is the disintegration of dry mycelium in a Sorvall Omnimixer followed by treatment on a Fritsch Pulverisette 6 ball-mill (17). The powdered mycelium is suspended in 1% solution of sodium dodecyl sulfate containing 0.02% of sodium azide and incubated with agitation overnight at 20°C. The cell wall material is collected by centrifugation at 4°C, washed with distilled water until free of cytoplasmic contamination, and, after two further washes with 50% and 100% alcohol, dried at 60°C in an aereated oven. In the middle of the washing process, narrow hyphae are subjected to ultrasonic treatment to facilitate cytoplasmic release.

C. Cell Wall Fractionation and Purification of Polysaccharides

Methods for cell wall fractionation are based on the solubility of polysaccharides in alkali or acid solutions (41). The method selected should preserve the integrity of the polysaccharides and avoid losses. Acid treatment may destroy most polysaccharides (3), and alkali extraction should therefore precede other treatments. The alkali-insoluble residue may be treated with acid solutions to obtain chitin or chitosan. The R-glucan (42), a branched β-(1 → 3)-glucan which forms a complex with chitin, can be isolated by treatment with nitrous acid or with chitinase (43).

Cellulose is dissolved with aqueous cupra-ammonium hydroxide (Schweizer reagent).

The polysaccharide material is precipitated from the extracts with methanol, ethanol, or acetone or by neutralization, and collected by centrifugation. The alkali or acid used in the extractions, and the solvents used in polysaccharide precipitation, should be removed by dialysis since these precipitates contain water-soluble polysaccharides (15–17) which are lost if they are washed with water. When polysaccharides are precipitated by neutralization, the water-soluble polysaccharides remain in the supernatant and are recovered by dialysis. The loss of wall components can be detected by determining the sugar composition of the cell wall and its recovery in the different fractions. If one or more sugars, detected in the cell wall, considerably decrease or disappear when the fractions are analyzed, the polysaccharides containing these sugars have been lost in the fractionation process. Polysaccharides containing certain sugars that are not detected in the analysis of the cell wall due to their low concentration, may appear in some fractions in which they are selectively extracted. Certain wall fractions represent about 5% of the dry cell wall (20–23) and may contain several polysaccharides; therefore extraction of 5 to 10 g of wall material will yield sufficient of these fractions for the purification and characterization of their polysaccharides. As an illustration, the extraction method used in our laboratory is shown in Figure 1.

Polysaccharides must be carefully purified before their chemical and structural characteristics can be determined. The selection of methods should depend on the physicochemical properties of a particular polysaccharide. Fractional purification, fractional solution, precipitation with ionic detergents or metallic ions, ultracentrifugation, electrophoresis, and ultrafiltration through graded membranes are commonly used. The application of gel filtration chromatography to polysaccharide purification has been reviewed (44), and the water-soluble polysaccharides of fraction F1S of fungal cell walls can be purified by gel filtration (15–17), by their different water solubilities (22) or by both methods sequentially.

D. Chemical and Structural Characterization of Polysaccharides

1. Chemical Analysis

The determination of the structure of a polysaccharide is a complicated task since it is necessary to know the identity and sequence of the monomers, size of the ring, position of the linkages, and anomeric configuration (45). Methods for determining all these factors have been developed and are described in the literature (46,47). Infrared spectroscopy of a polysaccharide provides general information on the predominant anomeric configuration of the glycosidic linkages and the presence of amino sugars and organic acids (48). The acid hydrolysis of polysac-

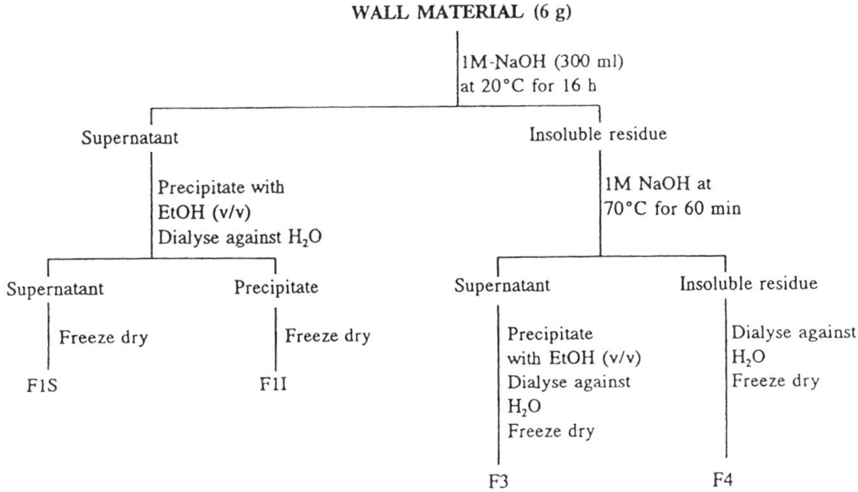

Figure 1 Fractionation of wall material.

charides gives information on monomer composition. Neutral polymers are generally hydrolyzed with variable concentrations of sulfuric, hydrochloric, formic, or trifluoracetic acids. If the polysaccharide contains furanoses, "soft" hydrolysis conditions are required because of the lability of these residues. The identification and quantification of the monosaccharides released can be determined either directly by high-performance liquid chromatography (HPLC) or after derivatization by gas-liquid chromatography (GLC).

The linkage types of the polymer are generally determined by methylation analysis. This method involves the complete methylation of a polysaccharide, its hydrolysis to give a mixture of partially methylated monosaccharides, and their conversion into partially methylated alditol acetates, which are analyzed by GLC and mass spectrometry (MS). Various methods involving different solvents and reagents have been described for complete methylation of polysaccharides. The Hakomori procedure (49) and modifications of this method (50,51) are the most widely used for structural analysis of polysaccharides. However, the anomeric configurations and the sequence of residues in the polysaccharide cannot be established from these data alone. Partial chemical degradation of polysaccharides using different methods has been successfully used for sequencing carbohydrates (52,53). The oligosaccharide fragments obtained can then be separated and characterized by GLC-MS, HPLC-MS, or NMR.

2. Nuclear Magnetic Resonance (NMR) Analysis

NMR spectroscopy provides a powerful and nondestructive tool for the investigation of the structure of complex carbohydrates, both in solution and in the solid state. The conventional, monodimensional (1D) ^{1}H- and/or ^{13}C-NMR spectra of oligo- and polysaccharides constitute identity cards or "fingerprints" of the carbohydrate polymers. These spectra can be used directly in fungal taxonomy for identification purposes, provided the primary structure of the polymers has been previously determined, or if dealing with a closely related structure, as shown in the pioneering papers on chemotaxonomy of yeasts by Gorin et al. (54). Even when the primary structure has not yet been resolved, the 1D NMR spectra can still be used as a demonstration of close or remote resemblance to other complex carbohydrates. We have found chemical similarities among wall polysaccharides from several *Penicillium*, *Eupenicillium*, and *Aspergillus* species by comparison of their ^{1}H and ^{13}C spectra (22). However, the ^{13}C spectra of the water-soluble polysaccharides were similar for most of the species of *Eupenicillium* (23). Similarly, four distinct polysaccharides have been extracted from the walls of several *Talaromyces* species (55). Additional studies on dermatophytes (20) and species of *Aphanoascus* (21) are also examples of the use of NMR spectroscopy in fungal chemotaxonomy.

When investigating an unknown biopolymer, the frequency (chemical shifts, δ, ppm) and pattern of signals, especially those isolated from others in the 1D ^{1}H-NMR spectra, are very useful for preliminary structural information. Among them, peaks corresponding to the anomeric protons (i.e., those appearing in the region 4.5 to 5.5 ppm), which provide information on the number, nature, proportion, and anomeric configuration of residues, are particularly relevant. The ^{13}C-NMR spectra can furnish additional information. Again, the anomeric region (95 to 110 ppm) reveals the number and nature of residues, and sometimes the presence of furanose rings in the biopolymer can be detected.

For the determination of the sequence and substitution positions of the different monosaccharides, a series of proton and carbon two- (2D) and three-dimensional (3D) NMR techniques can provide valuable information about the usually crowded regions of the conventional 1D spectra. Thus, integrated approaches including 1D or 2D ^{1}H-^{1}H homonuclear NMR shift-correlation experiments (DQF- and TQF-COSY, HOHAHA or TOCSY, NOESY, CAMELSPIN or ROESY), and ^{1}H-detected ^{1}H-^{13}C heteronuclear shift correlation (HMQC, HMBC) have proved to be extremely useful. Illustrative examples have been published on the scope and limitations of these techniques (56–58), and there are many recent examples of their use in the structural determination of oligo- and polysaccharides. The structure of the polysaccharides mentioned in this chapter has been investigated using such techniques (24,59–61). Theoretical conformational analyses, through the use of force field programs (MM2, MM3, HSEA,

AMBER, Monte Carlo, Molecular Dynamics, etc.) are complementary to NMR and allow the determination of the conformation of complex carbohydrates in solution (62).

III. NEW CHEMOTAXONOMIC MARKER POLYSACCHARIDES

A. β-(1 → 5)-Galactofuranan

The principal component of the alkali- and water-soluble fraction (F1S) of some species of *Eupenicillium*, *Penicillium*, and *Aspergillus*, is a β-galactofuranan (22,23) which amounts to 1.5% to 8.2% of dry cell wall material. Its chemical composition and structure were determined by GC-MS and NMR spectroscopy (24) revealing the repeating unit **1**. The NMR spectrum of this galactan is shown in Figure 2A.

$$[\rightarrow 5)\text{-}\beta\text{-D-Gal}f\text{-}(1\rightarrow]_n$$

1

In some species the water-insoluble fraction (F1I) contains an α-(1 → 3)-glucan which amounts to 14% to 37% of dry cell wall material, and in others there is a β-(1 → 3)-glucan which amounts to 1% to 10% (23). The species of *Penicillium* and *Eupenicillium* containing this galactan are listed in Table 1 grouped according to their glucans. The species of *Aspergillus* and related genera that contain the β-(1 → 5)-galactofuranan are listed in Table 2. In all these species, F1I is an α-(1 → 3)-glucan.

The presence of β-(1 → 5)-galactofuranan in all the species of *Eupenicillium* investigated shows a perfect correlation between wall chemistry and morphology. The galactans of *E. nepalense* CBS 203.84 (Fig. 2B) and *E. euglaucum* CBS 229.60 contain variable proportions of other linkage types in addition to β-(1 → 5)-linkages (23). Species of *Penicillium* containing β-(1 → 5)-galactofuranan should be considered anamorphs of *Eupenicillium*, and separated from species containing other polysaccharides. The different types of glucans, xylans, and other polysaccharides of fraction F1S may help to arrange species of *Eupenicillium* and *Penicillium* in smaller groups. The presence of β-(1 → 5)-galactofuranan in species of *Aspergillus* (22) and, in their perfect states, *Emericella*, *Eurotium*, *Fennelia*, *Hemicarpenteles*, *Hemisartorya*, *Neosartorya*, *Petromyces*, and *Warcupiella* (unpublished results from this laboratory), shows a closer relationship of most of the species of *Eupenicillium* and certain species of *Penicillium* with species of *Aspergillus* and their teleomorphs, than with species of *Talaromyces* and other species of *Penicillium*. The similarity in cell wall composition of the aspergilli was unexpected since in the penicillia there are several types

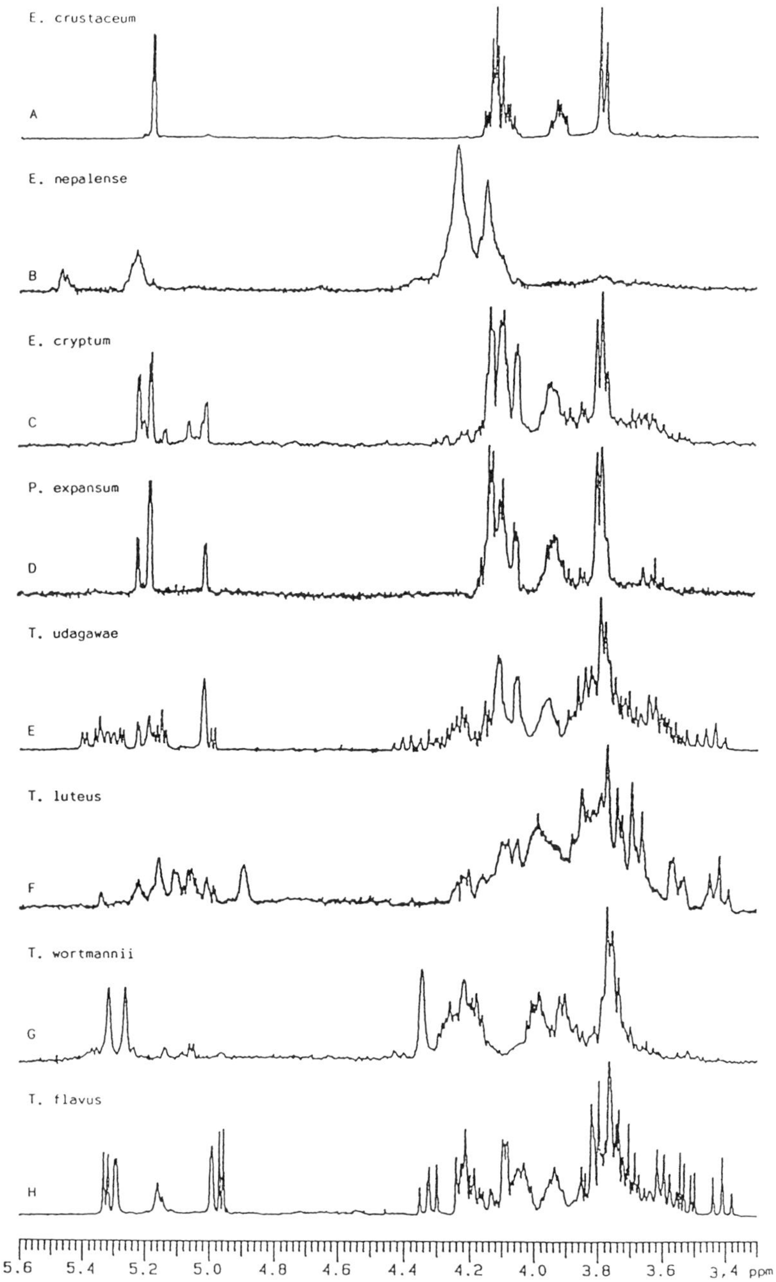

Figure 2 ^{1}H-NMR spectra of the main polysaccharide of fraction F1S obtained from the cell walls of species of the genus *Eupenicillium*, *Penicillium*, and *Talaromyces*. The structure of the polysaccharide and the species that contain a particular polysaccharide are described in the text.

Table 1 Species of *Penicillium* and *Eupenicillium* Containing β-(1-5)-Galactofuranan Grouped by Type of Glucan in Fraction F1I

Microorganisms	Strain	Microorganisms	Strain
α-(1-3)Glucan			
Eupenicillium		Penicillium	
E. angustiporcatum	CBS 202.84	*P. allii*	CBS 161.42
E. cinnamopurpureum	CBS 429.65	*P. atramentosum*	CBS 291.48
E. crustaceum	CBS 635.70	*P. aurantiogriseum*	CBS 123.14
E. lapidosum	CBS 343.48	*P. brevicompactum*	CBS 168.44
E. ochro-salmoneum	CBS 231.60	*P. camemberti*	CBS 122.08
E. pinetorum	CBS 235.60	*P. citrinum*	CECT 2268
E. shearii	CBS 290.48	*P. charmesinum*	CBS 305.48
E. sinaicum	CBS 279.82	*P. chrysogenum*	CBS 205.57
E. stolkiae	CBS 315.67	*P. coprophilus*	CBS 210.70
		P. crustosum	CBS 313.48
		P. decumbens	CBS 258.33
		P. frequentans	CBS 345.51
		P. hordei	CBS 701.68
		P. ingelheimense	CBS 107.66
		P. ingelheimense	CBS 163.42
		P. italicum	CECT 2294
		P. oxalicum	CECT 7531
		P. roqueforti	CBS 221.30
		P. solitum	CBS 288.36
		P. spinulosum	CBS 269.29
		P. thomii	CBS 381.48
β-(1-3)Glucan			
Eupenicillium		Penicillium	
E. baarnense	CBS 134.41	*P. erubescens*	CBS 253.35
E. catenatum	CBS 352.67	*P. charlesii*	CBS 304.48
E. inusitatum	CBS 351.67		
E. parvum	CBS 359.48		
E. terrenum	CBS 317.67		

of cell walls based on the water-insoluble glucans and the water-soluble polysaccharides.

B. β-(1 → 5)(1 → 6)-Galactofuranan

The main component of fraction F1S obtained from cell wall material of *Eupenicillium cryptum* CBS 271.89 (Fig. 2C). *Penicillium expansum* strains CBS 325.48, CECT 2278, and CECT 2275, and *P. digitatum* (Fig. 2D), is a galactofuranan, whose structure has the repeating unit **2** as determined by GC-MS and NMR (61).

Table 2 Species of *Aspergillus* and Species of Its Teleomorphic States Containing β-(1-5)-Galactofuranan

Microorganisms	Strain	Microorganisms	Strain
Aspergillus		Fennelia	
A. alliaceus	ATCC 58,470	*F. flavipes*	CBS 129.61
A. clavatus	CECT 2674	*F. nivea*	CBS 115.27
A. fumigatus	MR 53		
A. niger	CBS 120.49	Hemicarpenteles	
A. penicilloides	CBS 540.65	*H. acanthosporus*	CBS 558.71
A. ochraceus	CBS 385.67	*H. thaxteri*	CBS 105.25
A. ornatus	CBS 385.53		
		Hemisartorya	
Emericella		*H. maritima*	CBS 136.72
E. bicolor	CBS 425.77		
E. corrugata	CECT 2830	Neosartorya	
E. desertorum	CBS 635.73	*N. fischeri*	CBS 297.67
E. nidulans	CECT 2544		
E. striata	CBS 592.65	Petromyces	
		P. alliaceus	CBS 110.86
Eurotium			
E. herbariorum	CBS 529.65		
		Warcupiella	
		W. spinulosa	CBS 812.65

[→6)-β-D-Gal*f*-(1→5)-β-D-Gal*f*-(1→5)-β-D-Gal*f*-(1→5)[β-D-Gal*f*-(1→]

2

The repeating unit of the polysaccharide from *Neosartorya aureola* CBS 106.45, *N. quadricincta* CBS 135.52, *N. stramenia* CBS 498.65, *Aspergillus fumigatus* CBS 113.26, and *A. fumigatus* CBS 133.61 is similar to that shown above, but contains two residues of (1 → 5)-β-D-Gal*f* instead of three (63).

Penicillium allii, *P. atramentosum*, *P. aurantiogriseum*, *P. brevicompactum*, *P. camemberti*, *P. chrysogenum*, *P. crustosum*, *P. hordei*, *P. roqueforti*, and *P. solitum* included in subgenus *Penicillium* with *P. expansum* (64), contain the β-(1 → 5)-galactofuranan in their cell wall which might indicate that this subgenus is heterogeneous. It is interesting to note that other species of *Neosartorya*, and *Aspergillus* and its teleomorphic genera contain the β-(1 → 5) galactofuranan.

C. β-(1 → 2)(1 → 3)-Galactofuranan

The water-soluble polysaccharide obtained from the cell wall of *Talaromyces wortmannii* CBS 391.48, *T. rotundus* CBS 369.48, (55) *Penicillium allaha-*

badense CBS 304.63, and *P. zacinthae* IJFM 7232, is a galactofuranan with the repeating unit **3**. The ^{1}H-RMN spectrum is shown in Figure 2G.

$$\rightarrow 3)\text{-Gal}f\text{-}\beta\text{-}(1\rightarrow 2)\text{-Gal}f\text{-}\beta\text{-}(1\rightarrow$$

3

Although this polysaccharide has up to now been found in a limited number of species, the linkage types of its galactofuranosyl residues and its characteristic NMR spectrum are very useful for grouping certain penicillia and establishing their relationship with species of *Talaromyces*.

D. Complex Glucomannogalactan

The main component of fraction F1S of several species of *Talaromyces* and *Penicillium* is a heteropolysaccharide composed of galactose, mannose, and glucose in the molar ratio (4:1:1) (Table 3). This polysaccharide (Fig. 2H) is formed of two chains (60) with repeating units **4** and **5**.

Table 3 Species of *Penicillium* and *Talaromyces* Containing the Heteropolysaccharide **4** Plus **5** Grouped According to Type of Glucan in Fraction F1I

Microorganisms	Strain	Microorganisms	Strain
α-(1-3)Glucan			
Penicillium		Talaromyces	
P. erythromellis	IJFM 7284	*T. bacillisporus*	CBS 136.45
P. funiculosum	CECT 2276	*T. macrosporus*	CBS 117.72
P. islandicum	CBS 338.48	*T. derxii*	CBS 412.89
P. purpurogenum	CBS 365.48	*T. ucrainicus*	CBS 162.67
β-(1-3)Glucan			
Pencillium		Talaromyces	
P. aculeatum	CBS 289.48	*T. assiutensis*	CBS 645.80
P. allahabadense	CBS 762.48	*T. flavus*	CBS 352.72
P. dendriticum	CBS 660.80	*T. flavus*	CBS 310.38
P. diversum	CBS 320.48	*T. flavus*	W 36
P. pinophilum	CBS 439.89	*T. helicus*	CBS 335.48
P. verruculosum	Wa 30	*T. intermedius*	CBS 152.65
		T. mimosinum	CBS 659.80
		T. purpureus	CBS 475.71
		T. stipitatus	CBS 375.48
		T. wortmannii	CBS 387.67

→6)-β-D-Gal*f*-(1→5)-β-D-Gal*f*-(1→
2
↑
α-D-Glc*p*-(1→2)-α-D- Gal*f*

4

→6)-α-D-Man*p*-(1→
2
↑
α-D-Gal*f*

5

This polysaccharide may be used as chemotaxonomic marker for a large number of species of *Talaromyces*, and species of *Penicillium* which have *Talaromyces* as perfect state. It is interesting to note that all the species of *Penicillium* which contain this polysaccharide in their cell wall belong to subgenus *Biverticillium* (27).

The water-soluble polysaccharide from the cell wall of *T. udagawae* CBS 579.72 has features that resemble the above-described polysaccharide, but it is more complex and its structure has not been determined. Its ^{1}H-NMR spectrum is shown in Figure 2E. The main component of fraction F1S of the cell wall of *T. luteus* CBS 348.51 is a heteropolysaccharide, different from those previously described (Fig. 2F). Its structure is still unknown. These are the only two species of *Talaromyces* that produce ascospores with transverse to spiral ridges (65).

E. Complex Glucogalactomannans

The genus *Paecilomyces* is related to *Penicillium* with teleomorphs in *Talaromyces*, *Thermoascus*, and *Byssochlamys* (66). Cell wall composition, polysaccharide fractions, and the heteropolysaccharide of fraction F1S are similar for four strains of *P. variotii* (19). The soluble heteropolysaccharide of fraction F1S has also been obtained from the cell wall of other species of *Paecilomyces* and their teleomorphs (Table 4). The ^{1}H-NMR spectra of the polysaccharides are shown in Figure 3. The spectra of the two species of *Byssochlamys* (not shown), the four strains of *P. variotii* (Fig. 3F), and *T. byssochlamydoides* (Fig. 3A) resemble the spectrum of *P. expansum* with slight variations. The polysaccharide of these species is formed by the repeating unit **2** found in *P. expansum* attached to *O*-2 of some residues of the (1 → 6)-linked mannan backbone. The spectra of *P. viridis* (Fig. 3D) and *P. lilacinus* (Fig. 3E) have several signals in common. The spectrum of *Talaromyces leycettanus* (Fig. 3B), teleomorphic state of certain species of *Paecilomyces*, differs from the other species of *Talaromyces*. The spectra of four strains of *P. fumosoroseus* (Table 4) and two of *P. farinosus* are similar (Fig. 3G) and their spectra show the greatest differences from those of related species. The

Table 4 Species of *Paecilomyces*, *Talaromyces*, and *Byssochlamys* Containing Complex Glucogalactomannans: Percentages of Main Linkage Types Deduced from Methylation Analysis

Microorganisms	Glc*p*-(1→	Man*p*-(1→	Gal*p*-(1→	Gal*f*-(1→	→5)-Gal*f*-(1→	→6)-Gal*f*-(1→	→2)-Gal*f*-(1→	→5,6)-Gal*f*-(1→	→2)-Man*p*-(1→	→6)-Man*p*-(1→	→2,6)-Man*p*-(1→	→4,6)-Man*p*-(1→	Figure
P. variotii CBS 323.34, 990.73A, 339.51, 371.70	2	2	—	6	34	4	—	—	7	32	12	—	3F
T. byssochlamydoides CBS 533.71	—	9	—	7	44	—	—	—	9	20	12	—	3A
T. leycettanus CBS 275.70	2	16	—	6	16	—	—	—	12	21	26	—	3B
B. fulva CBS 604.71, 132.33	2	4	—	9	28	3	—	—	11	34	9	—	3A
B. nivea CBS 100.11	—	4	—	6	16	2	—	—	11	44	14	—	3A
P. lilacinus CBS 432.87	—	1	—	33	6	13	2	—	2	2	39	—	3E
P. viridis CBS 348.65	—	3	—	20	8	9	8	—	8	10	34	—	3D
P. fumosoroseus CBS 244.31, 375.70, 337.52, 339.54	7	—	29	1	8	—	3	—	4	9	4	33	3G
P. marquandii CBS 182.27	7	2	27	—	1	18	—	27	—	12	—	4	3C

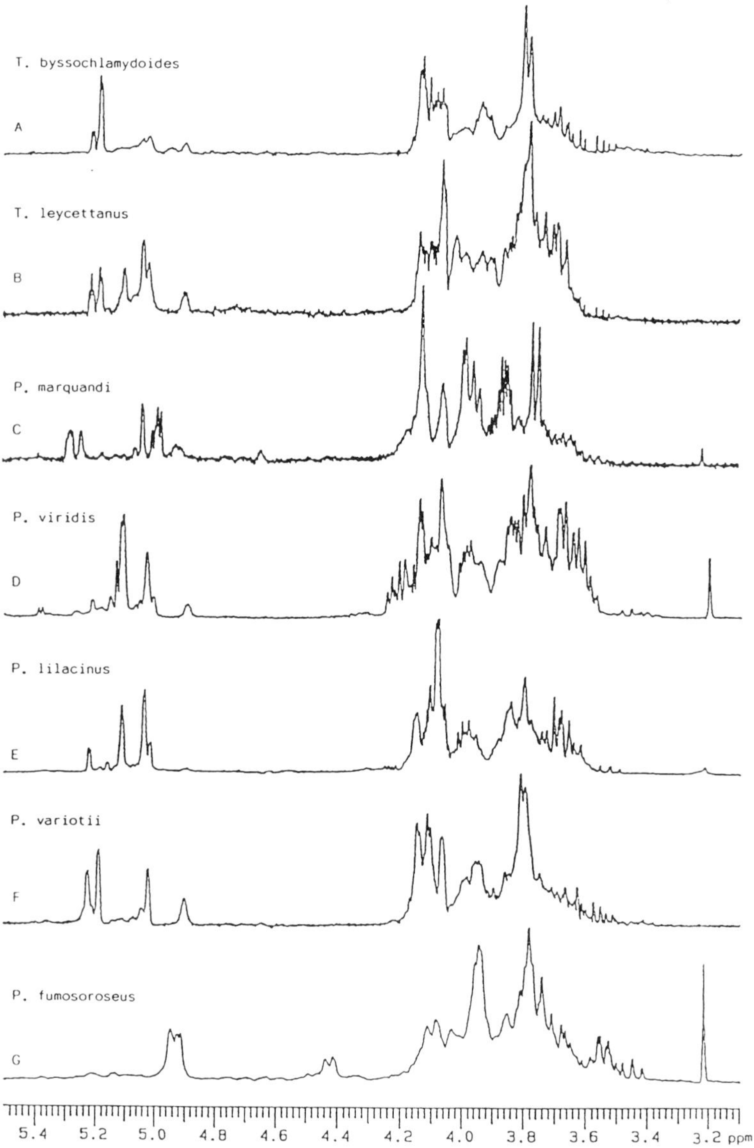

Figure 3 ^{1}H-NMR spectra of the main polysaccharide of fraction F1S obtained from the cell walls of species of the genus *Talaromyces* and *Paecilomyces*. The structure of the polysaccharide and the species that contain a particular polysaccharide are described in the text.

structure and antigenic properties of these polysaccharides have been recently described (67).

Although these polysaccharides have not been completely characterized, GC-MS analysis of their permethylated derivatives (Table 4) and the ^{1}H-NMR spectra show differences in the polysaccharides of the species investigated which confirms that the genus *Paecilomyces* is heterogeneous and that water-soluble polysaccharides may be used as chemotaxonomic markers for subdivision of the genus and for establishing relationships with other genera.

F. α-(1 → 2)-(1 → 6)-Mannans

The water-soluble polysaccharides obtained from the cell wall of different species of *Aphanoascus* are an α-(1 → 4)-glucan and an α-(1 → 2)-(1 → 6)-mannan (21). The species studied are listed in Table 5, and their ^{1}H-NMR spectrum is shown in Figure 4D. Further studies on the structure of this mannan (59) have shown the disaccharide **6** as a repeating unit.

```
→(-6-α-D-Manp-1)→
          2
          ↑
          1
    α-D-Manp
```

6

```
{→6)-α-D-Manp-(1→[6-)-α-D-Manp-(1→]₂}
        2
        ↑
        1
  α-D-Manp
```

7

The polysaccharide fractions of the cell wall of several species of *Microsporum*, *Epidermophyton*, *Trichophyton*, and *Keratinomyces* have been studied (20,26) and further species have since been studied (unpublished results). The main component of fraction F1S in all species is a mannan with the repeating unit **7**. The species investigated, the type of mannan, and the spectra are listed in Table 5 and Figure 4E, F, G.

G. Acidic Glucogalactans

The water-soluble polysaccharide of the cell wall of several species of *Fusarium* amounts to 10% to 15% of the cell wall dry weight. The ^{1}H-NMR spectra of the polysaccharide of *F. javanicum* CECT 2864, *F. solani* CECT 2199, and *F. proliferatum* CBS 240.64 (Fig. 4B) are similar but slightly different from the polysac-

Table 5 Microorganisms Containing Mannan as Main Component of Fraction F1S

Microorganisms	^{1}H-NMR	Repeating unit	Microorganisms	^{1}H-NMR	Repeating unit
Aphanoascus			Epidermophyton		
A. fulvescens FMR 3392	Fig. 4D	6	*E. stockdaleae* FMR 4006	Fig. 4E	7
A mephitalus FMR 2113	Fig. 4D	6	*E. floccosum* CCFVB 1001	Fig. 4F	7*
A. reticulisporus FMR2899	Fig. 4D	6	Keratinomyces		
A. saturnoideus FMR 2002	Fig. 4D	6	*K. ajelloi* FMR 4010	Fig. 4E	7
A. verrucosus FMR 2141	Fig. 4D	6	Microsporum		
Trichophyton			*M. cookei* FMR 4007	Fig. 4E	7
T. concentricum IFO 31068	Fig. 4E	7	*M. fulvum* FMR 4008	Fig. 4G	unknown
T. equinum IFO 31610	Fig. 4G	unknown	*M. gypseum* FMR 3314	Fig. 4E	7
T. rubrum IFO 5467	Fig. 4E	7	*M. nanum* FMR 4009	Fig. 4G	unknown
T. simii FMR 4011	Fig. 4F	7*	*M. persicolor* FMR 4010	Fig. 4G	unknown
T. soudanense P. 12931	Fig. 4E	7	*M. racemosum* FMR 3315	Fig. 4E	7
			M. audouinii IFO 8147	Fig. 4F	7*
			M. equinum P. 21979	Fig. 4F	7*
			M. canis FMR 3373	Fig. 4E	7

*Repeating unit 7, but more complex.

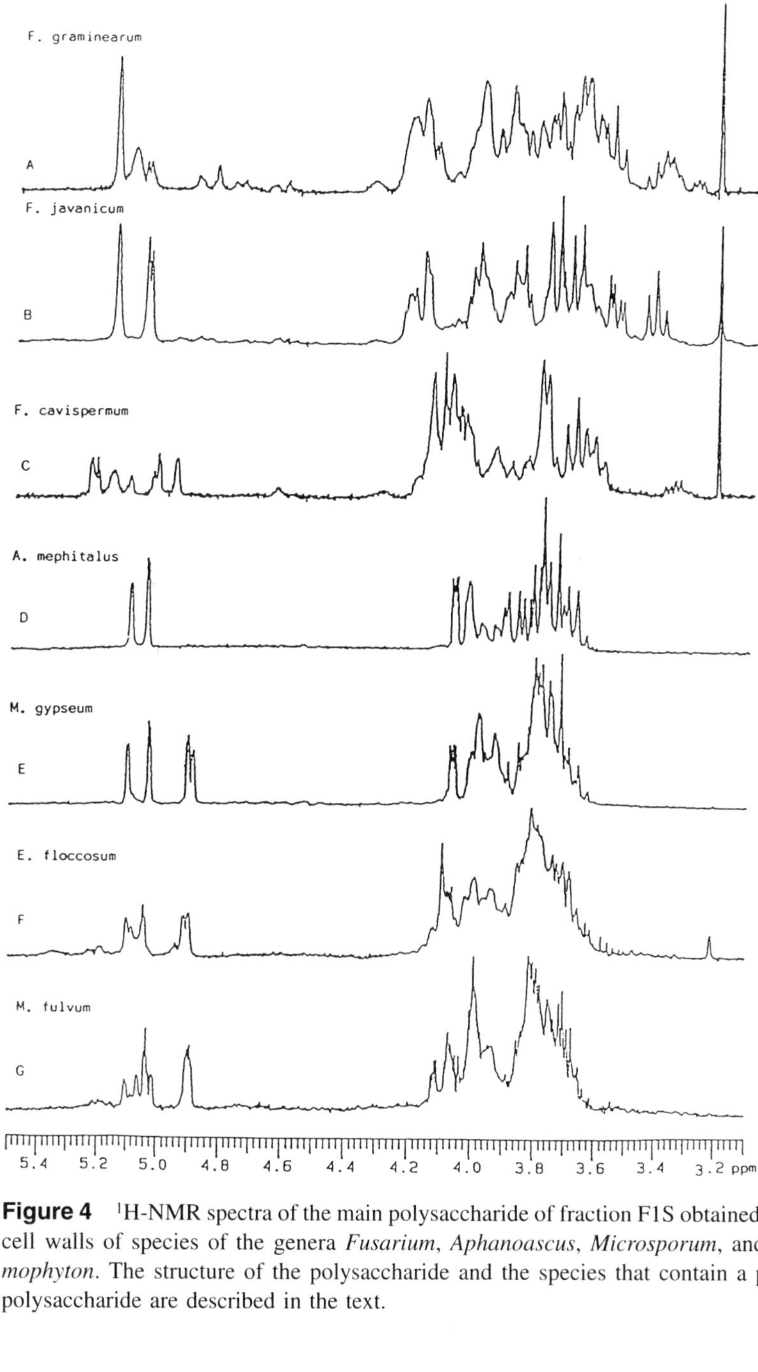

Figure 4 ^{1}H-NMR spectra of the main polysaccharide of fraction F1S obtained from the cell walls of species of the genera *Fusarium*, *Aphanoascus*, *Microsporum*, and *Epidermophyton*. The structure of the polysaccharide and the species that contain a particular polysaccharide are described in the text.

charide of *F. oxysporum* var. *lycopersici*, *F. culmorum* CECT 2148, *F. decemcellulare* CECT 2140, and *F. graminearum* CECT 2150 (Fig. 4A). These polysaccharides have the tetrasaccharide **8** as their main component. The proportion of glucopyranose and glucuronic acid vary according to the different species.

In addition, some units of glucuronic acids are substituted at position 4 by short chains of mannose (unpublished results). Similar structures have been found in glycoproteins isolated from mycelium of *Fusarium* sp. (68).

→6)-β-D-Gal*f*-(1→6)-β-D-Gal*f*-(1→
2 2
↑ ↑
1 1
α-D-Glc*p* α-D-Glc*p*A

8

→6)-α-D-Man*p*-(1→
2
↑
1
β-D-Gal*f*

9

The ^{1}H-NMR spectrum of the polysaccharide of *F. cavispermum* CBS 172.31 (Fig. 4C) is more complex, and its structure has not been determined. The polysaccharide with structure **8** might be considered a chemotaxonomic marker of the genus *Fusarium*.

H. Galactomannans

The polysaccharide of *Neurospora crassa* CECT 2255 and *Neurospora sitophila* CECT 2630 has the disaccharide **9** as their main repeating block (69). The ^{1}H-NMR spectrum of this polysaccharide is shown in Figure 5G.

I. Survey of Fungal Water-Soluble Polysaccharides

In an attempt to confirm that each genus may have at least a characteristic water-soluble polysaccharide, cell walls were prepared and fractionated from several species of different genera, chosen at random. The ^{1}H-NMR spectra of the different microorganisms studied are given in Figure 5. Although further work is required in the purification, analysis, and characterization of the structure of these polysaccharides, the spectra show that the polysaccharides differ from those previously described.

The spectra of the polysaccharides of *Rhizoctonia solani* CECT 2815 (Fig.

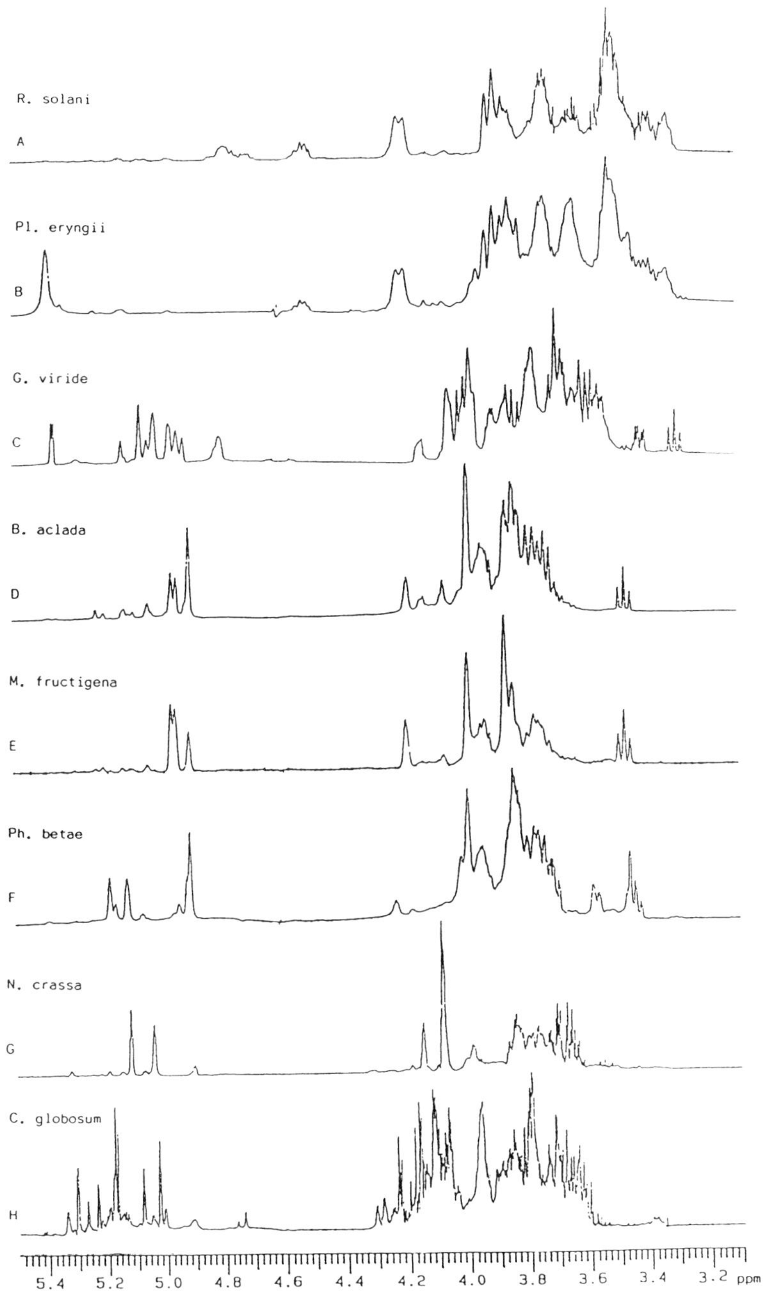

Figure 5 ^{1}H-NMR spectra of the main polysaccharide of fraction F1S obtained from the cell walls of (A) *Rhizoctonia solani*; (B) *Pleurotus eryngii*; (C) *Gliocladium viride*; (D) *Botrytis aclada*; (E) *Monilinia fructigena*; (F) *Phoma betae*; (G) *Neurospora crassa*; and (H) *Chaetomium globosum*.

5A) and *Pleurotus eryngii* A-169 (Fig. 5B) are very similar. *P. eryngii* has a signal at 5.35 ppm which may be due to a contaminating glucan. The main component of the polysaccharide of these two species is glucose (Table 6). The spectra of the polysaccharides of *Botrytis aclada* CECT 2851 (Fig. 5D) and *Monilia fructigena* CECT 2407 (Fig. 5E) are also similar. These two species belong to family Sclerotiniaceae. Both polysaccharides contain rhamnose, mannose, and glucose. *M. fructigena* contains a higher proportion of glucose, which may be due to an accompanying glucan (Table 6).

The spectrum of *Gliocladium viride* (Fig. 5C), partially characterized as an acidic glucogalactomannan, and the spectrum of *Chaetomium globosum* (Fig. 5H), a glucomannogalactan, are the most complex (Table 6). The spectrum of *Phoma betae* polysaccharide is shown in Figure 5F. Its main components are mannose and glucose.

J. Other Fungal Polysaccharides

Different polysaccharides and oligosaccharides have been extracted from mycelium, cell walls, and culture media of fungi. The need to obtain antigens that would allow the detection and characterization of fungal pathogens or food contaminants, the identification of the substances responsible for alergic reactions, and the search for antitumor polysaccharides have led to the characterization of a variety of these polysaccharidic substances. The chemotaxonomic application of these findings is irrelevant because in most cases single microorganisms were investigated. Nevertheless, they show that fungi produce further polysaccharides which might be used as additional chemotaxonomic characters. Water-soluble

Table 6 Neutral Sugars Released (%) from Fraction F1S of Different Fungi After Hydrolysis with 2M H_2SO_4 at 100°C During 5 Hours Determined as Alditol Acetates by Gas-Liquid Chromatography

Microorganisms	Rha	Xyl	Man	Gal	Glc	Recovery (%)
Chaetomium globosum CECT 2701	0.0	0.0	20.6	36.2	10.8	67.6
Rhizoctonia solani CECT 2815	0.0	1.0	3.2	3.1	47.7	54.0
Pleurotus eryngii A-169	0.0	0.0	2.4	4.1	73.0	79.5
Botrytis aclada CECT 2851	9.3	0.0	51.0	2.2	11.7	74.2
Monilinia fructigena CECT 2407	8.2	0.0	21.1	0.7	53.2	83.2
Phoma betae CECT 2358	0.0	0.0	32.8	1.2	40.8	74.8
Neurospora crassa CECT 2255	0.0	0.0	11.0	8.5	9.2	28.7
Trichoderma reeseii CECT 2414	0.0	0.0	21.5	26.0	34.1	81.6
Gliocladium viride A-122	0.0	0.0	16.9	12.3	45.2	74.4

heteropolysaccharides containing mannose, galactose, and fucose have been isolated from the fruit bodies of *Polyporaceae* (70–73).

Characteristic polysaccharides have also been found in the Mucorales. The hyphal walls of these fungi contain uronic acids, neutral sugars, and hexosamine. The polysaccharide fractions isolated from Mucorales species contain glucuronic acid and fucose (74–80). These heteropolysaccharides are typical of members of Mucorales and may be used as chemotaxonomic markers for this taxon.

The composition and structure of polysaccharides and oligosaccharides found in fungal cell wall glycoproteins and peptidopolysaccharides have been reviewed (81). Protein-containing galactomannans have been obtained from mycelium of *Cordyceps sinensis* (82) and *Cordyceps cicadae* (83). *Penicillium charlesii* produces an extracellular peptidophosphogalactomannan (84–86).

The production of extracellular polysaccharides (EPS) by fungi (87–92) is affected by the composition of the culture medium and by the growth conditions. Certain EPS are secreted during the growth phase of the microorganism, and their synthesis ceases when the nutrients are depleted. Generally, these polysaccharides are not found in the cell wall. Malonoglucans are produced by different species of *Penicillium* (93–98). Malonogalactan was isolated in a nuclease preparation from *Penicillium citrinum* (99–101). Malic acid-containing glucans are produced by *Aureobasidium pullulans* (102,103). Other strains of *A. pullulans* produce pullulan, an α-(1 $\rightarrow$ 3)(1 $\rightarrow$ 6)-glucan (104,105). N-acetylgalactosaminogalactans have been isolated from the liquid culture of certain species of *Penicillium* and *Aspergillus* (106–112). (1 $\rightarrow$ 3)-linked glucans are produced by different fungi (35,38,39,113–117). Extracellular acidic heteropolysaccharides have been isolated from *Cryptococcus laurentii* (118,119) and several species of *Tremella* (120), suggesting a possible taxonomic relationship between these two groups of microorganisms. The β-(1 $\rightarrow$ 5)-galactofuranosides of the extracellular polysaccharides of *Penicillium* and *Aspergillus* species are immunodominant (87–89). Mannose residues (1 $\rightarrow$ 2)- and/or (1 $\rightarrow$ 6)-linked are present in EPS of *Cladosporium herbarum* (90), *Alternaria solani*, *Fusarium solani* (91), and *Mucor hiemalis* (92).

Exocellular polysaccharides found in cultures after long periods of incubation might be components of the cell wall that are released during autolysis. Varianose was isolated from the cultures of *Penicillium varians* (121) after 8 weeks of incubation. Its structure (122) is similar to that of the complex polysaccharide **4** obtained in this laboratory from the cell walls of certain species of *Talaromyces* and *Penicillium*; varianose might therefore be a wall component released to the culture medium after cell wall degradation.

Glucans are important structural or skeletal constituents of the cell wall of filamentous fungi and yeast (123–125). Although the fungal cell wall contains several glucans, they have scarcely been used in chemotaxonomy because the isolation and characterization of their chemical structure present enormous difficulties, because the same glucans have been found in different taxa, and because they have not been systematically studied. Nevertheless, the differences in the structure of cell wall glucans may be useful in fungal chemotaxonomy based on the carbohydrate composition of the cell wall.

IV. CONCLUSIONS

Among the fungal cell wall components of several groups of fungi, the alkali-extractable and water-soluble polysaccharides show the greatest differences in composition and structure. Although the number of genera and species investigated are small and only a few water-soluble polysaccharides have been characterized to date, the results are encouraging and allow the following conclusions:

1. Alkali-extractable, water-soluble polysaccharides have a similar composition and structure in most species of a genus such as *Eupenicillium*, showing that it is homogeneous. Other genera (*Penicillium*, *Talaromyces*, and *Paecilomyces*) have several polysaccharides which are found separately in groups of species, indicating that these genera are heterogeneous. Polysaccharides can be chemotaxonomic markers for the genus *Eupenicillium*, and markers at subgeneric level in some other fungi.
2. Water-soluble polysaccharides may help to establish the relatedness of the species of anamorphic genera (*Penicillium*, *Paecilomyces*, etc.) to their teleomorphs (*Eupenicillium*, *Talaromyces*). This could improve the systematics of anamorphic genera where species could be grouped by the similarity of their wall polysaccharides and with those found in the teleomorphs.
3. The water-soluble polysaccharides and other cell wall polysaccharides may be useful in fungal systematics in combination with morphology, secondary metabolite analysis, genetic studies, and isoenzyme electrophoresis.

ACKNOWLEDGMENTS

Research from the authors' laboratories was supported by CICYT grants PB 87/0243 and PB 91/0054 from Dirección General de Investigación Cientifica y Técnica. The work from these laboratories is largely the result of the efforts of Drs. Gómez-Miranda, Jiménez-Barbero, Guerrero, Prieto, Rupérez, and Domenech, and Mr. Ahrazem. We are grateful for their assistance.

REFERENCES

1. Aronson JM. The cell wall. In: Ainsworth GC, Sussman AS, eds. The Fungi, Vol. I. The Fungal Cell. New York: Academic Press, 1965:49–76.
2. Bartnicki-García S. Cell wall chemistry, morphogenesis, and taxonomy of fungi. Annu Rev Microbiol 1968; 22:87–108.

3. Rosenberger RE. The cell wall. In: Smith JE, Berry DR, eds. The Filamentous Fungi, Vol. II. Biosynthesis and metabolism. London: Edward Arnold, 1976:328–344.
4. Burnett JH. Aspects of the structure and growth of hyphal walls. In: Burnett JH, Trinci APJ, eds. Fungal Walls and Hyphal Growth. Cambridge: Cambridge University Press, 1979:1–26.
5. Barker SA, Bourne EJ, Omant DM, Stacey M. Studies of *A. niger*. Part. VI. The separation and structures of oligosaccharides from nigeran. J Chem Soc 1957:2448–2454.
6. Bobbitt F, Nordin JH. Hyphal nigeran as a potential phylogenetic marker for *Aspergillus* and *Penicillium* species. Mycologia 1978; 70:1201–1211.
7. Gander JE, Jentoft NH, Drewes LR, Rick PD. The 5-O-β-D-galactofuranosyl-containing exocellular glycopeptide of *Penicillium charlesii*. Characterization of the phosphogalactomannan. J Biol Chem 1974; 249:2063–2072.
8. Matsunaga T, Okubo A, Fukami M, Yamazi S, Toda S. Identification of β-galactofuranosyl residues and their rapid internal motion in the *Penicillium ochrochloron* cell wall probed by ^{13}C-NMR. Biochem Biophys Res Commun 1981; 102: 502–530.
9. Azuma I, Kimura H, Hirao F, Tsubura E, Yamamura Y, Misaki A. Biological and immunological studies on *Aspergillus*. III. Chemical and immunological properties of the glycopeptide obtained from *Aspergillus fumigatus*. Jpn J Microbiol 1971; 15:237–246.
10. Bardalaye PC, Nordin JH. Galactosaminogalactan from cell walls of *Aspergillus niger*. J Bacteriol 1976; 125:655–669.
11. Leal JA, Moya A, Gómez-Miranda B, Rupérez P, Guerrero C. Differences in cell wall polysaccharides in some species of *Penicillium*. In: Nombela C, ed. Cell Wall Synthesis and Autolysis. Amsterdam: Elsevier Science, 1984:149–155.
12. Gómez-Miranda B, Moya A, Leal JA. Hyphal polysaccharides as potential phylogenetic markers for *Eupenicillium* species. Exp Mycol 1986; 10:184–189.
13. Rupérez P, Moya A, Leal JA. Cell wall polysaccharides from *Talaromyces* species. Arch Microbiol 1988; 146:250–255.
14. Gómez-Miranda B, Moya A, Leal JA. Differences in the cell wall composition in the type species of *Eupenicillium* and *Talaromyces*. Exp Mycol 1988; 12:258–263.
15. Rupérez P, Leal JA. Mannoglucogalactans from the cell walls of *Penicillium erythromellis*: isolation and partial characterization. Carbohydr Res 1987; 167: 269–278.
16. Prieto A, Rupérez P, Hernández-Barranco A, Leal JA. Partial characterisation of galactofuranose-containing heteropolysaccharides from the cell walls of *Talaromyces helicus*. Carbohydr Res 1988; 177:265–272.
17. Gómez-Miranda B, Prieto A, Leal JA. Chemical composition and characterization of a galactomannoglucan from *Gliocladium viride* wall material. FEMS Microbiol Lett 1990; 70:331–336.
18. Barreto-Bergter E, Travassos LR, Gorin PAJ. Chemical structure of the D-galacto-D-mannan component from hyphae of *Aspergillus niger* and other *Aspergillus* spp. Carbohydr Res 1980; 86:273–285.
19. Domenech J, Prieto A, Bernabé M, Leal JA. Cell wall polysaccharides of four strains of *Paecilomyces variotii*. Curr Microbiol 1994; 24:169–173.

20. Guarro J, Cano J, Leal JA, Gómez-Miranda B, Bernabé M. Composition of the cell wall polysaccharides in some geophilic dermatophytes. Mycopathologia 1993; 122:69–77.
21. Leal JA, Gómez-Miranda B, Bernabé M, Cano J, Guarro J. The chemical composition of the wall of six species of *Aphanoascus*: the taxonomic significance of the presence of α-(1-2)(1-6)mannan and α-(1-4)glucan. Mycol Res 1992; 96:363–368.
22. Leal JA, Guerrero C, Gómez-Miranda B, Prieto A, Bernabé M. Chemical and structural similarities in wall polysaccharides of some *Penicillium*, *Eupenicillium* and *Aspergillus* species. FEMS Microbiol Lett 1992; 90:165–168.
23. Leal JA, Gómez-Miranda B, Prieto A, Bernabé M. Differences in cell wall polysaccharides of several species of *Eupenicillium*. FEMS Microbiol Lett 1993; 108: 341–346.
24. Leal JA, Prieto A, Gómez-Miranda B, Jiménez-Barbero J, Bernabé M. Structure and conformational features of an alkali- and water-soluble galactofuranan from the cell walls of *Eupenicillium crustaceum*. Carbohydr Res 1993; 244:361–368.
25. Sasaki S, Uchida K. Isolation and characterization of cell wall polysaccharides from *Aspergillus oryzae*. Agric Biol Chem 1987; 51:2595–2596.
26. Takeda T, Kawarasaki I, Ogihara Y. Studies on the structure of a polysaccharide from *Epidermophyton floccosum* and approach to a synthesis of the basic trisaccharide repeating unit. Carbohydr Res 1981; 89:301–308.
27. Pitt JJ. The genus *Penicillium* and its teleomorphics states *Eupenicillium* and *Talaromyces*. London: Academic Press, 1979.
28. Corina DL, Munday KA. The metabolic stability of carbohydrates in walls of hyphae of *Aspergillus clavatus*. J Gen Microbiol 1971; 65:253–257.
29. Malowitz R, Pisano UA. Changes in cell wall carbohydrate composition of *Paecilomyces persicinus* P-10 M1 during growth and cephalosporin C production. Appl Environ Microbiol 1982; 43:916–923.
30. Gómez-Miranda B, Leal JA. Carbohydrate stability during ageing in *Penicillium expansum* cell wall. Microbiol Sem 1985; 1:67–75.
31. Gómez-Miranda B, Guerrero C, Leal JA. Effect of culture age on cell wall polysaccharides of *Penicillium allahabadense*. Exp Mycol 1984; 8:298–303.
32. Bartnicki-García S, Nickerson WS. Isolation, composition and structure of cell walls of filamentous and yeast-like forms of *Mucor rouxii*. Biochim Biophys Acta 1962; 58:102–119.
33. Kanetsuna F, Carbonell LM. Cell wall glucans of the yeast and mycelial forms of *Paracoccidioides brasiliensis*. J Bacteriol 1970; 101:675–680.
34. Martin SM, Adams GA. A survey of fungal polysaccharides. Can J Microbiol 1956; 2:715–721.
35. Buck KW, Chen AW, Dickerson AG, Chain EB. Formation and structure of extracellular glucans produced by *Claviceps* species. J Gen Microbiol 1968; 51:337–352.
36. Bull AT. Environmental factors influencing the synthesis and excretion of exocellular macromolecules. J Appl Chem Biotechnol 1972; 22:261–292.
37. Ono K, Yasuda N, Ueda S. Effect of pH on pullulan elaboration by *Aureobasidium pullulans*, S-1. Agric Biol Chem 1977; 41:2113–2118.
38. Santamaria F, Reyes F, Lahoz R. Extracellular glucan containing (1-3)-β and (1-6)-β linkages isolated from *Monilia fructigena*. J Gen Microbiol 1978; 109:287–293.

39. Leal JA. Rupérez P, Gómez-Miranda B. Extracellular glucan production by *Botrytis cinerea*. Trans Br Mycol Soc 1979; 72:172–176.
40. Taylor IFP, Cameron DS. Preparation and quantitative analysis of fungal cell walls. Strategy and tactics. Annu Rev Microbiol 1973; 27:243–260.
41. Mahadevan PR, Tatum EL. Relationship of the major constituents of the *Neurospora crassa* cell wall to wild-type and colonial morphology. J Bacteriol 1956; 90:1073–1081.
42. Wessels JGH, Kreger DR, Marchant R, Regensburg BA, De Vries BHH. Chemical and morphological characterization of the hyphal wall surface of the Basidiomycete *Schizophyllum commune*. Biochim Biophys Acta 1972; 273:346–358.
43. Sietsma JH, Wessels JGH. Evidence for covalent linkages between chitin and β-glucan in fungal wall. J Gen Microbiol 1979; 114:99–108.
44. Whistler RL, Anisuzzaman AKM. Gel permeation chromatography. Carbohydr Chem 1980; 8:45–54.
45. Lee CK, Gray GR. A general strategy for the chemical sequencing of polysaccharides. J Am Chem Soc 1988; 110:1292–1293.
46. McNeil M, Darvill AG, Aman P, Franzen LE, Albersheim P. Structural analysis of complex carbohydrates using high-performance liquid chromatography, gas chromatography and mass spectrometry. Methods Enzimol 1982; 83:3–45.
47. Fournet B. Chemical methods to study the structure of fungal polysaccharide antigens. In: Drouet E, Cole GT, de Repentigny L, Latgé JP, Dupont B, eds. Fungal Antigens. New York: Plenum Press, pp. 41–56.
48. Barker SA, Bourne EJ, Whiffen DH. Use of infrared analysis in the determination of carbohydrate structure. Methods Biochem Anal 1956; 3:213–245.
49. Hakomori SI. A rapid permethylation of glycolipid and polysaccharide catalysed by methylsulfinyl carbanion in dimethyl sulfoxide. J Biochem (Tokio) 1964; 55: 205–208.
50. Jansson P-E, Kenne L, Liedgren H, Lindberg B, Lönngren J. A practical guide to the methylation analysis of carbohydrates. Chem Commun Univ Stockholm 1976; 8.
51. Paz-Parente J, Cardon P, Leroy Y, Montreuil J, Fournet B, Ricart G. A convenient method for methylation of glycoprotein glycans in small amounts by using lithium methyl-sulfinyl carbanion. Carbohydr Res 1985; 141:41–47.
52. Lindberg B, Lönngren J, Svensson S. Specific degradation of polysaccharides. Adv Carbohydr Chem Biochem 1975; 31:185–240.
53. Svensson S. Degradation of polysaccharides by oxidation and β-elimination. Methods Enzymol 1978; 50:33–38.
54. Gorin PAJ, Spencer JFT. Proton magnetic resonance spectroscopy—an aid in identification and chemotaxonomy of yeast. Adv Appl Microbiol 1970; 13:25–89.
55. Prieto A, Bernabé M, Leal JA. Isolation, purification and chemical characterization of alkali-soluble polysaccharides from the cell walls of *Talaromyces* species. Mycol Res 1995; 99:69–75.
56. Dabrowski J. Two-dimensional proton magnetic resonance spectroscopy. Methods Enzymol 1989; 179:122–156.
57. Van Haalbeck H. NMR of complex carbohydrates. In: Live D, Armitage IM, Patel D, eds. Frontiers of NMR in Molecular Biology. UCLA Symposia Series Vol. 109. New York: Alan R. Liss, 1990:195–213.

58. Van Haalbeck H. Structural characterization of the carbohydrate moieties of glycoproteins by high-resolution ^{1}H-NMR spectroscopy. Methods Mol Biol 1993; 17: 115–148.
59. Jiménez-Barbero J, Bernabé M, Leal JA, Prieto A, Gómez-Miranda B. Chemical structure and conformational features of cell-wall polysaccharides isolated from *Aphanoascus mephitalus* and related species. Carbohydr Res 1993; 250:289–299.
60. Parra E, Jiménez-Barbero J, Bernabé M, Leal JA, Prieto A, Gómez-Miranda B. Structural studies of fungal cell-wall polysaccharides from two strains of *Talaromyces flavus*. Carbohydr Res 1994; 251:315–325.
61. Parra E, Jiménez-Barbero J, Bernabé M, Leal JA, Prieto A, Gómez-Miranda B. Structural investigation on two cell wall polysaccharides of *Penicillium expansum* strains. Carbohydr Res 1994; 257:239–248.
62. Meyer B. Conformational aspects of oligosaccharides. Top Curr Chem 1990; 154: 141–208.
63. Leal JA, Jiménez-Barbero J, Gómez-Miranda B, Parra E, Prieto A, Bernabé M. Structural investigation of cell-wall polysaccharides from *Neosartorya*: relationships with their putative anamorphs of *Aspergillus*. Carbohydr Res 1995; 273: 255–262.
64. Pitt JI, Cruickshank RH. Speciation and synonymy in *Penicillium* subgenus *Penicillium*—towards a definitive taxonomy. In: Samson RA, Pitt JI, eds. Modern Concepts in *Penicillium* and *Aspergillus* classification. Vol. 185. New York: Plenum Press, 1990:103–119.
65. Stolk AC, Samson RA. The genus *Talaromyces*. Studies on *Talaromyces* and related genera II. In: Studies in Mycology. No. 2, Centraalbureau voor Schimmelcultures (Baarn); 1972:1–65.
66. Samson RA. *Paecilomyces* and some allied hyphomycetes. In Studies in Mycology, No. 6, Centraalbureau voor Schimmelcultures (Baarn). 1974; 1–119.
67. Domenech J, Barasoin I, Prieto A, Gómez-Miranda B, Bernabé M, Leal JA. An antigenic water-soluble glucogalactomannan extracted from cell walls of *Paecilomyces fumosoroseus* and *Paecilomyces farinosus*. Microbiology 1996; 142:3497–3503.
68. Jikibara T, Tkegawa K, Iwahara S. Studies on the uronic acid-containing glycoproteins of *Fusarium* sp. M7. III. The primary structures of the acidic polysaccharides of the glycoproteins. J Biochem 1992; 111:236–243.
69. Leal JA, Jiménez-Barbero J, Gómez-Miranda B, Prieto A, Domenech J, Bernabé M. Structural investigation of a cell-wall galactomannan from *Neurospora crassa* and *N. sitophila*. Carbohydr Res 1996; 283:215–222.
70. Bhavanandan VP, Bouveng MD, Lindberg B. Polysaccharides from *Polyporus giganteus*. Acta Chim Scand 1964; 18:504–512.
71. Fraser RN, Karaesoni S, Lindberg B. Polysaccharides elaborated by *Polyporus pinicola* (Fr.). Acta Chim Scand 1967; 21:1783–1789.
72. Björndal H, Wagstrom B. A heterogalactan elaborated by *Polyporus squamosus* (Huds.). Acta Chim Scand 1969; 23:3313–3320.
73. Usui T, Iwasaki Y, Mizuno T. Isolation and characterization of two kinds of heterogalactans from the fruit bodies of *Ganoderma applanatum* by employing a column of concanavalin A-sepharose 4B. Carbohydr Res 1981; 92:103–114.

74. Bartnicki-García S, Reyes F. Polyuronides in the cell wall of *Mucor rouxii*. Biochim Biophys Acta 1968; 170:54–62.
75. Bartnicki-García S, Lindberg B. Partial characterization of mucoran: the glucuronomannan component. Carbohydr Res 1972; 23:75–85.
76. Miyazaki T, Irino T. Acidic polysaccharides from the cell wall of *Absidia cylindrospora*, *Mucor mucedo* and *Rhizopus nigricans*. Chem Pharm Bull 1970; 18:1930–1931.
77. Miyazaki T, Irino T. Studies on fungal polysaccharides. IX. The acidic polysaccharide from the cell wall of *Rhizopus nigricans*. Chem Pharm Bull 1971; 19:2545–2550.
78. Datema R, Van der Ende H, Wessels JGH. The hyphal wall of *Mucor mucedo*. 1. Polyanionic polymers. Environ J Biochem 1977; 80:611–619.
79. Dow JM, Rubery PH. Chemical fractionation of the cell walls of mycelial and yeast-like forms of *Mucor rouxii*. J Bacteriol 1977; 155:1088–1093.
80. Tsuchihashi H, Yadomae T, Miyazaki T. Structural analysis of the cell-wall D-glucuronans from the fungi *Absidia cylindrospora*, *Mucor mucedo* and *Rhizopus nigricans*. Carbohydr Res 1983; 111:330–335.
81. Gander JE. Fungal cell wall glycoproteins and peptido-polysaccharides. Annu Rev Microbiol 1974; 28:103–119.
82. Kiho T, Tabata H, Ukai S, Hara C. A minor, protein-containing galactomannan from a sodium carbonate extract of *Cordyceps sinensis*. Carbohydr Res 1986; 156: 189–197.
83. Kiho T, Miyamoto I, Nagal K, Ukai S, Hara C. Minor, protein-containing galactomannans from the insect-body portion of the fungal preparation chan hua. Carbohydr Res 1988; 181:207–215.
84. Preston JF, Gander JE. Isolation and partial characterization of the extracellular polysaccharides of *Penicillium charlesii*. Arch Biochem Biophys 1968; 124: 504–512.
85. Rick PD, Drewes LR, Gander JE. The 5-O-β-D-galactofuranosyl-containing exocellular glycopeptide of *Penicillium charlesii*. Occurrence of ethanolamine and partial characterization of the peptide portion and the carbohydrate-peptide linkage. J Biol Chem 1974; 249:2073–2078.
86. Unkefer CJ, Jackson C, Gander JE. The 5-O-β-D-galactofuranosyl-containing glycopeptide from *Penicillium charlesii*. J Biol Chem 1982; 257:2491–2497.
87. Notermans S, Veeneman GH, van Zuylen CWEM, Hoogerhout P, van Boom JH. (1-5)-Linked β-D-galactofuranosides are immunodominant in extracellular polysaccharides of *Penicillium* and *Aspergillus* species. Mol Immunol 1988; 25: 975–979.
88. Van Bruggen–van der Lugt AW, Kamphuis HJ, de Ruiter GA, Miscnick P, van Boom JH, Rombouts FM. New structural features of the antigenic extracellular polysaccharides of *Penicillium* and *Aspergillus* species revealed with exo-β-D-galactofuranosidase. J Bacteriol 1992; 174:6096–6102.
89. Kamphuis HJ, De Ruiter GA, Veeneman GH, van Boom JH, Rombouts FM, Notermans SHW. Detection of *Aspergillus* and *Penicillium* extracellular polysaccharides (EPS) by ELISA: using antibodies raised against acid hydrolysed EPS. Antonie van Leeuwen 1992; 61:323–332.

90. Miyazaki T, Naoi Y. Extracellular polysaccharide of *Cladosporium herbarum*. Studies on fungal polysaccharide XIII. Chem Pharm Bull 1974; 22:1360–1365.
91. Miyazaki T, Naoi Y. Chemical structures of main extracellular polysaccharides of *Alternaria solani* and *Fusarium solani*. Studies on fungal polysaccharides XVIII. Chem Pharm Bull 1975; 23:1752–1758.
92. Miyazaki T, Hayashi O, Oshima Y, Yadomae T. Studies on fungal polysaccharides. The immunological determinant of the serologically active substances from *Absidia cylindrospora*, *Mucor hiemalis* and *Rhizopus nigricans*. J Gen Microbiol 1979; 111:417–422.
93. Raistrick H, Rintoul ML. Studies in the biochemistry of micro-organisms. XIII. On a new type of mucilaginous material, luteic acid, produced from glucose by *Penicillium luteum* Zukal. Trans R Soc (Lond) 1931; 220:255–267.
94. Anderson CG, Howart WN, Raistrick H, Stacey M. XXXIII. Polysaccharides synthesized by microorganisms. IX. The molecular constitution of luteose. Biochem J 1939; 33:272–279.
95. Lloyd PF, Pon MG, Stacey M. Polysaccharides from *Penicillium luteum* Zukal. Chem Ind (Lond) 1956:172–173.
96. Baddiley JJ, Buchanan G, Thain EM. The polysaccharide of *Penicillium islandicum* Sopp. J Chem Soc 1953:1944–1946.
97. Rupérez P, Gómez-Miranda B, Leal JA. Extracellular β-malonoglucan from *Penicillium erythromellis*. Trans Br Mycol Soc 1983; 80:313–318.
98. Rupérez P, Gómez-Miranda B, Leal JA. Acidic exopolysaccharide from *Penicillium allahabadense*. Can J Microbiol 1984; 30:1157–1162.
99. Fujimoto M, Kuninaka A, Yonei S, Kohama T, Yoshino H. Occurrence of malonogalactan in the nuclease preparation from *Penicillium citrinum*. Agric Biol Chem 1969; 33:1666–1668.
100. Kohama T, Fujimoto M, Kuninoka A, Yoshino H. Structure of malonogalactan an acidic polysaccharide of *Penicillium citrinum*. Agric Biol Chem 1974; 38:127–134.
101. Ogura M, Kohama T, Fujimoto M, Kuninaka A, Yoshino H, Sugiyama H. Structure of malonogalactan: carbon-13 nuclear magnetic resonance spectra of malonogalactan. Agric Biol Chem 1974; 38:2563–2564.
102. Leal-Serrano G, Rupérez P, Leal JA. Acidic polysaccharide from *Aureobasidium pullulans*. Trans Br Mycol Soc 1980; 75:57–62.
103. Promma K, Miki E, Marsuda M, Okutani K. Malic acid-containing polysaccharide produced by the fungus *Aureobasidium pullulans* of the marine origin. Nippon Suisan Gakkaishi 1993; 59:529–534.
104. Bouveng HO, Kiessling H, Lindberg B, Mckay J. Polysaccharides elaborated by *Pullularia pullulans*. III. Polysaccharides synthesised from xylose solutions. Acta Chim Scand 1963; 17:1351–1356.
105. Catley BJ. The extracellular polysaccharide, pullulan, produced by *Aureobasidium pullulans*: a relationship between elaboration rate and morphology. J Gen Microbiol 1980; 120:265–268.
106. Distler JJ, Roseman S. Galactosamine polymers produced by *Aspergillus parasiticus*. J Biol Chem 1960; 235:2538–2541.
107. Gorin PAJ, Eveleigh PE. Extracellular 2-acetamido-2-deoxy-D-galacto-D-galactan from *Aspergillus nidulans*. Biochemistry 1970; 9:5023–5027.

108. Bardalaye PC, Nordin JH. Chemical structure of the galactomannan from the cell wall of *Aspergillus niger*. J Biol Chem 1977; 252:2584–2591.
109. Leal JA, Rupérez P. Extracellular polysaccharide production by *Aspergillus nidulans*. Trans Br Mycol Soc 1978; 70:115–120.
110. Rupérez P, Leal JA. Extracellular galactosaminogalactan from *Aspergillus parasiticus*. Trans Br Mycol Soc 1981; 77:621–625.
111. Gómez-Miranda B, Leal JA. Extracellular and cell wall polysaccharides of *Aspergillus alliaceus*. Trans Br Mycol Soc 1981; 76:249–253.
112. Guerrero C, Prieto A, Leal JA. Extracellular galactosaminogalactan from *Penicillium frequentans*. Microbiol Sem 1988; 4:39–46.
113. Dubourdieu D, Fournet B, Bertran A, Ribéreau-Gayon P. Identification du glucane sécrété dans la baie de Raisin par *Botrytis cinerea*. C R Acad Sci Paris 1978; 286: 229–231.
114. Dubourdieu D, Pucheu-Planté B, Mercier M, Ribéreau-Gayon P. Structure, rôle et localisation du glucane exo-cellulaire sécrété par *Botrytis cinerea* dans la baie de Raisin. C R Acad Sci Paris 1978; 287:571–573.
115. Kritzman G, Chet I, Henis Y. Isolation of extracellular polysaccharides from *Sclerotium rolfsii*. Can J Bot 1979; 57:1855–1859.
116. Latgé JP, Boucias DG, Fournet B. Structure of the exocellular polysaccharide produced by the fungus *Numuraea rileyi*. Carbohydr Res 1988; 181:282–286.
117. Stasinopoulos SJ, Seveour RJ. Exopolysaccharide formation by isolates of *Cephalosporium* and *Acremonium*. Mycol Res 1989; 92:55–60.
118. Abercrombie MJ, Jones JKN, Lock MV, Perry MB, Stoodley RJ. The polysaccharides of *Cryptococcus laurentii* (NRRL Y-1401). Can J Chem 1960; 38:1617–1624.
119. Jeannes A, Pittsley JE, Watson PR. Exocellular polysaccharide production from glucose by *Cryptococcus laurentii* var. *flavescens*. NRRL Y-1401: chemical and physical characterization. J Appl Polymer Sci 1964; 8:2775–2787.
120. Slodki ME, Wickerham LJ, Bandoni RJ. Extracellular heteropolysaccharides from *Cryptococcus* and *Tremella*: a possible taxonomic relationship. Can J Microbiol 1966; 12:489–494.
121. Haworth WN, Raistrick H, Stacey M. CCCXVI. Polysaccharides synthesised by microorganisms. II. The molecular structure of varianose produced from glucose by *Penicillium varians* G. Smith. Biochem J 1935; 29:2668–2678.
122. Jansson P, Lindberg B. Structural studies of varianose. Carbohydr Res 1980; 82: 97–102.
123. Gorin PAJ, Spencer JET. Structural chemistry of fungal polysaccharides. Adv Carbohydr Chem 1968; 23:367–417.
124. Fleet GH, Phaff HJ. Fungal glucans—structure and metabolism. In: Tanner W, Loewus FA, eds. Enciclopedia of Plant Physiology. New Series Vol. 13B. Berlin: Springer-Verlag, 1981; 416–440.
125. Barreto-Bergter E, Gorin PAJ. Structural chemistry of polysaccharides from fungi and lichens. Adv Carbohydr Chem Biochem 1983; 41:67–103.

8

Chemotaxonomy of Fungi by Unsaponifiable Lipids

R. Russell Monteith Paterson
International Mycological Institute, Egham, Surrey, United Kingdom

I. INTRODUCTION

Lipids, as a subject for chemical studies, are often overlooked in fungi, whereas other compounds such as nucleic acids, polysaccharides, proteins, enzymes, and secondary metabolites have been given more attention. They cannot be deemed a single chemical entity except in the most basic of terms, unlike the other major macromolecules. The situation is compounded when chemotaxonomic studies are considered, when even less meaningful data are available. The current state of the field is hindered by a lack of studies which take a standardized approach and/or are based on inappropriate methods. The fundamental definition of lipids is of compounds with high solubility in organic solvents (e.g. chloroform, alcohols, hydrocarbons, ethers) and slight solubility in water. These can be further divided as saponifiable and unsaponifiable. The former produce water-soluble soaps on heating with alkali. They are, or contain, fatty acids and are dealt with elsewhere in this book. The latter are soluble in lipid solvents but are not saponified by alkali. However, all these compounds are biosynthetically related, being derived from condensation of acetate (i.e., isoprenoids).

The isoprenoid classification followed here is of Stoffel (1), and concerns ubiquinones, steroids, carotenoids, and vitamins A, D, E, and K. Only the ubiquinones have been investigated in a standardized manner and predominantly by one group of researchers. Most information is available for fungal sterols within the steroids, but few standardized studies are available. Virtually none have been published for carotenoids, and none for the vitamins. Lipids that are secondary metabolites (e.g., some carotenoids) are not considered. Emphasis has been placed here on taxonomic studies using standard methods, and little attempt has been made to compare studies by diverse research groups. Only filamentous fungi will be considered in this chapter. The section on fungal ubiquinones represents the first review of these data.

II. UBIQUINONES

Ubiquinones are a chemical class of terpenoid lipids which are used in microbial chemotaxonomy because of the structural variation observed between some taxa (2). The length of the isoprenoid chains, and degree of saturation, can vary depending on from which biological system the ubiquinones were isolated. They are constituents of the eukaryotic mitochondrial plasma membranes and are important in electron transport, oxidative phosphorylation, and possibly active transport. Transfer of electrons is by alternate reduction and oxidation of the two quinone/phenolic groups on the benzoquinone ring (Fig. 1).

Theoretically, fungal ubiquinones can provide greater discriminating power than bacterial ones, as both partially saturated and saturated structures are found, whereas only the latter have been detected from bacteria. Interestingly, phenolic and quinone precursors of ubiquinones have been reported from fungi, which may form the basis of some additional characters (3). The momentum for considering ubiquinones as characters for the taxonomy of fungi comes from the perception of existing ones being inadequate (4). Lechevalier and Lechevalier (5) state that "the general lack of interest in using chemical and physiological characters is due to the comparative ease with which fungi can be classified on the basis of morphological characters"! This apparent contradiction may reflect perspectives from different disciplines (e.g., mycologists vs. bacteriologists).

The ambiguities inherent in employing different methods for ubiquinone analysis in fungal taxonomy have been alluded to (6), and concerns about the validity of some previous approaches are relevant to this issue. Various publica-

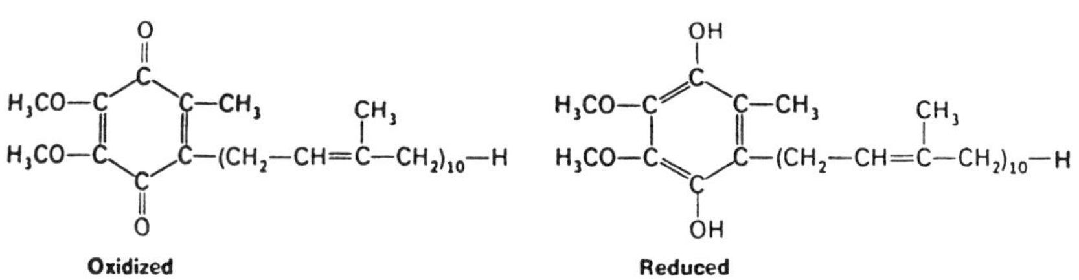

Figure 1 Oxidized and reduced forms of ubiquinone Q10.

tions have indicated they are useful in classifying filamentous fungi (7–16). However, there are some potential problems. Authors specify rigorous extraction by saponification (7–9,14). Direct extraction methods are less likely to alter ubiquinones chemically because they are mild and rapid (2). In most reports on fungi (15,18) the method of Yamada and Kondo (17) is cited which is based on saponification; in other reports, alcoholic pyrogallol is included to avoid quinone oxidation (7–14). However, Mitchell and Fallon (19) compared a direct method with a pyrogallol saponification extraction and found higher Q10 levels in one in 10 strains of bacteria extracted by the direct method. On the other hand, a method devised for the safe handling of pathogenic fungi (20) indicated various sterilization regimes did not affect ubiquinone properties, and so they may not be especially unstable. Unfortunately, the effect of saponification was not ascertained. Two direct extractions for fungi have now been reported (21,22), and a novel direct extraction method for bacterial ubiquinones which employs two-dimensional thin-layer chromatography (TLC) may also be suitable for fungi (23). The choice of extraction procedure has additional significance when "cut-off" values are used to decide if concentrations are sufficiently high to be considered (7–9) and the concentrations are used to decide what is minor appear to be arbitrary (7–9,14,17).

As mentioned previously, many of the above methods were based on Yamada and Kondo (17) which involve removing ubiquinones from the middle section of normal phase TLC plates. It is unclear whether all ubiquinones would be removed, as they may have a wide range of Rf values. Rf values from low to high were observed for various ubiquinones from reversed-phase high-performance TLC (19,24) and by paper chromatography (17).

In many studies which use high-performance liquid chromatography (HPLC), quantification is by comparing peak areas (10–13). By implication, the same method is used by Kuraishi et al. (7–9), although this is not stated. However, these approaches are inaccurate in terms of concentrations due to differences in molar extinction coefficients for each ubiquinone, unless the peak areas for known concentrations of each ubiquinone were obtained. A sample with the same molar concentrations of different ubiquinones would not give the same peak areas for each in the resulting chromatogram. Finally, too few strains of each species have been examined, and repeat analyses of the same strains are not generally carried out, as far as can be determined.

Some of the above factors may explain why ubiquinones are most useful at the genus level in fungi as any further discriminatory power is reduced. Strains with basically the same ubiquinones could be considered as different because concentrations of some ubiquinones were below the cut off value chosen, and vice versa. Therefore, some caution is recommended when using these schemes (16). The methods are time-consuming and use HPLC, which can be experimentally demanding. The studies are reviewed here in some depth as a wide range of fungi has been analyzed by standard techniques, and they have been compared to results from other characters in some cases.

A. Taxonomic Applications

1. Studies Across Groups

Kuraishi et al. (7) analyzed 220 samples representing 193 strains and 27 fruitbody specimens assigned to 218 species, 195 teleomorph and 41 anamorph genera, 67 families, and four anamorph sections in higher fungi (Tables 1–3), and concluded that classification from ubiquinones was consistent with generic circumscriptions based on other characters. Zygomycotina and Basidiomycotina (particularly homobasidiomycetes) were considered to have predominantly Q9. However, the data for the Zygomycotina were not included, and some representatives of the Basidiomycotina had Q10 listed in the report (Table 2). Major amounts of hydrogenated Q10 were not found in the Basidiomycotina. The ubiquinones of *Aspergillus*, *Penicillium*, and *Paecilomyces* were reported to be heterogeneous, and it was suggested that these genera may be polyphyletic. However, the teleomorphs were usually homogeneous. Hydrogenated ubiquinones were present in the Ascomycetes. Fungi can be divided into nine main ubiquinone systems; most higher fungi possess either Q9, 10, or 10(H_2) as the major ubiquinones according to this investigation. On the other hand, Q5, 6, 7, 8, 9, 10, 10(H_2), 10(H_4) have been reported from fungi (12). It should be pointed out that there was no indication of repeats on the same strains, that only one strain of each species was analyzed, and that very few representatives of each genus were tested (17). Kuraishi et al. (8,9), remedied the third problem but not the first, and insufficient strains were tested of the same species in some cases.

In a different study, *Sporobolomyces* was separated from *Bensingtonia* by the main ubiquinones being Q10 (or 10(H_2)) and Q9, respectively, in a revision of ballistoconidia-forming yeast and fungi (25). Also the ballistoconidia-forming genera *Tilletiopsis* and *Tilletiaria* were found to contain Q10, whereas *Itersonilia* contained Q9 (15).

Although not a taxonomic study per se, the isoprenoid quinone and phenol precursors of ubiquinone and dihydroquinones from certain fungi were determined (3) as mentioned previously. Ten fungi examined were of three types: ubiquinones only (*Aspergillus fumigatus* and *Penicillium brevicompactum*) or ubiquinones (H_2) (*Alternaria solani*, *Claviceps purpurea*, and *Talaromyces stipitatus*); 5′ demethoxyubiquinones and ubiquinones (*Agaricus campestris*, *Aspergillus niger*, *Phycomyces blakesleeanus*) or 5′ demethoxyubiquinones (H_2) and ubiquinones (H_2) (*Emericella quadrilineata*); and *A. flavus*, which had 2-decaprenyl (H_2) phenol, 6-methoxy-2-decaprenyl (H_2) phenol, 6-methoxy-2-decaprenyl (X-H_2)-1, 4-benzoquinone, 5-demethoxyubiquinone-10 (X-H_2), and ubiquinones (H_2). Therefore, there is a greater diversity than simply ubiquinones and hydrogenated ubiquinones, which may represent a source of new characters for fungal chemotaxonomy.

2. Hyphomycetes

a. Penicillium *and Related Genera.* Living isolates (335) assigned to 118 species of *Penicillium* and related teleomorph and anamorph genera have been examined (9) (Table 4). Q9 was detected in members of the 21 species of the genus *Eupenicillium*. Isolates from the subgenera *Aspergilloides*, *Furcatum*, and *Penicillium* from the genus *Penicillium* were considered to be of Q9, except for *P. megasporum* (Q10) and *P. asperosporum* (Q10(H_2)), which were placed in series *Megaspora*.

Species in section *Coremigena* had Q9 in subgenus *Biverticillium*, but section *Simplicia* had Q10(H_2) as did the genus *Geosmithia* except for *G. argillacea* (Q10). Nearly all *Talaromyces* had Q10(H_2), but some also had Q10(H_4). It is not recorded if those defined as Q10(H_2) also had Q10(H_4) but at a percentage lower than the cut-off value. *T. thermophilus*, *T. leycettanus*, *T. avellaneus*, and *T. striatus* had Q10. A great deal of strain variation was recorded in *T. trachyspermus*, and means and standard deviations were 44% (20.8) and 56% (20.1) for 10(H_2) and 10(H_4), respectively, assuming 10(H_4) was 90% for one strain where the actual figure was not provided (i.e., strain "IFO 317557"). Similar variation was observed for *T. byssochlamydoides* and *T. emersonii*. *Eladia*, *Penicilliopsis*, and *Dendrosphaera* had Q9. *Trichocoma* had both Q10(H_2) and Q10(H_4). It was concluded the ubiquinone systems in *Penicillium*, its teleomorph, and related anamorph genera are heterogeneous. The taxonomic position of those species with different ubiquinones and morphological characters should be revised.

Taylor and LoBuglio (16) considered the ubiquinone data of Kuraishi et al. (9) to be particularly useful in determining relationships in penicillia with clustered conidiophores (*P. duclauxii*, *P. vulpinum*, and *P. clavigerum*) in a phylogenetic study in *Penicillium*, *Talaromyces*, and *Eupenicillium*. However, only a few strains of each species were analyzed for ubiquinones so the conclusion cannot be considered definitive. LoBuglio et al. (26) concluded inter alia, that *P. duclauxii* and *P. clavigerum* represented two species by rDNA sequence data, and referred to the recommendation of classifying *P. duclauxii* with Q9 as *P. clavigerum* (9). The situation was summarized by assigning Q9 to almost all species of *Eupenicillium* and to *Penicillium* subgenera *Aspergilloides*, *Furcatum*, and *Penicillium* (27). *Talaromyces* and *Penicillium* subgenus *Biverticillium* possessed Q10 or Q10(H_2) systems. A few anomalies with some of the original classifications (28) were mentioned. Another study on *Penicillium* was undertaken of 38 species (41 strains) of *Penicillium sensu lato* and related teleomorphic genera (14). In general, results were consistent with those of Kuraishi et al. (9). In eight of the 10 sections, Q9 was found to be predominant, whereas for *Geosmithia* Q10 was highest with only trace amounts of Q9.

Table 1 Principal Ubiquinone Systems in the Ascomycotina

Family	Teleomorph (Anamorph)	Ubiquinone
Hemiascomycetes		
Endomycetaceae	*Endomyces geotrichum*	9
	(*Geotrichum candidum*)	9
Plectomycetes		
Gymnoascaceae	*Amauroascus reticulatus*	$10(H_2)$
	Apinisia graminicola	9
	Arachniotus ruber	$10(H_2)$
	Arachnotheca glomerata	9
	Arthroderma benhamiae	9
	Ctenomyces serratus	9
	Glymnoascus reessii	$10(H_2)$
	Myxotrichum cancellatum	$10(H_2)$
	(*Geomyces asperulatus*)	$10(H_2)$
	Nannizzia gypsea	9
	Pseudogymnoascus roseus	$10(H_2)$
	Rollandina capitata	$10(H_2)$
Onygenaceae	*Anixiopsis fulvescens*	10
	Aphanoascus cinnabarinus	$10(H_2)$
	Arachnomyces minimus	9
	Dichotomyces cejpii	10
	Onygena corvina	9
	Xylogone sphaerospora	$10(H_2)$
Monascaceae	*Monascus ruber*	10
Thermoascaceae	*Thermoascus aurantiacus*	9
Trichocomaceae	*Chaetosartorya stromatoides*	9
	Edyuillia athecia	9
	Emericella nidulans	$10(H_2)$
	Eurotium halophilicum	9
	Fennellia nivea	$10(H_2)$
	Hemicarpenteles acanthosporus	10
	Hemicarpenteles paradoxus	9
	Neosartorya fischeri	$10(H_2)$
	Petromyces alliaceus	10
	Sclerocleista ornata	9
	Warcupiella spinulosa	10
	(*Aspergillus flavus*	$10(H_2)$
	(*Aspergillus fumigatus*)	10
	(*Aspergillus niger*)	9
	Dendrosphaera eberhardtii	9
	Eupenicillium crustaceum	9
	Hamigera avellanea	10
	Penicilliopsis clavariiformis	9
	Talaromyces flavus	$10(H_2)$
	Trichocoma paradoxa	$10(H_2)(66\%)+10(H_4)(34\%)$

Table 1 Continued

Family	Teleomorph (Anamorph)	Ubiquinone
Trichocomaceae (cont.)	(*Eladia saccula*)	9
	(*Penicillium islandicum*)	$10(H_2)$
	(*Penicillium megasporum*)	10
	(*Penicillium thomii*)	9
	(*Byssochlamys nivea*)	9
	(*Paecilomyces farinosus*)	$10(H_2)$
	(*Paecilomyces variotii*)	9
	(*Paecilomyces variotii*)	10
Pseudeurotiaceae	*Albertiniella polyporicola*	$10(H_2)$
	Emericellopsis terricola	$10(H_2)$
	Pseudeurotium zonatum	$10(H_2)$
Ascosphaeraceae	*Ascosphaera apis*	9
	Bettsia alvei	10
Pyrenomycetes		
Ophiostomataceae	*Ceratocystis piceae*	$10(H_2)$
Melanosporaceae	*Achaetomium globosum*	$10(H_2)$
	Ascotricha lusitanica	$10(H_2)$
	Boothiella tetraspora	$10(H_2)$
	Chaetomium funicola	$10(H_4)$
	Chaetomium globosum	$10(H_2)$
	Chaetomium indicum	$10(H_2)(82\%)+10(H_4)(16\%)$
	(*Scopulariopsis brevicaulis*)	$10(H_2)$
	Kernia nitida	$10(H_4)$
	Kernia ovata	$10(H_2)$
	Kernia pachypleura	$10(H_2)(78\%)+10(H_4)(18\%)$
	Thielavia basicola	$10(H_2)$
Sphaeriaceae	(*Chloridium chlamydosporis*)	$10(H_2)$
	(*Codinaea simplex*)	$10(H_2)$
Hypocreaceae	*Calonectria rigidiuscula*	$10(H_2)$
	Gibberella pulicaris	$10(H_2)$
	Hypocrea rufa	$10(H_2)$
	Micronectriella nivalis	$10(H_2)$
	Nectria cinnabarina	$10(H_2)$
	Nectriopsis solani	$10(H_2)$
	Neocosmospora vasinfecta	$10(H_2)$
	Peloronectriella sasae	$10(H_2)$
	Pseudohypocrea citrinella	$10(H_2)$
	Pseudonectria rousselliana	$10(H_2)$
	(*Acremonium strictum*)	$10(H_2)$
	(*Fusarium oxysporum*)	$10(H_2)$
	(*Gliocladium virens*)	$10(H_2)$
	(*Myrothecium verrucaria*)	$10(H_2)$
	(*Trichoderma longibrachiatum*)	$10(H_2)$
	(*Verticillium lateritium*)	$10(H_2)$

Table 1 Continued

Family	Teleomorph (Anamorph)	Ubiquinone
Polystigmataceae	*Glomerella fusarioides*	$10(H_2)$
Pyronemataceae	*Ascodesmis nigricans*	9
	Thelebolus crustaceus	10
	Ascophanus carneus	9
	Trichophaea abundans	9
	Anthracobia melaloma	9
Stictidaceae	*Apostemidium leptospora*	$10(H_2)$
Geoglossaceae	*Spathularia clavata*	$10(H_4)$
	Spathularia flavida	$10(H_2)$
	Spathularia velutipes	$10(H_4)$
Sclerotiniaceae	*Botryotinia arachidis*	9
	Botryotinia fuckeliana	$10(H_2)(60\%)+10(H_4)(40\%)$
	Sclerotinia sclerotiorum	$10(29\%)+10(H_2)(70\%)$
	Sclerotinia tuberosa	$10(H_2)$
	Ciborinia allii	$10(H_2)$
	Monilinia fructicola	$10(23\%)+10(H_2)(75\%)$
	Monilinia fructigena	$10(H_2)$
	(*Botrytis squamosa*)	$10(H_2)(45\%)+10(H_4)(50\%)$
	(*Cristulariella moricola*)	$10(H_2)$
	(*Sclerotium hydrophilum*)	9
	(*Sclerotium cepivorum*)	$10(68\%)+10(H_2)(31\%)$
	Phaeosclerotinia phaeospora	$10(H_2)$
	Ovulinia azaleae	$10(H_2)$
	Scleromitrula shiraiana	$10(H_2)$
	Stromatinia cepivorum	$10(38\%)+10(H_2)(61\%)$
	Stromatinia gladioli	$10(H_2)$
	Lambertella corni-maris	$10(H_2)$
	Rustroemia cuniculi	$10(H_2)$
	Ciboria amentacea	$10(21\%)+10(H_4)(54\%)$ $10(H_2)(25\%)$
	C. americana	$10(42\%)+10(H_2)(47\%)$
	C. carunculoides	$10(H_2)$
	Orbilia xanthostigma	$10(H_2)$
Dermataceae	*Tapesia hydrophila*	$10(H_2)$
	Belonopsis ericae	$10(H_2)$
	Chlorosplenium aeruginosum	$10(H_2)$
	Leptotrochila trifolii	$10(H_2)$
	Diplocarpon rosae	$10(H_2)$
Hyaloscyphaceae	*Dasyscyphus ciliaris*	$10(H_2)$
	Helotium herbarum	$10(H_2)$
Leotiaceae	*Bulgaria inquinans*	$10(H_2)$
	Neocudoniella jezoensis	$10(H_2)$
	Leotia lubrica	10

Table 1 Continued

Family	Teleomorph (Anamorph)	Ubiquinone
Leotiaceae (cont.)	*Coryne sarcoides*	10(H_2)
	Cudonia constrictospora	10(H_2)
	Hymenoscyphus varicosporoides	10(H_2)
Sordariaceae	*Apiosordaria verruculosa*	10(H_2)
	Coniochaeta tetraspora	10(H_2)
Diaporthaceae	*Diaporthe citri*	10(H_2)
Xylariaceae	*Xylaria hypoxylon*	10(H_2)
Clavicipitaceae	*Claviceps purpurea*	10(H_2)
	Cordyceps militaris	10(H_2)
Hypomycetaceae	*Hypomyces aurantius*	10(H_2)
	(*Trichothecium roseum*)	10(H_2)
Loculoascomycetes		
Botryosphaeriaceae	*Botryosphaeria dothidea*	10(H_2)
Pleosporaceae	*Cochliobolus heterostrosphus*	10(H_2)
	(*Curvularia lunata*)	10(H_2)
	(*Drechslera australiensis*)	10(H_2)
	Phaeosphaeria eustoma	10(H_2)
Sporormiaceae	*Preussia isomera*	10(H_2)
	Westerdykella ornata	10(H_2)
Dothioraceae	*Leptosphaeria doliolum*	10(H_2)
	(*Dothichiza ferruginea*)	10(H_2)
Hysteriaceae	*Gloniopsis constricta*	10(H_2)
	Glonium lineare	10
	Hysterographium minus	10(H_2)
Hierarchical position unknown	*Phaerocreopsis hypoxyloides*	9(36%)+10(61%)
Discomycetes		
Terfeziaceae	*Terfezia gigantea*	8(50%)+9(50%)
Sarcosomataceae	*Chorioactis geaster*	9
	Desmazierella acicola	9
	Urnula craterium	9
Sarcoscyphaceae	*Wynnea gigantea*	9
	Phillipsia domingensis	9
Ascobolaceae	*Ascobolus denudatus*	9
Pezizaceae	*Aleuria aurantia*	9
	Peziza ostracoderma	9
	Patella scutellata	9
	Galactinia micropus	9
Morchellaceae	*Morchella esculenta*	8(67%)+9(32%)
Helvellaceae	*Helvella elastica*	10
	Macropodia macropus	10

Source: Kuraishi et al. (7).

Table 2 Principal Ubiquinone Systems in the Basidiomycotina

Hymenomycetes		
Tremellaceae	*Phlogiotis helvelloides*	9
	Pseudohydnum gelatinosum	9
Sirobasidiaceae	*Fibulobasidium inconspicuum*	9
Auriculariaceae	*Auricularia mesenterica*	9
	Helicobasidium monpa	10
Exobasidiaceae	*Exobasidium gracile*	10
Carcinomycetaceae	*Christiansenia pallida*	10
	Christiansenia effibulata	9
Dacrymycetaceae	*Dacrymyces*	10
Ceratobasidiaceae	*Thanatephorus cucumeris*	9
	Protodaedalea hispida	9
Thelephoraceae	*Hericium laciniatum*	9
Clavariaceae	*Clavulinopsis miyabeana*	9
Echinodontiaceae	*Mycoleptodonoides pergameneus*	9
Hymenochaetaceae	*Coltricia cinnamomea*	9
Ganodermataceae	*Ganoderma lucidum*	9
Polyporaceae	*Fomes fomentarius*	9
Stereaceae	*Stereum ostrea*	9
Boletaceae	*Boletus edulis*	9
Trichlomataceae	*Laccaria laccata*	9
	Lentinus edodes	9
Amanitaceae	*Amanita muscaria*	9
Bolbitiaceae	*Conocybe tenera*	9
Lepiotaceae	*Chlorophyllum molybdites*	9
Agaricaceae	*Agaricus campestris*	9
Strophariaceae	*Pholiota aurivella*	9
Coprinaceae	*Coprinus comatus*	9
Russulaceae	*Lactarius volemus*	9
Gasteromycetes		
Clathraceae	*Linderia biolumnata*	9
Phallaceae	*Mutinus caninus*	9
Lycoperdaceae	*Calostoma japonica*	9
Nidulariaceae	*Crucibulum* sp.	9

Source: Ref. 7.

The rapid method of Mitchell and Fallon (19) for Legionellaceae was adapted for fungi by Paterson and Buddie (22) for *Penicillium*. The method involved mild and direct extractions. A more limited range of spots were considered as ubiquinones and characters by Mitchell and Fallon than was the case in Paterson (24), where all the spots detected (i.e., probable ubiquinones and other

Table 3 Principal Ubiquinones of the Deuteromycotina

Hyphomycetes		
Series Phialosporae	*Caldariomyces fumago*	9
	Clonostachys cylindrospora	10(H_2)
	Gliomastix murorum	10(H_2)
	Metarhizium anisopliae	10(H_2)
		10(H_2)
Series Aleuriosporae	*Beverwykella pulmonaria*	10(H_2)
	Epicocum humicola	10(H_2)
	Doratomyces stemonitis	10(H_2)
	Gilmaniella humicola	10(H_2)
Series Sympodiosporae	*Dactylaria purpurella*	10(H_2)
	Diplorhinotrichum ampulliforme	10(H_2)
Series Arthrosporae	*Moniliella acetobutans*	9
	Oidiodendron citrinum	10(H_2)
	Oidiodendron tenuissimum	9
	Oidiodendron truncatum	10
	Trichosporonoides spathulata	9
Miscellaneous	*Wallemia sebi*	9
Coelomycetes		
Sphaeropsidaceae	*Chaetomella oblonga*	10
	Coniothyrium hellebori	10(H_2)
	Macrophomina phaseolina	10(H_2)
	Phoma sorghina	10(H_2)
Melanconiaceae	*Colletotrichum lindemuthianum*	10(H_2)

Source: Ref. 7.

lipids) were considered as potential characters on the basis of all evidence being "grist for the taxonomic mill" (29). This provided sufficient data for numerical analysis. Four terverticillate penicillia strains from four species or varieties clustered separately and independently of growth period except in one case where the growth period was long (25 days). *Fusarium*, *Chaetomium*, *Colletotrichum*, and *Metarhizium* have also been analyzed by this method (Paterson, unpublished results). However, it is still experimental and requires further assessment.

b. Aspergillus *and Teleomorphs.* The distribution of the ubiquinone systems in *Aspergillus* and teleomorphs was examined in relation to the taxonomic systems of Raper and Fennell compared to that of Gams et al. (Table 5) (8). Q9 was considered to be the major ubiquinone in subgenus *Aspergillus* sections *Aspergillus* and *Restricti*. Atypically, approximately equal amounts of Q8 and 9 were reported from one isolate of *A. penicilloides*. Data are presented on glucose-6-phosphate dehydrogenase, malate dehydrogenase, fumarase, alcohol dehydrogenase, and glutamate dehydrogenase electrophoresis patterns which di-

Table 4 Major Ubiquinones in *Penicillium* and Related Genera

Eupencillium		
Series		
Alutacea	*E. alutaceum*	9
	E. anatolicum	9
	E. gracilentum	9
	E. cinnamopurpureum	9
	E. meridianum	9
Erubescentia	*E. erubescens*	9
	E. hirayamae	9
Fracta	*E. catenatum*	9
	E. fractum	9
	E. ochrosalmoneum	9
	E. ornatum	9
Tularensia	*E. lassenii*	9
	E. tularense	9
Pinetora	*E. pinetorum*	9
Javanica	*E. brefeldianum*	9
	E. javanicum	9
	E. ludwigii	9
	E. zonatum	9
Lapidosa	*E. lapidosum*	9
Crustaceum	*E. crustaceum* (4 strains)	9
	(1 strain)	8(19%)+9(81%)
	E. molle	9
Penicillium		
Subgenus *Aspergilloides*		
Section *Aspergilloides*		
Glabra	*P. donkii*	9
	P. glabrum	9
	P. lividum	9
	P. sclerotiorum	9
	P. spinulosum	9
	P. thomii	9
Implicata	*P. bilaii*	9
	P. chermesinum	9
	P. implicatum	9
	P. montanense	9
	P. quercetorum	9
Section *Exilicaulis*		
Restricta	*P. capsulatum*	9
	P. restrictum	9
Citreonigra	*P. citreonigrum*	9
	P. decumbens	9
	P. sublateritium	9

Table 4 Continued

Subgenus *Furcatum*		
Section *Divaricatum*		
Janthinella	*P. janthinellum*	9
	P. ochrochloron	9
Canescentia	*P. canescens*	9
	P. daleae	9
	P. janczewskii	9
	P. melinii	9
Fellutana	*P. fellutanum*	9
	P. waksmanii	9
Section *Furcatum*		
Oxalica	*P. oxalicum*	9
	P. raistrickii	9
	P. rolfsii	9
	P. simplicissimum	9
Citrina	*P. citrinum*	9
	P. corylophilum	9
	P. herquei	9
	P. miczynskii (4 strains)	9
	(1 strain)	8(29%)+9(71%)
	P. novae-zeelandiae	9
Megaspora	*P. megasporum*	10
	P. asperosporum	10(H_2)
Subgenus *Penicillium*		
Section *Cylindrosporum*		
Italica	*P. digitatum*	9
	P. italicum	9
Section *Inordinate*		
Arenicola	*P. arenicola*	9
Section *Coronatum*		
Olsonii	*P. olsonii*	9
Section *Penicillium*		
Camembertii	*P. camembertii*	9
Expansa	*P. atramentosum*	9
	P. chrysogenum	9
	P. expansum	9
Viridicata	*P. aurantiogriseum*	9
	P. commune	9
	P. crustosum	9
	P. echinulatum	9
	P. hirsutum	9
	P. roquefortii	9
	P. solitum	9
	P. viridicatum	9

Table 4 Continued

Section *Penicillium*		
Urticicola	*P. brevicompactum*	9
	P. glandicola	9
	P. griseofulvum	9
	P. verrucosum	9
Subgenus *Biverticillium*		
Section *Coremigena*		
Duclauxii	*P. vulpinum*	9
	P. duclauxii (3 strains)	10(H_2)
	P. duclauxii (2 strains)	9
Section *Simplicia*		
Miniolutea	*P. diversum*	10(H_2)
	P. funiculosum	10(H_2)
	P. minioluteum	10(H_2)
	P. pinophilum	10(H_2)
	P. purpurogenum	10(H_2)
	P. verrucolosum	10(H_2)
Islandica	*P. brunneum*	10(H_2)
	P. islandicum	10(H_2)
	P. rugulosum	10(H_2)
	P. variabile	10(H_2)
Geosmithia		
	G. argillacea	10
	G. cylindrospora	10(H_2)
	(1 strain)	10(11%)+10(H_2)(89%)
	G. lavendula	10(H_2)
	G. namyslowskii	10(H_2)
	G. putterillii	10(H_2)
Eladia		
	E. saccula	9
Talaromyces		
Section *Talaromyces*		
Flavi	*T. flavus*	10(H_2)
	T. helicus	10(H_2)
	T. stipitatus	10(H_2)
	T. avellaneus	10
	T. striatus	10
Lutei	*T. assiutensis*	10(H_2)(75%)+10(H_4)(25%)
	T. luteus	10(H_2)
	T. rotundus	10(H_2)
	T. wortmannii	10(H_2)
Trachyspermi	*T. galapagensis*	10(H_4)
	T. gossypii	10(H_2)

Table 4 Continued

Section *Talaromyces*		
Trachyspermi	*T. intermedius*	10(H_2)
	T. mimosinus	10(H_2)
		10(H_2)
	T. ohiensis	
	T. trachyspermus	
	ATCC 52507	10(H_2)(48%)+10(H_4)(52%)
	IFO 6440	10(H_2)(67%)+10(H_4)(33%)
	IFO 8890	10(H_2)(62%)+10(H_4)(38%)
	IFO 9861	10(H_2)(35%)+10(H_4)(65%)
	IFO 30066	10(H_2)(71%)+10(H_4)(29%)
	IFO 31360	10(H_2)(19%)+10(H_4)(81%)
	IFO 31757	10(H_4)
	IFO 31907	10(H_2)(40%)+10(H_4)(60%)
	Mean (SD)	44%(20.8)+56%(20.1)
Series unknown	*T. derxii*	10(H_2)(89%)+10(H_4)(11%)
Section *Purpureus*		
Purpurei	*T. purpureus*	10(H_2)
Section *Thermophilus*		
Thermophilii	*T. thermophilus*	10
Section *Emersonii*	*T. byssochlamydoides*	
	CBS 413.71	10(31%)+10(H_2)(69%)
	CBS 533.71	10(87%)+10(H_2)(13%)
	CBS 150.75	10(77%)+10(H_2)(23%)
	IAM 13445	10(86%)+10(H_2)(14%)
	IFO 31900	10(53%)+10(H_2)(47%)
	IFO 31069, IFO 31150	10(H_2)
	T. leycettanus	10
	T. bacillisporus	10(H_2)
	T. emersonii	
	IFO 9734	10(39%)+10(H_2)(61%)
	IFO 9860	10(12%)+10(H_2)(88%)
	IFO 31070, IFO 31127, IFO 31159, IFO 31232, IFO 31852, IFO 31901	10(H_2)
Pencilliopsis	*P. clavariiformis*	9
Trichocoma	*T. paradoxa*	
	IFO 6765	10(H_2)(67%)+10(H_4)(33%)
	IFO 9685	10(H_2)(69%)+10(H_4)(31%)
	IFO 30659	10(H_2)(68%)+10(H_4)(32%)
Dendrosphaera	*Penicillium* anamorph (*P. eberhardtii*)	9

Source: Ref. 9.

Table 5 Principal Ubiquinone Systems in *Aspergillus* Anamorphs and Teleomorphs

Subgenus *Aspergillus*	
Section *Aspergillus*	
Eurotium amstelodami	9
E. chevalieri	9
E. chavalieri var. *intermedium*	9
E. echinulatum	9
E. halophilicum	9
E. herbariorum	9
E. leucocarpum	9
E. pseudoglaucum	9
E. repens	9
E. rubrum	9
E. tonophilum	9
E. umbrosum	9
E. xerophilum	9
Edyuillia athecia	9
Aspergillus proliferans	9
Section *Restrici*	
A. caesiellus	9
A. conicus	9
A. gracilis	9
A. penicilloides	9
1 strain	8(49%) 9(51%)
A. restrictus	9
Subgenus *Fumigati*	
Section *Fumigati*	
Neosartorya aurata	10
N. aureola	10
N. fennelliae	10
N. fischeri	10
N. fischeri var. *glabra*	10
N. fischeri var. *spinosa*	10
N. fischeri var. *verrucosa*	10
N. quadricincta	10
N. stramenia	10
A. brevipes	10
A. fumigatus	10
A. unilateralis	10
Section *Cervini*	
A. cervinus	9
A. kanagawaensis	9
A. nutans	9
A. bisporus	$10(H_2)$

Table 5 Continued

Subgenus *Ornati*	
Hermicarpenteles acanthosporus	10
H. paradoxus	9
Sclerocleista ornatus	9
S. thaxteri	9
Warcupiella spinulosa	10
A. brunneo-uniseriatus	9
A. brunneo-uniseriatus var. *nanus*	10(H_2)
A. apicalis	10
A. ivoriensis	10(H_2)
A. raperi	10(H_2)
Subgenus *Clavati*, Section *Clavati*	
A. clavatus	10
A. clavato-nanica	10
A. giganteus	10
A. longivesica	9(49%)+10(46%)
Subgenus *Nidulantes*, Section *Nidulantes*	
Emericella aurantiobrunnea	10(64%)+10(H_2)(36%)
E. bicolor	10(H_2)
E. cleistominuta	10(H_2)
E. desertorum	10(H_2)
E. foveolata	10(H_2)
E. fruticulosa	10(52%)+10(H_2)(48%)
E. heterothallica	10(H_2)
E. navahoensis	10(H_2)
E. nidulans	10(H_2)
E. nidulans var. *acristata*	10(H_2)
E. nidulans var. *dentata*	10(H_2)
E. nidulans var. *echinulata*	10(H_2)
E. nidulans var. *lata*	10(H_2)
E. nivea	10(H_2)
E. parvathecia	10(85%)+10(H_2)(15%)
E. purpurea	10(H_2)
E. quadrilineata (1 strain)	10(34%)+10(H_2)(66%)
(4 strains)	10(H_2)
E. rugulosa (8 strains)	10(H_2)
(1 strain)	10(12%)+10(H_2)(88%)
E. spectabilis	10(H_2)
E. striata	10(H_2)
E. sublata	10(12%)+10(H_2)(88%)
E. unguis	10(H_2)
E. variecolor	10(H_2)
E. violacea	10(H_2)
A. multicolor	10(H_2)

Table 5 Continued

Section *Versicolores*	
A. asperescens	$10(H_2)$
A. caespitosus	$10(H_2)$
A. janus	$10(H_2)$
A. silvaticus	$10(H_2)$
A. sydowi	$10(H_2)$
A. sydowii var. *achlamidosporus*	$10(H_2)$
A. sydowii var. *inaequalis*	$10(H_2)$
A. varians	$10(H_2)$
A. versicolor	$10(H_2)$
Section *Usti*	
A. deflectus	$10(H_2)$
A. puniceus	$10(H_2)$
A. ustus	$10(H_2)$
Section *Terrei*	
A. terreus	$10(H_2)$
A. terreus var. *africanus*	$10(H_2)$
A. terreus var. *aureus*	$10(H_2)$
Section *Flavipedes*	
Fennellia flavipes	$10(H_2)$
F. nivea	$10(H_2)$
Aspergillus carneus	$10(H_2)$
A. iizukae	$10(H_2)$
A. niveus	$10(H_2)$
Subgenus *Circumdata*	
Section *Wentii*	
A. terricola	$10(H_2)$
A. terricola var. *americana*	$10(H_2)$
A. thomii	$10(H_2)$
A. wentii	9
A. wentii var. *minimus*	9
Section *Flavi*	
A. avenaceus	10
A. flavus	$10(H_2)$
A. flavus var. *asper*	10
A. flavus var. *columnaris*	$10(H_2)$
A. oryzae	$10(H_2)$
A. parasiticus	$10(H_2)$
A. sojae	$10(H_2)$
A. tamarii	$10(H_2)$
A. toxicarius	$10(H_2)$
A. zonatus	$10(H_2)$

Table 5 Continued

Section *Nigri*	
A. aculeatus	9
A. awamori	9
A. awamori var. *fumeus*	9
A. awamori var. *fuscus*	9
A. awamori var. *minimus*	9
A. awamori var. *piceus*	9
A. carbonarius	9
A. ficuum	9
A. foetidus	9
A. helicothrix	9
A. heteromorphus	9
A. japonicus	9
A. japonicus var. *aculeatus*	9
A. niger	9
A. niger var. *awamori*	9
A. niger var. *phoenicis*	9
A. phoenicis	9
A. pulverulentus	9
A. saitoi	9
A. saitoi var. *kagoshimaensis*	9
A. usamii	9
Section *Circumdati*	
Petromyces alliaceus	10
A. dimorphicus	9
A. lanosus	10
A. auricomus	10(H_2)
A. bridgeri	10(H_2)
A. campestris	10(H_2)
A. elegans	10(H_2)
A. insulicola	10(H_2)
A. melleus	10(H_2)
A. ochraceus	10(H_2)
A. ochraceus var. *microsporus*	10(H_2)
A. ochraceoroseus	10(H_2)
A. ostianus	10(H_2)
A. petrakii	10(H_2)
A. robustus	10(H_2)
A. sclerotiorum	10(H_2)
A. sulphureus	10(H_2)
Section *Candidi*	
A. candidus	9

Table 5 Continued

Section *Cremei*	
Chaetosartorya chrysella	9
C. cremea	9
C. stromatoides	9
A. itaconicus	9
Section *Sparsi*	
A. gorakhpurensis	9
A. biplanus	$10(H_2)$
A. diversus	$10(H_2)$
A. funiculosus	$10(H_2)$
A. sparsus	$10(H_2)$

Source: Ref. 8.

vide the two sections into three clusters (30). The first is divided into seven subclusters and the second into two subclusters, so there are considerable electrophoretic differences despite them all having the Q9 system. *Neosartorya* and *Aspergillus* had Q10 as the major compound in subgenus *Fumigati* sect. *Fumigati*. Ubiquinone and electrophoresis analysis was applied to clinical isolates of *A. fumigatus* and some other *Aspergillus* species (31). Strains from clinical and nonclinical sources within the same species were considered to have identical ubiquinones and similar electrophoresis patterns.

Three of four species in sect. *Cervini* had Q9 as the dominant structure, whereas the other had $Q10(H_2)$ (8); however, more strains need to be examined before a definitive statement can be made. The authors proposed two groups for subgenus *Ornati* with Q9 or Q10 the major structure, and to exclude those with $Q10(H_2)$, uniseriate conidial heads, and Hülle cells. However, an electrophoretic study of five enzymes (glucose-6-phosphate dehydrogenase, malate dehydrogenase, fumarase, alcohol dehydrogenase, and glutamate dehydrogenase) for *Ornati*, including sect. *Cremei* (all Q9), was undertaken (32). The strains were divided into five groups, although similarities were low for some subclusters. There was little evidence to suggest splitting *Ornati* into groups on the basis of the ubiquinone data, but there were indications of more than two groups. Eleven isolates of subgenus *Clavati* were examined (8), and 10 possessed Q10, but one of *A. longivesica* also had approximately 50% of Q9. Four of the five sections of subgenus *Nidulantes* only had the $Q10(H_2)$ system (i.e., *Versicolores*, *Usti*, *Terrei*, and *Flavipedes*), but five *Nidulantes* isolates also had high Q10 levels.

Nine isolates in subgenus *Circumdati* sect. *Wentii* had the Q9 system, but single strains from two species had $Q10(H_2)$. $Q10(H_2)$ was usually found in sect. *Flavi*, but Q10 was detected in *A. avenaceus* and *A. flavus* var. *asper*. The authors conclude the Q10 species are probably incorrectly classified in sect. *Flavi*. In relation to this, 41 strains from sect. *Flavi* were examined by electrophoresis of eight enzymes (6-phosophogluconate dehydrogenase, malate dehydrogenase, phosphoglucomutase, glucose-6-phosphate dehydrogenase, fumarase, alcohol dehydrogenase, lactate dehydrogenase, and glutamate dehydrogenase), and the results compared to the major ubiquinone of a representative 27 strains from the same section (33). As before, *A. avenaceus* and *A. flavus* var. *asper* were classified as Q10, as was *A. leporis*. However, the electrophoretic analysis indicated that *A. flavus* var. *asper* was typical of the other Q10(H_2) strains, although the Q10 species were different. So, although *A. flavus* var. *asper* is similar morphologically and electrophoretically to the *Flavi* section, its ubiquinone system is considered different by these authors. Subsect. *Circumdati* can be divided into two subgroups I and II consisting of the teleomorph *Petromyces* (Q10) and others (Q10(H_2)) (8). The authors summarize by stating species from subgenera *Aspergillus*, *Fumigati*, *Ornati*, and *Clavati* produce Q9 or 10, although a few produce Q10(H_2). Species from subgenus *Nidulantes* produce Q10(H_2), and the ubiquinone systems in *Circumdati* were more complex than those from other subgenera. Representative species of the various subgenus sections have the following ubiquinone systems (30): *Eurotium repens* (9); *Aspergillus fumigatus* (10); *Sclerocleista ornata* (9); *Hemicarpenteles paradoxus* (9); *Warcupiella spinulosa* (10); *Aspergillus raperi* (10(H_2)); *Emercella nidulans* (10(H_2)); *A. flavus* (10(H_2)); *Chaetosartorya cremea* (9).

c. Coccidioides. The work of Fukushima et al. (34) is interesting as data are presented on the analysis of 11 strains of *Coccidioides immitis* which give an insight into the degree of variation within strains of the same species with implications for the use of cut-off values (see also *T. trachysperma* in Ref. 9). The means and standard deviations are: Q10 88% (2.7), Q9 11% (3.8), and Q8 1% (1.9). Obviously if values lower than 10% were ignored as in Reference 9, some strains would not be considered to have Q9 whereas others would. However, the claim that the method could be used to identify the fungus is unrealistic, as the data are very similar to those for other fungi.

d. Sporothrix. Suzuki and Nakase (10) state the ubiquinones of *Sporothrix* are heterogeneous. However, only 13 strains of *Sporothrix* and one of *Stephanoascus* were examined; more need to be analyzed to support some of the conclusions made. The authors used molar ratios to determine the relative amounts of ubiquinones, which is more appropriate than simply using peak areas, as discussed previously. The range of structures observed was Q8, 9, 9(H_2), 10, and 10(H_2).

e. Paecilomyces. The ubiquinones in *Paecilomyces* and related genera have been reported (35). However, the small publication was in Japanese, and a translation was not available to the present reviewer for comment.

f. More than One Genus. *Penicillium*, *Eupenicillium*, *Geotrichum*, and *Sporothrix* were analyzed by Kreisel and Schubert (36). Q9 was considered the main structure in all strains except *Sporothrix nivea*, in which Q10(H_2) was detected (10). The ubiquinone systems of the causative pathogens of a serious deep mycosis, *Paracoccidioides brasiliensis*, and *Blastomyces dermatitidis* have been examined (13). These dimorphic fungi are similar except for the budding processes. The authors determined Q10 was the major component with minor amounts of Q9, so the ubiquinones were also very similar. However, there may be significant quantitative differences between the two sets of data.

3. Urediniomycetes *and* Ustilaginomycetes

The ubiquinone system of smut and rust fungi as determined by the methods described previously have been reported (11). *Sporosporium*, *Sphacelotheca*, *Tolyposporium*, and *Ustilago* in the Ustilaginaceae; *Tilletia* in the Tilletiaceae; and *Graphiola* in the Graphiolaceae had Q10. Q9 was present but in either "trace or small" amounts (up to 12%). Q9 was prominent in *Pucciniastrum*, *Melampsora*, *Gymnosporangium*, and *Puccinia*, but these also had Q8 and 10 in amounts which gave smaller HPLC peaks.

4. Ascomycotina

It was encouraging to note a direct extraction method for *Corollospora*, and that major and minor components were considered (21). However, only two strains of each of only 13 species were examined, so the degree of variation cannot be determined. *C. angusta*, *C. cinnamomea*, *C. colossa*, *C. filiformis*, *C. fisca*, *C. gracilis*, *C. intermedia*, *C. luteola*, *C. maritima*, *C. pseudopulchella*, and *C. quinqueseptata* had Q10(H_2) as the major structure. *C. lacera*, and *C. pulchella* had Q10(H_4) although the former also had a substantial proportion of Q10(H_2) (34%).

5. Mastigomycotina

Ubiquinone systems in the Oomycetes were examined for the first time (18), although the methods were also based on Yamada and Kondo (17). All strains were of Q9, and *Saprolegnia ferax* and *S. hypogyna* contained traces of Q6 and 8, respectively.

III. STEROIDS

Steroids have not been studied adequately as taxonomic characters. However, they have been advocated as an area for taxonomic research in zoosporic fungi (37). Steroids are important in fungi, as with other eukaryotes, because they interact with lipid acyl chains by condensing and strengthening the lipid-lipid bilayers which form all membranes. A generalized clinical structure of sterols is provided

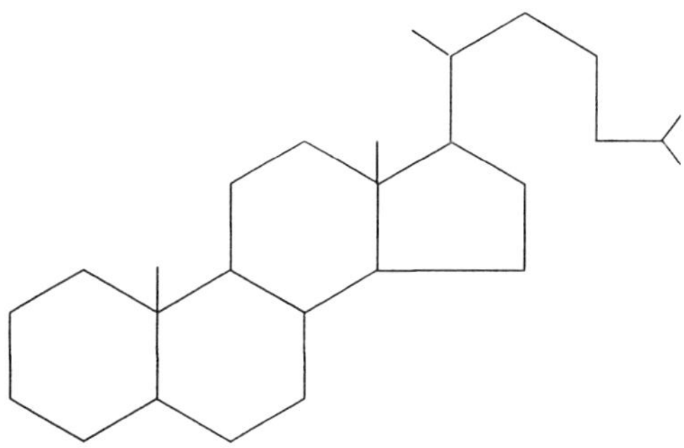

Figure 2 Generalized structure of sterols.

in Figure 2. Generally, "true" fungi possess 24-methyl sterols such as ergosterol (Fig. 3), although some may have ethyl sterols of the stigmastine series, and occasionally 24-desalkyl ones of the cholestane series. In contrast, the Oomycetes and Hyphochytriomycetes contain cholesterol and 24-alkylidene sterols (predominantly ucosterol). However, sterols were not detected from some Oomycetes (e.g., *Lagenidium giganteum*) and members of the Peronosporales, which depend on an exogenous supply for growth. The presence of cholesterol in the Phycomycetes (or Mastigomycotina and Zygomycotina) has been confirmed. Ergosterol has also been detected in a *Mucor* species (Zygomycotina), and in other Phycomycetes. Also, in one of two *Mucor* species cholesterol was not detected (1). So the situation with respect to the sterols of true fungi is somewhat confused from a taxonomic standpoint and may require additional studies.

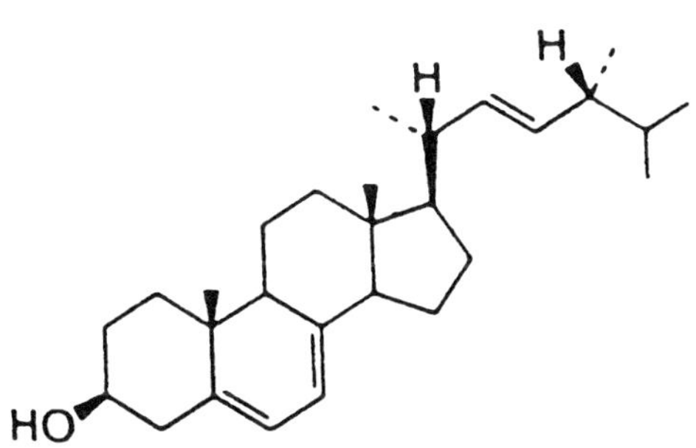

Ergosterol

Figure 3 Chemical structure of ergosterol.

It is affirmed adequately by Wassef (38) in a review of fungal lipids that "too few studies on sterol composition are available to provide adequate conclusions on the distribution within fungal taxa." C28 sterols are produced by most species, and ergosterol is common. Fungisterol is present in most fungi again usually with a high concentration of ergosterol. Most fungi have sterols unsaturated at the Δ^7 position, but some lower classes of Phycomycetes have C27, C28, and C29 sterols predominantly unsaturated at Δ^5. Unique C29 sterols are present in oils of spores from rust fungi. Some Oomycetes apparently do not produce sterols. Ergosterol is not produced by the aquatic Oomycetes or the rust fungi. Since there is little information on lichen fungi it may be worth adding that the isolated lichenized fungus *Xanthoria parientina* was found to contain C28 sterols such as ergosterol, and minor amounts of 24β sterols (39) (see also Table 6).

Lösel (40) tabulated quantitative and qualitative data obtained by other authors on the presence of sterols in fungi according to taxonomic considerations. A summary of this is presented in Table 6; the original can be referred to for complete information. The author asserts chemotaxonomic patterns emerge; however, in the present author's view, the use of data from a large number of studies by various authors can lead to spurious conclusions. For example, two different sterol profiles are listed for *Physarum polycephalum* and similarly for *Claviceps purpurae*. The review could form the basis of new work if this was thought to be desirable and a numerical analysis of the data presented in Reference 40 may be useful. Some of the problems of this area of research are supported by Lösel's conclusion:

> Any satisfaction gained from more complete records of the diversity of metabolites and the possibility of greater insight into the significance to chemotaxonomy in the biochemistry of fungi is tempered by the perception of new problems requiring greater refinement of data than previously available.

However, some interesting studies have emerged more recently.

The sterol composition of 14 species of dermatophyte and two nondermatophytes have been examined by gas chromatography and mass spectroscopy (GCMS) (41). Data were subjected to statistical analysis to determine chemotaxonomic relationships. Cholesterol, brassicasterol, ergosterol, fecosterol, campesterol, episterol, and sitosterol were detected, although campesterol and sitosterol were not found in *Trichophyton terrestre*, and campesterol was absent in *Microsporum andouinii* and *M. ferrugineum*. A dimethylzymosterol-like sterol was absent in *M. ferrugineum* and *T. verrucosum*, but it was present in the other fungi. Campesterol appeared to be strain-dependent. However, it was found in species from all three dermatophyte genera. The dermatophytes were not differentiated by this method, which reflects the close taxonomic relationships in this group. *Nattrassia mangiferae* (syn. *Hendersonula toruloides*) and *Scytalidium hyalinum* had similar sterols which also reinforces their similarity; however, *N. mangiferae*

Table 6 Sterol Profiles of Fungi[a]

	Sterol																							
	a	b	c	d	e	f	g	h	i	j	k	l	m	n	o	p	q	r	s	t	u	v	w	x
Myxomycota																								
Dictyostelium discoideum																						1		
Physarum polycephalum	1		1														1	1	1	1				
P. polycephalum	1					1											1			1				
P. flavicomum						1											1			1				
Labyrinthula minuta			1																					
Plasmodiophora brassicae			1		1													1	1					
	a	b	c	d	e	f	g	h	i	j	k	l	m	n	o	p	q	r	s	t	u	v	w	x
Eumycota																								
Mastigomycotina																								
Chytridiomycetes																								
Allomyces macrogynus			1		1													1						
Allomyces sp.			1		1																			
Blastocladia ramosa			1																					
Monoblepharella sp.			1		1													1						
Rhizophlyctis rosea			1																					
Hyphochytriomycetes																								
Hyphochytrium catenoides					1													1				1		
H. catenoides															1									
Rhizidiomyces apophysatus					1													1						
Rhizidiomyces sp.						1									1				1					
Oomycetes																								
Saprolegniales																								
	a	b	c	d	e	f	g	h	i	j	k	l	m	n	o	p	q	r	s	t	u	v	w	x
Achyla bisexualis			1										1								1			
A. caroliniana			1		1								1											
A. hypogyna			1	1									1						1		1			
with cycloartenol			1	1									1								1			
with lanosterol	1		1	1									1								1			
Aphanomyces astaci			1										1								1			
Aplanopsis terrestris			1										1						1		1			
A. terrestris			1										1								1			
Atkinsiella dubia			1	1									1								1			
Haliphthoros milfordense				1									1								1			
H. milfordense						1							1						1					
with cycloartenol						1							1						1					
with lanosterol	1			1		1																		
Geolegnia inflata																					1			
Leptolegnia caudata			1	1									1								1			
Pythiopsis cymosa			1	1																	1			

Table 6 Sterol Profiles of Fungi[a]

	Sterol																							
	a	b	c	d	e	f	g	h	i	j	k	l	m	n	o	p	q	r	s	t	u	v	w	x
Oomycetes (cont).																								
Saprolegniales (cont.)																								
Saprolegnia mygasperma			1	1									1								1			
S. ferex			1	1									1								1			
	a	b	c	d	e	f	g	h	i	j	k	l	m	n	o	p	q	r	s	t	u	v	w	x
Lagenidiales																								
Lagenidium giganteum																								
with cycloartenol			1																		1			
Leptomitales																								
Apodachlyella completa			1										1								1			
A. completa				1									1		1						1			
with cycloartenol			1			1							1						1		1			
with lanosterol	1		1										1						1					
A. brachynema			1	1									1								1			
A. minima			1	1									1								1			
Peronosporales																								
Zoophagus insidians	1									1					1									
Phytophthora cactorum																								
with cycloartenol	1		1												1				1					
with lanosterol	1		1			1									1									
with fucosterol			1	1			1														1			
Ph. cinnamomi																								
with cycloartenol	1		1	1																	1			
with lanosterol	1														1									
with fucosterol			1			1	1								1						1			
Ph. infestans																								
with cycloartenol	1		1																					
with lanosterol	1		1																					
with fucosterol			1	1			1														1			
Peronophythora litchii																								
with cycloartenol	1		1	1											1									
with lanosterol	1		1												1									
with fucosterol						1	1														1			
Pythium debaryanum																								
with cycloartenol	1		1	1																				
with lanosterol	1			1											1									
with fucosterol			1	1			1														1			

Table 6 Continued

	Sterol																							
	a	b	c	d	e	f	g	h	i	j	k	l	m	n	o	p	q	r	s	t	u	v	w	x
Zygomycotina																								
Zygomycetes																								
Absidia glauca																								
+												1			1									
−			1									1			1									
Linderina pennispora							1																	
Mucor hiemalis																								
+			1									1			1									
−			1									1			1									
M. dispersus												1			1									
Rhizopus arrhizus							1		1						1									
R. stolonifer												1			1									
Phycomyces blakesleeanus																								
+										1		1			1									
−												1			1									
Ascomycotina																								
Hemiascomycetes																								
	a	b	c	d	e	f	g	h	i	j	k	l	m	n	o	p	q	r	s	t	u	v	w	x
Protomyces											1													
Taphrina deformans											1													
Plectomycetes																								
Aspergillus niger spores															1									
A. flavus			1									1			1									
A. fennelliae			1												1									
A. fumigatus		1													1									
A. oryzae free sterols		1	1						1	1	1							1						
	a	b	c	d	e	f	g	h	i	j	k	l	m	n	o	p	q	r	s	t	u	v	w	x
sterol esters			1						1	1	1													
A. parasiticus			1									1			1									
A. nidulans	1														1									
Penicillium claviforme							1								1									
P. expansum					1		1								1									
Pyrenomycetes																								
	a	b	c	d	e	f	g	h	i	j	k	l	m	n	o	p	q	r	s	t	u	v	w	x
Claviceps purpurea							1		1						1									
C. purpurea							1		1					1										
Monilinia fructigena		1													1									
Nectria galligena			1						1						1	1		1						
Neurospora crassa			1						1						1									

Table 6 Continued

	Sterol																							
	a	b	c	d	e	f	g	h	i	j	k	l	m	n	o	p	q	r	s	t	u	v	w	x
Pyrenomycetes (cont.)																								
Discomycetes																								
Terfezia sp.											1				1									
Tuber brumale											1				1									
Tu. melanosporum											1				1									
(Lichens)																								
Xanthoria parietina																								
mycobiont total sterol										1	1			1	1	1								
Lobaria pulmonaria			1		1			1		1	1				1	1		1		1				
L. scrobiculata			1		1			1		1	1				1	1		1		1				
Usnea longissima			1		1			1		1	1				1	1		1		1				
Deuteromycotina																								
Alternaria alternata															1									
A. kikuchiana	1	1													1									
Fusarium oxysporum															1			1						
F. roseum											1				1									
Pullularia pullulans															1				1					
Spicaria elegans															1									
Trichophyton rubrum											1				1									
	a	b	c	d	e	f	g	h	i	j	k	l	m	n	o	p	q	r	s	t	u	v	w	x
Basidiomycotina																								
Hymenomycetes																								
Agaricus bisporus							1					1			1									
A. campestris	1	1								1					1									
Amanita caesarea												1			1									
Clitocybe illudens															1									
Coprinus atramentarius											1				1									
Flammulina velutipes									1			1			1	1								
Hygrocybe punica												1			1									
Lampteromyces japonicus												1			1									
Lentinus edodes												1			1									
Leucopaxillus giganteus												1			1									
Russula foetens												1			1									
R. nigricans												1			1									
R. senecis							1								1									
Coriolus heteromorphus	1	1							1			1												
C. pergamenus	1	1					1								1									
C. versicolor							1		1						1									
Fomitopsis cytisina		1							1						1									
F. pinicola	1	1					1								1									
Gloeophyllum saepiarum							1		1						1									

Table 6 Sterol Profiles of Fungi[a]

	Sterol																						
	a	b	c	d	e	f	g	h	i	j	k	l	m	n	o	p	q	r	s	t	u	v	w	x
Basidiomycotina (cont.)																								
Hymenomycetes (cont.)																								
Grifola frondosa	1	1					1								1									
Lenzites trabaea		1					1								1									
Microporus flabelliformis	1	1										1			1									
Piptoporus betulinus										1		1												
Schizophyllum commune															1									
Cryptoderma citrinum	1	1					1		1						1									
Daedalea quercina									1						1									
Gasteromycetes																								
Calvatia gigantea															1									
Scleroderma aurantium															1									
Ustilaginomycetes																								
Ustilago nuda							1								1									
U. zeae			1				1								1									
Urediniomycetes																								
Cronartium fusiforme							1																	
C. fusiforme							1																	
Puccinia graminis							1																	
Uromyces phaseoli																							1	1
	a	b	c	d	e	f	g	h	i	j	k	l	m	n	o	p	q	r	s	t	u	v	w	x

[a]This is an abbreviated version of the table in (40), where full details can be obtained.
Key: a = lanosterol, b = 24-methylenedihydrolanosterol, c = cholesterol, d = desmosterol, e = campesterol, f = ergost-5-enol, g = fungisterol, h = fecosterol, i = 5-dihydroergosterol, j = episterol, k = brassicasterol, l = 22-dihydroergosterol, m = 24-methylenecholesterol, n = ergosta-7,22,24(28)-trienol, o = ergosterol, p = lichesterol, q = clionasterol, r = β-sitosterol, s = stigmasterol, t = poriferasterol, u = fucosterol, v = stigmast-22-enol, w = stigmast-7-enol, x = stigmasta-7,24(28)-dienol. 1 = sterol detected.

form 3 had different sterols to forms 1 and 2. Ergosterol was the major lipid in all the species in a study of the Agaricales (42). Dihydroergosterol, fungisterol and lanosterol were also detected as minor components.

Sterols and fatty acids were included as taxonomic characters in 42 strains belonging to 16 species and 11 genera within the Phycomycetes, Ascomycetes, and Basidiomycetes (29) in an interesting and novel study. The sterols were detected in extracts prepared for fatty-acid analysis simply by extending the length of time of the gas chromatographic analysis. Profiles were evaluated by multivariate discriminant analysis. Only some genera could be separated on the fatty-acid profiles alone, but improvements were obtained when the sterols were included.

Differentiation of the S and P type "sibling" species of *Heterobasidium annosum* was possible, and ergosterol and ergosta-7,22-dien-3-ol were especially useful for these fungi.

This work was extended to the combined profiles of intersterility groups S, P, and F and used more strains, including a wider range from different geographical locations (43). Interesting data on the large variation of fatty acids with cultivation are presented, but this feature does not fall within the context of this chapter. Principal-component analysis did not cluster the intersterility groups separately, although outliers were observed. However, three nonoverlapping clusters were produced, each corresponding to an intersterility group, when the strains were combined, a priori, according to intersterility and geographical origin. Separation into geographical origin within a single sterility group was not observed. Among the sterols that gave the highest loading by discriminant analysis were ergostenol, ergostadienol, and ergosterol for cultures on modified orange-serum agar, and ergosterol for potato tomato soytone agar. Type S strains had higher levels of sterols (and fatty acids) than type P, and type F were characterized by high total amounts of extractives.

The above method was applied to 66, 17, and 13 strains of *Ascocoryne cylichium*, *Nectria fuckeliana*, and *Neobulgaria premnophilia*, respectively, obtained from 16 inoculated and four reference trees of Norway spruce (44). Preliminary identifications based on conventional methods were confirmed by discriminant analysis of combined fatty acids and sterols (ergostatrienol, ergostenon, and ergosterol). The sterols and fatty acids were important for effective separations; ergosterol was particularly useful. However, the intraspecies variation of fatty acids and sterols was high. A higher degree of variation was observed in strains from inoculated spruce compared to uninoculated trees.

Finally, ergosterol (Fig. 3) is increasingly being considered as a measure of fungal growth or content in solid substrates (45). It is generally considered to be unique to fungi and hence more valid than, for example, the chitin assay. Chitin forms the major part of the insect cuticle, so analysis for fungal chitin could be contaminated with that from insects. However, ergosterol constitutes part of the sterols of insects (46,47), which also could interfere. It is also useful in determining whether an unknown organism is of fungal origin, although not all fungi produce it (Table 6), and it is not unique to fungi, as mentioned previously. In the

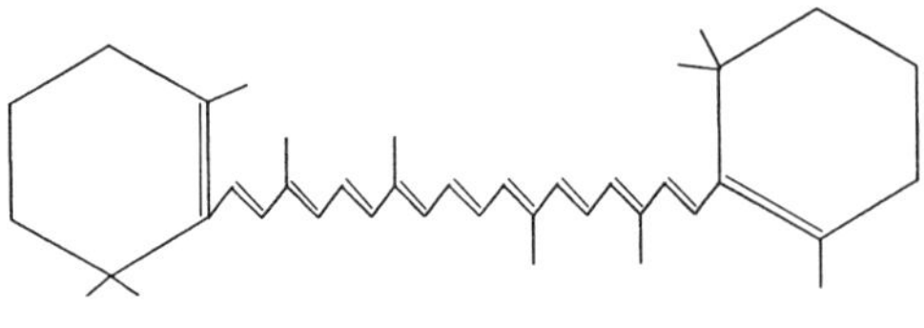

Figure 4 Chemical structure of β-carotene.

present reviewer's opinion ergosterol is still a useful indicator of both parameters. However, lipid analysis of the opportunistic pathogen *Pneumocystis carinii* was considered to provide further evidence of its being closely related to fungi despite ergosterol not being detected (48).

IV. CAROTENOIDS

Carotenoids are not as significant in fungi as they are in plants because fungi are obviously not photosynthetic. However, the mating cultures of *Blakeslea trispora* produce copious amounts of β-carotene which led to the discovery of the pheromones of trisporic acid and precursors (49). As many as 60% of all fungi examined contain them, and distribution appears capricious (50). The main fungal carotenoid is β-carotene (Fig. 4) and is present in all classes. However, it is absent in some Chytridiomycetes where only χ-carotene is present, and which is also widely distributed in fungi. As indicated above, species of the Zygomycetes order Mucorales have β-carotene as the predominant carotenoid, whereas species of Chytridiales and Blastocladiales have α-carotene. Members of the lower Ascomycetes *Protomyces* and *Taphrina* can be separated by the former containing β-carotene and the latter producing none although they have similar cultural characteristics. Also, the presence of carotenoids has been used in the recognition of a new family of Discomycetes—the Aleuriaceae (1). Xanthophylls have a wide distribution in higher fungi and become predominant in Discomycetes (e.g., Pezizales but not Helotiales) and some Hymenomycetes.

V. CONCLUSIONS

Fungi have more distinctive morphological characters than, for example, most bacteria, so careful consideration is necessary to decide whether a chemotaxonomic approach is required. However, there are areas where the conventional methods are inadequate and where chemotaxonomy can play a role (e.g., morphologically similar species with functional differences). There is a requirement for chemotaxonomic methods to be inexpensive and experimentally undemanding for the nonchemist/biochemist. Consideration also has to be given to the procedures being employed and their potential benefits and limitations. It is necessary to be critical of previous methods which may have been superseded by new approaches. There is a scientific requirement for a multidisciplinary approach, coupled with numerical analysis, to combine these different approaches (51).

At the risk of being overly pedantic, it may be worthwhile to mention the inherent variation involved in any form of measurement. This applies to lipids and other biomolecular analyses. There is little value in testing a fungal strain only once, yet this is the situation in some of the reports I have seen. The variation of the method should at first be assessed. I would suggest the following protocol. Assess the variation in:

1. The analytical method by using standard purified chemicals. This would involve extracting and analyzing the pure compounds in an identical, or as near identical, way as would be carried out with strains.

2. Five cultures of each of three strains with time (3 months). Freeze-dried ampules rather than cultures may be better for this purpose.

3. Different strains of the same species.

Ten strains of three species would be sufficient as an initial assessment. I recommend that the data obtained from these types of preliminary studies be included in subsequent publications. Of course, repeat analyses would still have to be included when the actual taxonomic study begins, and I suggest that each analysis be carried out in triplicate. The above are suggestions; protocols could be established by committees of the systematic societies. Essentially, this is simply giving sufficient thought to experimental design and will help to ensure valid taxonomic results. Time and effort are required to assess and control variation.

In the case of the ubiquinones, future studies must employ direct extraction methods and avoid saponification. Cut-off values have to be used with care, especially where complete profiles would be more appropriate. Also, greater consideration has to be given to the significance of different side-chain lengths: Is it really significant that some strains have predominantly Q10 while others have Q9?

A large-scale study on the sterols of certain fungi, which employs standard methods, would be interesting. The combined fatty-acid and sterol method described above would appear to have considerable potential and could be extended. It is not appropriate to consider other lipids at this stage when there are such significant gaps of knowledge in the cases of the ubiquinones and steroids. In future studies, much more thought has to be given to experimental design, including the inherent variation of the particular method, strain variation, and the overall validity of the technique. .

REFERENCES

1. Weete JD. Lipid Biochemistry of Fungi and Other Organisms. New York: Plenum Press, 1980:224–255.
2. Collins MD. Isoprenoid quinone analysis in bacterial classification and identification. In: Goodfellow M, Minnikin DE, eds. Chemical Methods in Bacterial Systematics. Society for Applied Microbiology. Technical Series 20. London: Academic Press, 1985:267–287.
3. Law AH, Threlfall DR, Whishtance GR. Isoprenoid phenol and quinone precursors of ubiquinones and dihydroubiquinones [ubiquinones (H_2)] in fungi. Biochem J 1971; 123:331–339.
4. Samson RS, Gams W. The taxonomic situation in the hyphomycete genera *Penicillium, Aspergillus*, and *Fusarium*. Antonie von Leeuwenhoek 1984; 50:815–824.

5. Lechavalier H, Lechavelier MP. Chemotaxonomic use of lipids—an overview. In: Ratledge C, Wilkinson SG, eds. Microbial Lipids. Vol 1. London: Academic Press, 1988:869–898.
6. Samson RA. Problems caused by new approaches in fungal taxonomy. Mycopathology 1991; 116:149–150.
7. Kuraishi H, Katayama-Fujimura U, Sugiyama J, Yokoyama T. Ubiquinone systems in fungi. I. Distribution of ubiquinones in the major families of Ascomycetes, Basidiomycetes, and Deuteromycetes, and their taxonomic implications. Trans Mycol Soc Jpn 1985; 26:383–395.
8. Kuraishi H, Itoh M, Tsusaki N, Katayama Y, Yokoyama T, Sugiyama J. The ubiquinone system as a taxonomic aid in *Aspergillus* and its teleomorphs. In: Samson RA, Pitt JI, eds. Modern Concepts in *Penicillium* and *Aspergillus* Classification. New York: Plenum Press, 1990:407–421.
9. Kuraishi A, Aoki M, Itoh M, Katayama Y, Sugiyama J, Pitt JI. Distribution of ubiquinones in *Penicillium* and related genera. Mycol Res 1991; 95:705–711.
10. Suzuki M, Nakase T. Heterogeniety of ubiquinone systems in the genus *Sporothrix*. J Gen Appl Microbiol 1986; 32:165–168.
11. Sugiyama J, Itoh M, Katayama Y, et al. Ubiquinone systems in fungi. II. Distribution of ubiquinones in smut and rust fungi. Mycologia 1988; 80:115–120.
12. Fukushima K, Takeo K, Takizawa K, Nishimura K, Miyaji M. Reevaluation of the teleomorph of the genus *Histoplasma* by ubiquinone systems. Mycopathology 1991; 116:151–154.
13. Fukushima K, Nishimura K, Takizawa K, et al. Ubiquinone systems of *Paracoccidioides brasiliensis* and *Blastomyces dermatitidis*. Jpn J Med Mycol 1991; 32:1–4.
14. Schubert M, Kreisel H. Ubiquinones in selected species of *Penicillium* and related teleomorph genera. Persoonia 1991; 14:341–346.
15. Boekhout T, Yamada Y, Weijman ACM, Roeymans HJ, Batenburg–van der Vegte WH. The significance of coenzyme Q, carbohydrate composition and septal ultrastructure for the taxonomy of ballistoconidia-forming yeasts and fungi. Syst Appl Microbiol 1992; 15:1–10.
16. Taylor JW, LoBuglio KF. Ascomycete phylogenetics: Morphology and molecules. Mycoscience 1994; 35:109–112.
17. Yamada Y, Kondo K. Coenzyme Q system in the classification of the yeast genera *Rhodotorula* and *Cryptococcus*, and the yeast-like genera *Sporobolomyces* and *Rhodosporidium*. J Gen Appl Microbiol 1973; 19:59–77.
18. Nakamura K, Yuasa K, Sinmuk S, Hatai K, Hara N. The ubiquinone system in Oomycetes. Mycoscience 1995; 36:121–123.
19. Mitchell K, Fallon RJ. The determination of ubiquinone profiles by reversed-phase high-performance thin-layer chromatography as an aid to the specification of Legionellaceae. J Gen Microbiol 1990; 136:2035–2041.
20. Fukushima K, Takizawa K, Okada K, Maebayashi Y, Nishaimura K, Kiyaji M. Suitability of sterilization methods for ubiquinones analysis of pathogenic fungi. Transact Mycol Soc Jpn 1993; 34:473–480.
21. Nakagiri A. Coenzyme Q systems in the genus *Corollospora* and allied marine fungi. Inst Ferment Res Commun (Osaka) 1991; 15:97–104.
22. Paterson RRM, Buddie A. Rapid determination of ubiquinone profiles in *Penicillium*

by reversed phase high performance thin-layer chromatography. Lett Appl Microbiol 1991; 13:133–136.

23. Hiraishi A, Shin YK, Sugiyama J. Rapid profiling of bacterial quinones by two-dimensional thin-layer chromatography. Lett Appl Microbiol 1992; 14:170–173.
24. Paterson RRM. Effects of growth on taxonomically-useful ubiquinone-lipid profiles from *Penicillium*. Mycol Res 1993; 97:173–178.
25. Boekhout T. A revision of ballistoconidia-forming yeasts and fungi. Stud Mycol 1991; 33, 1 Oct.
26. LoBuglio KF, Pitt JI, Taylor JW. Independent origins of the synnematous *Penicillium* species *P. duclauxii*, *P. clavigerum*, and *P. vulpinum*, as assessed by two ribosomal regions. Mycol Res 1994; 98:250–256.
27. Pitt JI. Chemotaxonomy of *Penicillium* and related teleomorphs. Jpn J Med Mycol 1991; 32(suppl 2):31–38.
28. Pitt JI. The Genus *Penicillium* and Its Teleomorphic States *Eupenicillium* and *Talaromyces*. London: Academic Press, 1979.
29. Müller MM, Kontola R, Kitunen V. Combining sterol and fatty acid profiles for the characterization of fungi. Mycol Res 1994; 98:593–603.
30. Sugiyama J, Rahayu ES, Chang J, Oyaizu H. Chemotaxonomy of *Aspergillus* and associated teleomorphs. Jpn J Med Mycol 1991; 32(suppl 2):39–60.
31. Matsuda H, Kohno S, Maesaki S, et al. Application of ubiquinone systems and electrophoresis comparisons of enzymes to identify clinical isolates of *Aspergillus fumigatus* and several other species of *Aspergillus*. J Clin Microbiol 1992; 30:1999–2005.
32. Sugiyama J, Yamatoya K. Electrophoretic comparison of enzymes as a chemotaxonomic aid among *Aspergillus* taxa: (1) *Aspergillus* sects, *Ornati* and *Cremei*. In: Samson RA, Pitt JI, eds. Modern Concepts in *Penicillium* and *Aspergillus* Classification. New York: Plenum Press, 1990:385–393.
33. Yamatoya K, Sugiyama J, Kuraishi H. Electrophoretic comparisons of enzymes as a chemotaxonomic aid among *Aspergillus* taxa: (2) *Aspergillus* sect. *flavi*. In: Samson RA, Pitt JI, eds. Modern Concepts in *Penicillium* and *Aspergillus* Classification. New York: Plenum Press, 1990:395–405.
34. Fukushima K, Takizawa K, Okada K, et al. Ubiquinone systems of *Coccidioides immitis*, the causative agent of coccidioidomycosis. FEMS Microbiol Lett 1993; 108:243–245.
35. Kuraishi H, Banno I, Sugiyama J, Samson RA. Distribution of ubiquinones in *Paecilomyces* and related genera. Proceedings of the 35th Anniversary Meeting of the Mycological Society of Japan 1991; Abstracts of submitted papers, 52.
36. Kreisel H, Schubert M. Ubiquinones in some filamentous fungi. Zentralbl Mikrobiol 1990; 145:91–94.
37. Warner SA, Domnas AJ. Biochemical characterization of zoosporic fungi: the utility of sterol metabolism as an indicator of taxonomic affinity. In: Fuller MS, Jaworski A, eds. Zoosporic Fungi in Teaching and Research. Georgia: Southeastern Publishing Corporation, 1987:202–208.
38. Wassef MK. Fungal lipids. In: Paoletti R, Kritchevsky D, eds. Advances in Lipid Research, Vol. 15. New York: Academic Press, 1977:159–231.
39. Nes WR. The biochemistry of plant sterols. In: Paoletti R, Kritchevsky D, eds. Advances in Lipid Research. Vol. 15. New York: Academic Press, 1977:233–324.

40. Lösel DM. Fungal lipids. In: Ratledge C, Wilkinson SG, eds. Microbial Lipids. London: Academic Press, 1988:699–806.
41. Howell SA, Moore MK, Mallet AI, Noble WC. Sterols of fungi responsible for superficial skin and nail infection. J Gen Microbiol 1990; 136:241–247.
42. Solberg Y. A literature review of the lipid constituents of higher fungi. New investigations of *Agaricales* species. Int J Mycol Lichenol 1989; 4:137–154.
43. Müller MM, Kantola R, Korhonen K, Uotila J. Combined fatty acid and sterol profiles of *Heterbasidion annosum* intersterility groups S, P and F. Mycol Res 1995; 99:1025–1033.
44. Müller MM, Hallaksela AM. Variation in combined fatty acid and sterol profiles of *Ascocoryne*, *Nectria*, and *Neobulgaria* strains isolated from Norway spruce. Eur J For Pathol 1994; 24:11–20.
45. Gao Y, Chen T, Breuil C. Ergosterol—a measure of fungal growth in wood for staining and pitch control fungi. Biotechnol Techniques 1993; 7:621–626.
46. Maurer P, Roger C, Mauchamp B, Porcheron P, Debieu D, Riba G. Occurrence of 28- and 27-carbon ecdysteroids and sterols in developing worker purpose of the leaf-cutting ant *Acromyrmex octospinosus* (Reich) (Hymenoptera, Formicidae: Attini). Arch Insect Biochem Physiol 1991; 16:1–9.
47. Maurer P, Debieu D, Malosse C, Leroux P, Riba G. Sterols and symbiosis in the leaf-cutting ant *Acromyrmex octospinosus* (Reich) (Hymenoptera, Formicidae: Attini). Arch Insect Biochem Physiol 1992; 20:13–21.
48. Furlong ST, Samia JA, Rose RM, Fishman JA. Phytosterols are present in *Pneumocystis carinii*. Antimicrob Agents Chemother 1994; 38:2534–2540.
49. Bu'Lock JD. Hormones in Fungi. In: Smith JE, Berry DR, eds. The Filamentous Fungi. Vol. 2. Biosynthesis and Metabolism. London: Edward Arnold, 1976:345–368.
50. Goodwin TW, Britton G. Distribution and analysis of carotenoids. In: Goodwin TW, ed. Plant Pigments. London: Academia Press, 1988:61–132.
51. Bridge PD, Hawksworth DL, Kozakiewicz Z, Onions AHS, Paterson RRM. A reappraisal of terverticillate penicillia using biochemical, physiological, and morphological features. Numerical taxonomy. J Gen Microbiol 1989; 135:2941–2966.

9

Fatty Acids in Fungal Taxonomy

J. L. F. Kock and A. Botha
University of the Orange Free State, Bloemfontein, South Africa

I. INTRODUCTION

Lipids are defined as being sparingly soluble in water but readily soluble in organic solvents such as chloroform, hydrocarbons, alcohols, ethers, and esters (1). Lipids can be loosely divided into two groups: the first contains compounds based on long-chain fatty acids (FAs), and the other is characterized by compounds derived from an isoprene unit including the terpenoid lipids. Since the aim of this chapter is to highlight the value of FA composition in fungal taxonomy, more emphasis will be placed on FA-based lipids.

The FA-based lipids include relatively simple compounds such as FAs as well as more complex molecules, including phospholipids (mainly associated with the cell membrane system), glyco and sphingolipids (associated with membranes and cell walls), and the triacylglycerols, which are usually present as lipid droplets in fungal cells. Fungi are known to possess both the ω3 and ω6 series of FAs (1). The ω3 polyunsaturated fatty acids (PUFAs) include α-linolenic acid (18:3 ω3) which in humans is transformed via eicosapentaenoic acid (20:5) to the 3 series eicosanoids (prostaglandins). In contrast, the ω6 series of PUFAs, including gamma linolenic acid (18:3 ω6), di homo gamma linolenic acid (20:3 ω6), and arachidonic acid (20:4 ω6), which occur in animals and some fungi and are eventually transformed to the 1, 2, and 3 series of eicosanoids in man. Both the ω3 and ω6 PUFA series are derived from linoleic acid (18:2 ω6) by the participation

---

This work is dedicated to Professor P. M. Lategan, former head and institutor of the Department of Microbiology at the University of the Orange Free State (South Africa), who initiated this lipid research program.

of different desaturase and elongase enzymes (2,3). Hydroxy FAs occur widely in nature as metabolic intermediates or secondary metabolites of FAs (4). The presence of these substances in fungi is well established (5,6), and the hydroxy FAs are produced via hydroxylation of PUFAs by lipoxygenase, cyclooxygenase, and P-450 pathways (7,8).

Various FAs including some of doubtful standing have been identified in fungi (9). The latter is ascribed to the fact that most authors working on fungal FAs used only the most rudimentary techniques. It is now generally accepted that FAs with chain lengths of C16 and C18 predominate in fungi while unsaturated FAs are normally present in the *cis* configuration. The most abundant FAs are 16:0 (palmitic acid), 16:1 (palmitoleic acid), 18:0 (stearic acid), 18:1 (oleic acid), 18:2 (linoleic acid), and 18:3 (linolenic acid). It is important to note that the FA composition of fungal cells is very susceptible to growth rate, culture age, oxygen availability, temperature, pH, and the composition of the growth medium; these should be taken into account when comparisons of FA compositions in fungi are made. For extensive reviews on this matter, the reader is referred to Erwin (10), Rattray et al. (11), and Rattray (6).

To combat fungi where necessary or to utilize them for the benefit of man, it is necessary to develop rapid and reliable techniques to identify these organisms. The classification of fungi is performed mainly according to morphological criteria. These techniques often pose problems, especially when closely related sporulating and nonsporulating fungi are studied. Consequently, several chemotaxonomic approaches have been attempted. Among these are proton magnetic resonance spectra of cell wall mannans, classification of isoprenoid quinones in the electron transport system, electrophoretic enzyme patterns, genome comparisons, DNA fingerprinting, and of course lipid profiling, which has shown much promise in yeast taxonomy. To avoid an exhaustive list of references the reader is referred to the following dissertations and theses: Augustyn (9), Botha (14), Coetzee (15), Cottrell (16,17), Jansen van Rensburg (18), Miller (19), Muller (20), M. S. Smit (2), E. J. Smit (21), Tredoux (22), Van der Berg (23), Van der Westhuizen (3), and Viljoen (24,25). These dissertations and theses are a compilation of work performed since 1982 in the laboratory of the first author. In this chapter significance of FA composition in the taxonomy of fungi will be reviewed. Special emphasis will be placed on the value of this phenotypic characteristic in fungal phylogeny and identification.

II. DISTRIBUTION OF FATTY ACIDS IN FUNGI

Interesting patterns are observed when the distribution of the predominant FA families i.e. ω3 and ω6 in fungi is compared to the phylogenetic scheme proposed for fungi by Kendrick (26) (Fig. 1). In this scheme, the Chytridiomycota, Hypho-

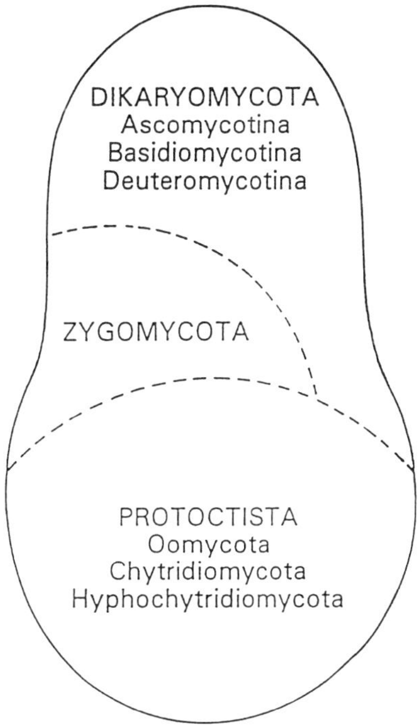

Figure 1 Schematic presentation of the higher classification of fungi proposed by Kendrick (26). *Source*: Van der Westhuizen (3).

chytridiomycota, and Oomycota (all members of the Protoctista) which produce flagellate zoospores are considered "ancestors" of the true fungi. From these fungi the mainly terrestrial true fungi were formed which comprise two phyla of which the representatives produce nonmotile cells (Zygomycota and Dikaryomycota). The Zygomycota seem to have a close affinity with the protoctistan fungi since they produce sporangia as well as coenocytic hyphae, while the Dikaryomycota do not produce these types of sporangia and have septate hyphae (26–28).

By comparing literature concerning the distribution of the ω3 and ω6 series of FAs in fungi, the following patterns can be observed (Table 1):

1. Fungi representing the Protoctista are in general characterized by the presence of the ω6 series of PUFAs comprising 18 (C18) and 20 carbons (C20).
2. Fungi of the Zygomycota also contain the ω6 series of PUFAs although most representatives analyzed only produce C18 PUFAs and not C20 PUFAs.
3. In general, representatives of the Dikaryomycota (i.e., Ascomycotina and Basidiomycotina) and affiliated anamorphs do not produce the ω6 series of

Table 1 Distribution of ω3 and ω6 Long-Chain Fatty Acids in Protoctistan and True Fungi

Unsaturated fatty acids		Omega 3 series (PUFAs)	Omega 6 series (PUFAs)	
18:1	18:2	18:3	C18	C20
	True fungi		Protoctista	
	Dikaryomycota		Chytridiomycota	
	Basidiomycotina			*Allomyces javanicus*
			Allomyces macrogynus	
	Phragmobasidiomycetes			*Blastocladia ramosa*
			Monoblepharis sp.	*Blastocladiella emersonii*
		Auricularia auriculajudae		
		Stilbum zacalloxanthium	*Rhizophlyctis rosea*	*Dermocystidium* sp.
	Homobasidiomycetes			*Phlyctochytrium punctatum*
	Agaricales			*Schizochytrium aggregatum*
				Synchytrium endobioticum
	Agaricus (2 species)	*Boletus edulis*		
	Amanita (2 species)	*Clitocybe tabescens*		*Thraustochytrium* (2 species)
	Armillaria mellea	*Collybla velutipes*		
	Tricholoma portentosum	*Coprinus comatus*		
		Exobasidium vexans		
		Hypholoma sublateritum	Hyphochytridiomycota	
		Lactarius (2 species)		
		Psathyrella candolleana		*Rhizidiomyces apophysatus*
		Suillus (2 species)		
		Tricholoma terreum		
	Polyporales		Oomycota	
Athelia bombycina	*Athelia* (2 species)	*Coriolus versicolor*		
	Cantherellus cibarius	*Corticium solani*		*Achlya americana*

Clavaria argillacea *Coriolus pergamenus* *Cryptoderma citrinum* *Hydnum rufescens* *Polyporus ramosissimus*	*Fomitopsis* (2 species) *Gloeophyllum saepiarium* *Grifola frondosa* *Hydnum repandum* *Microporus flabelliformis* *Piptoporus betulinus* *Polyporus* (2 species) *Ramaria flava* *Rhizoctonia* (2 species)		*Phytophthora* (3 species) *Pythium* (2 species) *Saprolegnia* (2 species)
		True fungi	
		Zygomycota	
Gasteromycetes	*Calvatia* (2 species) *Geastrum triplex* *Lycoperdon* sp.	*Basidiobolus* (3 species) *Blakeslea trispora* *Choanephora cucurbitarum* *Cunninghamella blakesleeana* *Helicostylum elegans* *Mucor* (4 species) *Phycomyces blakesleeanus* *Piptocephalis virginiana* *Rhizopus* (2 species) *Thamnidium elegans*	*Conidiobolus* (15 species) *Entomophthora* (12 species)
Ustilaginomycetes *Tilletia foetens* *Ustilago zeae*	*Tilletia* (2 species) *Ustilago scitaminea*		

Table 1 Continued

Unsaturated fatty acids		Omega 3 series (PUFs)	Omega 6 series (PUFAs)	
18:1	18:2	18:3	C18	C20
	Urediniomycetes			
		Cronartium (2 species)		
		Gymnosporangium clavipes		
		Melampsora lini		
		Puccinia graminis		
		Uromyces phaseoli		
	Ascomycotina			
	Hemiascomycetes			
		Taphrina deformans		
	Plectomycetes			
		Aspergillus niger		
		Erysiphe graminis		
		Penicillium (2 species)		
		Sphaerotheca humuli		
	Pyrenomycetes			
		Ceratocystis (2 species)		
		Chaetomium gibbosum		
		Claviceps (2 species)		
		Fusarium (2 species)		
		Myriococcum albomyces		
		Nectria ochroleuca		
		Neurospora crassa		
		Ophiostoma (8 species)		
		Thielavia thermophila		

	Discomycetes	
	Pyrtonema domesticum	*Botrytis* (2 species)
		Sclerotinia scleroliorum
	Loculoascomycetes	
		Alternaria dauci
		Leptosphaeria typhae
	Deuteromycotina	
	Dactylaria gallopava	*Beauveria bassiana*
	Epidermophyton floccosum	
		Hirsutella gigantea
		Isaria farinosa
		Metarrhizium anisopliae
	Microsporum (5 species)	*Microsporum canis*
	Scolecobasidium terreum	*Papulospora thermophila*
		Pullularia pullulans
		Sclerotium (2 species)
		Sporothrix schenckii
	Trichophyton (9 species)	*Trichoderma viride*
		Trichophyton (5 species)
	Yeasts	
	Ascomycotina	
		Ambrosiozyma monospora
		Clavispora lusitaniae
Arxiozyma telluris	*Botryoascus synnaedendrus*	*Dipodascopsis* (2 species)

Table 1 Continued

Unsaturated fatty acids		Omega 3 series (PUFs)	Omega 6 series (PUFAs)	
18:1	18:2	18:3	C18	C20
	Yeasts (cont.)			
	Ascomycotina (cont.)			
Hanseniaspora (2 species)	*Dipodascus* (8 species)	*Kluyveromyces* (2 species)		
Kluyveromyces (7 species)	*Endomyces fibuliger*	*Lipomyces* (4 species)		
Octosporomyces octosporus	*Nematospora coryli*	*Lodderomyces elongisporus*		
Pachytichospora transvaalensis	*Pichia* (6 species)	*Metschnikowia* (2 species)		
Saccharomyces (6 species)	*Torulaspora delbrueckii*	*Pichia* (75 species)		
Saccharomycodes ludwigii	*Zygosaccharomyces rouxii*	*Saccharomycopsis capsularis*		
Schizosaccharomyces (2 species)	*Zygozyma oligophaga*	*Schwanniomyces occidentalis*		
		Waltiozyma mucosa		
		Williopsis saturnus		
		Wingea robertsiae		
		Zygosaccharomyces bailii		
	Basidiomycotina			
		Cystofilobasidium (2 species)		
	Filobasidium floriforme	*Filobasidiella neoformans*		
		Filobasidium (3 species)		
		Rhodosporidium (7 species)		
	Deuteromycotina			
Candida (11 species)	*Candida* (10 species)	*Candida* (45 species)		
	Geotrichum (5 species)	*Cryptococcus* (16 species)		
	Myxozyma (2 species)	*Rhodotorula* (9 species)		
		Tremella (4 species)		

[a]Key: 18:1 = oleic acid; 18:2 = linoleic acid; 18:3 = linolenic acid; C18 = fatty acids with 18 carbons; C20 = fatty acids with 20 carbons; PUFAs = polyunsaturated fatty acids.

PUFAs. Some of these fungi are characterized by the presence of 18:3(ω3), and others are capable of producing FAs only up to 18:2(ω6).

4. Certain yeasts, which are considered reduced fungi (29), do not produce the ω3 and ω6 FAs. In fact, these fungi are incapable of producing unsaturated FAs greater than C18 monoenoic FAs.

With generalizations of this kind, exceptions to the rule are always evident. For instance, some protoctistan fungi (*Blastocladiella emersonii*, *Phlyctochytrium punctatum*, *Thraustochytrium aureum*, and *Thraustochytrium roseum*) also form 18:3(ω3) in combination with 18:3(ω6) (5). *Hyphochytrium catenoides* is reported to produce only up to 18:2(ω6) fatty acids (5). *Thamnidium elegans* of the Zygomycota produces 18:3(ω3) as well as 18:3(ω6) (30).

In many studies concerning the determination of FA profiles in fungi, different cultivation procedures were used, which according to our experience may have a significant influence on their relative FA compositions (31–33). Combining FA results from different sources in order to construct a database for identifying fungi is therefore risky, especially when relative FA compositions are used. The presence of the ω3 and ω6 series of PUFAs in fungi seems, however, to be conserved, and they are not influenced by different cultivation procedures (Table 1).

III. FATTY-ACID PROFILES AND IDENTIFICATION

A. Filamentous Fungi

The fatty acid composition of filamentous fungi is well established. For detailed reviews the reader is referred to Lösel (5) and Augustyn (9). Unfortunately, in some of these studies no standard cultivation procedures were used, and these are known to influence the FA profiles (relative amounts of FAs) present in fungi (31,34). In certain studies, fungi were cultivated in complex media known to contain FA-based lipids which can be incorporated into the lipids of fungi, thereby influencing their FA profiles (35). When utilizing FA profiles for identification purposes it is of crucial importance that reproducible profiles, which are produced de novo, are obtained from fungi before any identification is attempted. It is also essential that, when working within species consisting of large populations of strains, as many strains as possible from different locations be included to obtain an indication of the variation in FA profiles within species. Promising results regarding the use of FAs for identification of filamentous fungi have been reported. For instance, Blomquist et al. (36) found that with the aid of FA profiles it was possible to differentiate between various *Aspergillus*, *Mucor*, and *Penicillium* species, while Ferreira and Augustyn (37) reported on the differentiation of *Eutypa lata* and *Cryptovalsa cf ampelina* by means of FA analysis. Unfortunately, only a limited number of strains and species have been included in most of these studies, which makes the evaluation of this method as an identification parameter difficult. Van der Westhuizen (3) has attempted to evaluate FA profiles as an identification tool in fungi by cultivating these organisms under standard conditions in a defined medium containing mainly yeast nitrogen base (YNB) (38) and glucose. He started by determining the influence of culture age on FA composition of filamentous fungi to determine at what stage of growth the most reproducible results were obtained. For this purpose changes in cellular FA composition (FA of total lipids) during growth of *Conidiobolus coronatus*, *Rhizomucor pusillus* (Fig. 2), *Cunninghamella elegans*, *Microsporum canis*, *Phialophora verrucosa*, *Sporothrix schenkii*, and *Wangiella dermatitidis* were obtained by cultivating these organisms under similar conditions. In this study highly reproducible results were reported for the FA analyzed at different time intervals, and FA profiles reached stability during stationary growth phase for all seven species tested.

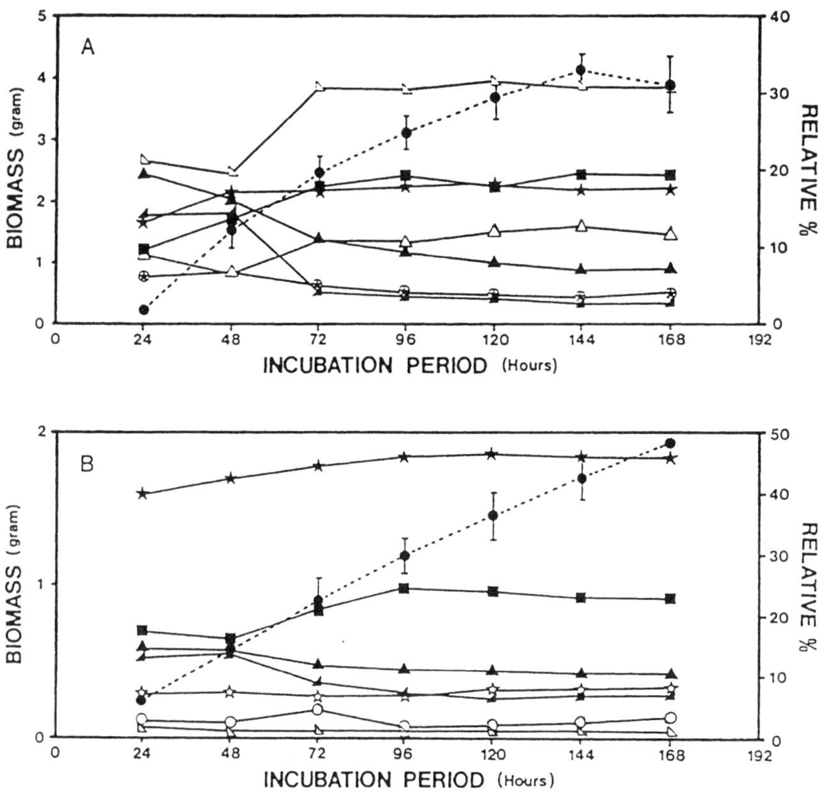

Figure 2 Changes in cellular long-chain fatty acid composition, with culture age, of (A) *Conidiobolus coronatus* CBS 176.55 and (B) *Rhizomucor pusillus* CBS 183.67. ● = biomass (g/400 ml medium); △ = tridecanoic acid; ◺ = myristic acid; ■ = palmitic acid; ☆ = palmitoleic acid; ○ = stearic acid; ★ = oleic acid; ▲ = linoleic acid; ◢ = linolenic acid; ⊛ = arachidonic acid. *Source*: Van der Westhuizen (3).

This stabilization of FA profiles, especially during the later phases of growth, may be due to the fact that FA synthesis as well as desaturation mechanisms decrease as the fungi enter stationary phase. Usually, in oleagenous fungi, most of the lipids are presented as neutral lipids at this stage (1). Having set growth conditions as well as culture growth period, Van der Westhuizen (3) determined the FA profiles of 266 fungi representing the Protoctista, Zygomycota, and Dikaryomycota (with affiliated anamorphs). Reproducible results were obtained (SE $< 5\%$) for all strains, tested and the following conclusions were drawn from this study:

1. Similar results were obtained concerning the distribution of the ω3 and ω6 series of FAs as reported in literature. Fungi of the Protoctista, i.e., *Allomyces*, are all characterized by the presence of the ω6 series of FAs and include C18 and C20 PUFAs. The zygomycete fungi studied also produced the ω6 series of FAs, although most representatives, excluding *Conidiobolus*, *Entomophaga*, *Entomophthora*, and *Mortierella*, only produced the C18 PUFAs (ω6) and not C20 PUFAs (ω6). Fungi representing the Ascomycotina and Basidiomycotina did not produce the ω6 series of PUFAs. Some are characterized by 18:3 (ω3); others can only produce FAs up to 18:2.
2. In general, some genera can be differentiated on the basis of the presence of ω3 and ω6 FAs. In the Zygomycota, *Basidiobolus* and *Cunninghamella* are differentiated by the presence of 18:3 (ω6) while *Absidia*, *Blakeslea*, *Choanephora*, *Gilbertella*, *Helicostylum*, *Mucor*, *Parasitella*, *Phycomyces*, *Pilaira*, *Piptocephalis*, *Rhizomucor*, *Rhizopus*, *Saksenaea*, *Syncephalastrum*, *Syncephalis*, *Thamnidium*, and *Zygorhynchus* are distinguished by the presence of 18:3 (ω6) as well as 20:0 and/or 20:1 FAs. The genera *Conidiobolus*, *Entomophaga*, *Entomophthora*, and *Mortierella* are unique in this group and produce 18:3 (ω6) as well as C20 PUFAs (ω6).

Genera representing the Deuteromycotina could be divided into groups on the basis of the presence of 18:2 and 18:3 (ω3). The hyphomycete genera *Cladophialophora*, *Cladosporium*, *Lecythophora*, and *Sporothrix* are all characterized by the presence of 18:2 and 18:3 (ω3) FAs. This group can be distinguished from the hyphomycete genera *Epidermophyton*, *Microsporum*, and *Trichophyton*, which are characterized by the presence of 18:2 and the absence of 18:3 (ω3). In this study, Van der Westhuizen concluded that further differentiation between species should only be attempted when more strains, preferably from different locations, are analyzed.

B. Yeasts

1. FA Composition of the Endomycetalean Families

An extensive survey on the cellular long-chain FA composition of yeasts in the Endomycetales, produced a scheme (Fig. 3) which compared the cellular FA composition and coenzyme Q systems of the endomycete families (39–58). From

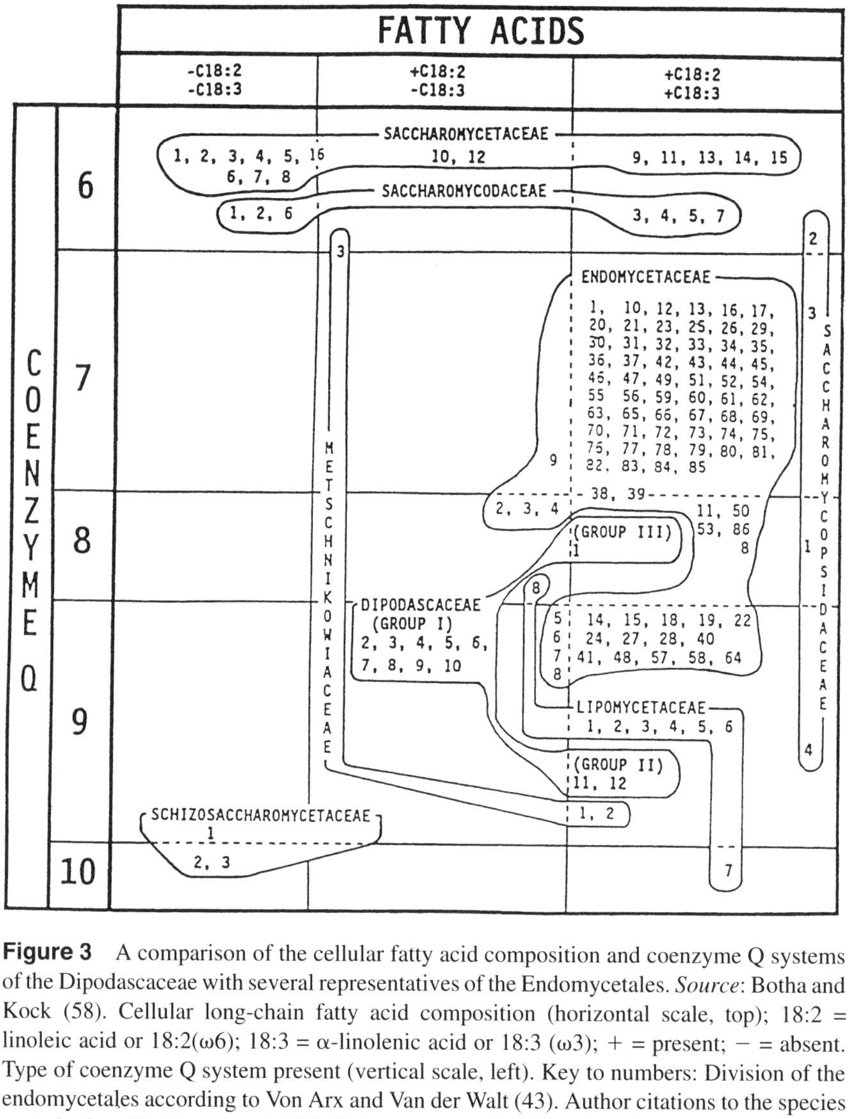

Figure 3 A comparison of the cellular fatty acid composition and coenzyme Q systems of the Dipodascaceae with several representatives of the Endomycetales. *Source*: Botha and Kock (58). Cellular long-chain fatty acid composition (horizontal scale, top); 18:2 = linoleic acid or 18:2(ω6); 18:3 = α-linolenic acid or 18:3 (ω3); + = present; − = absent. Type of coenzyme Q system present (vertical scale, left). Key to numbers: Division of the endomycetales according to Von Arx and Van der Walt (43). Author citations to the species are to be found in Barnett et al. (44). Endomycetaceae J. Schröt. 1 *Ambrosiozyma monospora*, 2 *Botryoascus synnaedendrus*, 3 *Endomyces fibuliger*, 4 *Pichia capsulata*, 5 *P. farinosa*, 6 *P. sorbitophila*, 7 *P. acaciae*, 8 *P. haplophila*, 9 *P. finlandica*, 10 *P. bovis*, 11 *P. burtonii*, 12 *P. henricii*, 13 *P. ciferrii*, 14 *P. philogaea*, 15 *P. media*, 16 *P. naganishii*, 17 *P. philodendra*, 18 *P. besseyi*, 19 *P. nakazawae*, 20 *P. subpelliculosa*, 21 *P. angusta*, 22 *P. ohmeri*, 23 *P. cactophila*, 24 *P. segobiensis*, 25 *P. onychis*, 26 *P. sydowiorum*, 27 *P. spartinae*, 28 *P. stipitis*, 29 *P. muscicola*, 30 *P. pijperi*, 31 *P. jadinii*, 32 *P. rabaulensis*, 33 *P. alni*, 34 *P. strasburgensis*, 35 *P. minuta*, 36 *P. canadensis*, 37 *P. wingei*, 38 *P. cellobiosa*,

these data (Fig. 3), it is obvious that the cellular long-chain FA compositions of the different families overlap. The presence of linoleic acid (18:2 ω6) and α-linolenic acid (18:3 ω3) can therefore not be used as a single criterion to delimit these families. The same is true of the terpenoid lipid, coenzyme Q. The Schizosaccharomycetaceae, with CoQ systems CoQ9 and CoQ10 and no 18:2 or 18:3, however, seem to occupy a rather isolated position. These yeasts are also unique since they are capable of producing extremely high relative amounts of oleic acid (18:1) (59). Discrepancies regarding cellular long-chain FA composition are also observed within genera (Fig. 3). The presence of 18:2 or 18:3 is therefore not necessarily correlated with a criterion such as ascospore morphology, considered a conserved character within the yeast genera (60).

2. *Cellular FA Profiles to Differentiate Yeast Taxa*

Over the past 7 years the cellular FA composition of yeasts has been repeatedly examined as a criterion to differentiate species. Van der Westhuizen et al. (33) examined the cellular long-chain FA composition of species of *Rhodosporidium*,

---

39 *P. methanolica*, 40 *P. castillae*, 41 *P. inositovora*, 42 *P. euphorbiae*, 43 *P. amethionina*, 44 *P. bimundalis*, 45 *P. deserticola*, 46 *P. fermentans*, 47 *P. euphorbiaphila*, 48 *P. mexicana*, 49 *P. wickerhamii*, 50 *P. holstii*, 51 *P. fabianii*, 52 *P. bispora*, 53 *P. heimii*, 54 *P. thermotolerans*, 55 *P. glucozyma*, 56 *P. petersonii*, 57 *P. triangularis*, 58 *P. guilliermondii*, 59 *P. populi*, 60 *P. rhodanensis*, 61 *P. salictaria*, 62 *P. anomala*, 63 *P. lynferdii*, 64 *P. scolyti*, 65 *P. mississippiensis*, 66 *P. quercuum*, 67 *P. opuntiae*, 68 *P. veronae*, 69 *P. nakasei*, 70 *P. trehalophila*, 71 *P. chambardii*, 72 *P. toletana*, 73 *P. kluyveri*, 74 *P. delftensis*, 75 *P. silvicola*, 76 *P. americana*, 77 *P. membranaefaciens*, 78 *P. norvegensis*, 79 *P. meyerae*, 80 *P. dryadoides*, 81 *P. heedii*, 82 *P. pini*, 83 *P. amylophila*, 84 *P. pseudocactophila*, 85 *P. kodamae*, 86 *Hyphopichia burtonii* (44–47). Dipodascaceae Gäumann. 1 *Clavispora lusitaniae*, 2 *Dipodascus aggregatus*, 3 *D. albidus*, 4 *D. ambrosiae*, 5 *D. australiensis*, 6 *D. magnusii*, 7 *Dipodascus ovetensis*, 8 *D. spicifer*, 9 *D. tetrasperma*, 10 *Galactomyces geotrichum*, 11 *Lodderomyces elongisporus*, 12 *Schwanniomyces occidentalis* (44,48). Lipomycetaceae Novak et Zsolt. 1 *Dipodascopsis uninucleata*, 2 *D. tothii*, 3 *Lipomyces anomalus*, 4 *L. kononenkoae*, 5 *L. starkeyi*, 6 *L. tetrasporus*, 7 *Waltomyces lipofer*, 8 *Zygozyma oligophaga* (43,49). Metschnikowiaceae Kamienski. 1 *Metschnikowia pulcherrima*, 2 *M. reukaufii*, 3 *Nematospora coryli* (44,50). Saccharomycodaceae Kudrjavzev. 1 *Hanseniaspora uvarum*, 2 *H. valbyensis*, 3 *Nadsonia commutata*, 4 *N. elongata*, 5 *N. fulvescens*, 6 *Saccharomycodes ludwigii*, 7 *Wickerhamia fluorescens* (44,50,51). Saccharomycetaceae Winter. 1 *Arxiozyma telluris*, 2 *Pachytichospora transvaalensis*, 3 *Saccharomyces castellii*, 4 *S. cerevisiae*, 5 *S. dairensis*, 6 *S. exiguus*, 7 *S. servazzii*, 8 *S. unisporus*, 9 *S. kluyveri*, 10 *Torulaspora delbreuckii*, 11 *Zygosaccharomyces bailii*, 12 *Z. rouxii*, 14 *Kluyveromyces marxianus*, 15 *K. thermotolerans*, 16 *K. africanus* (44,50,52–56). Saccharomycopsidaceae v. Arx et v.d. Walt. 1 *Saccharomycopsis capsularis*, 2 *Waltiozyma mucosa*, 3 *Williopsis saturnus*, 4 *Wingea robertsii* (44,50,56,57). Schizosaccharomycetaceae Beijerinck. 1 *Octosporomyces octosporus*, 2 *Schizosaccharomyces malidevorans*, 3 *S. pombe* (44,50).

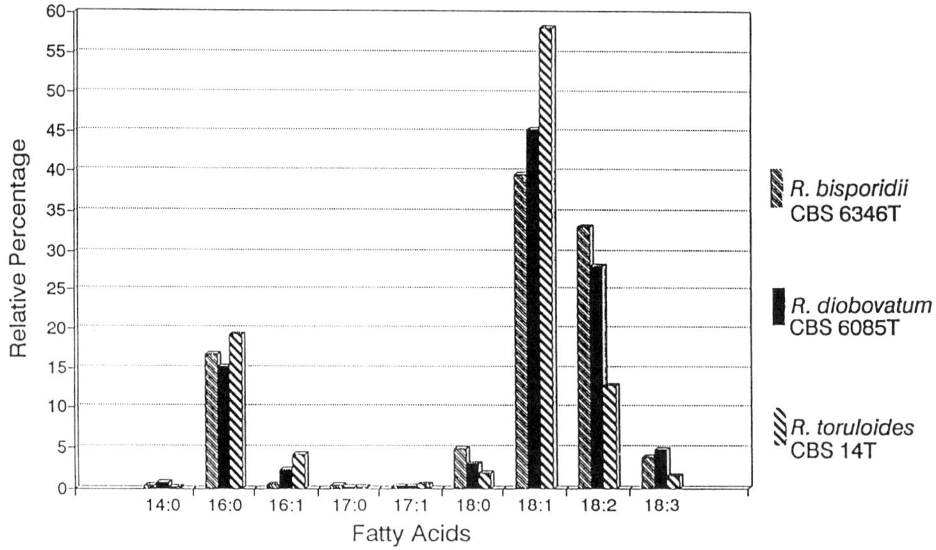

Figure 4 Variation of FA profiles within the genus *Rhodosporidium*. *Source*: Botha and Kock (61).

and suggested that FA profiles can be used for the rapid differentiation between species in this genus (Fig. 4) (61). Tredoux (62) obtained evidence that species of the genus *Saccharomyces* may be identified by their FA profiles. Successive studies on the cellular FA composition of yeasts from various taxa (47,49,50,63) have shown that many yeast strains show characteristic FA profiles within species.

This variation within species demonstrates that as many representative strains as possible must be examined to obtain a true representative FA profile of a particular yeast species. Nevertheless, FA profiles, in conjunction with other phenotypic characters, were used with success by various yeast taxonomists to differentiate a number of taxa. In 1987, the genus *Schizosaccharomyces* was subdivided with the proposal of a new genus, *Hasegawaea* (64). The latter genus differs from *Schizosaccharomyces* on account of its smooth ascospores, the absence of CoQ and the presence of linoleic acid (Fig. 5). Golubev et al. (51) revised the genus *Nadsonia* using cellular FA composition as one of the criteria to differentiate between *Nadsonia commutata* and *Nadsonia fulvescens* (Fig. 6).

3. Hydroxylated FAs and Yeast Taxonomy

Hydroxy FAs occur widely in nature as metabolic intermediates or secondary metabolites (4). The production of hydroxy FAs by fungi has been reported in various studies. For a review the reader is referred to Smit (2). Using radio-

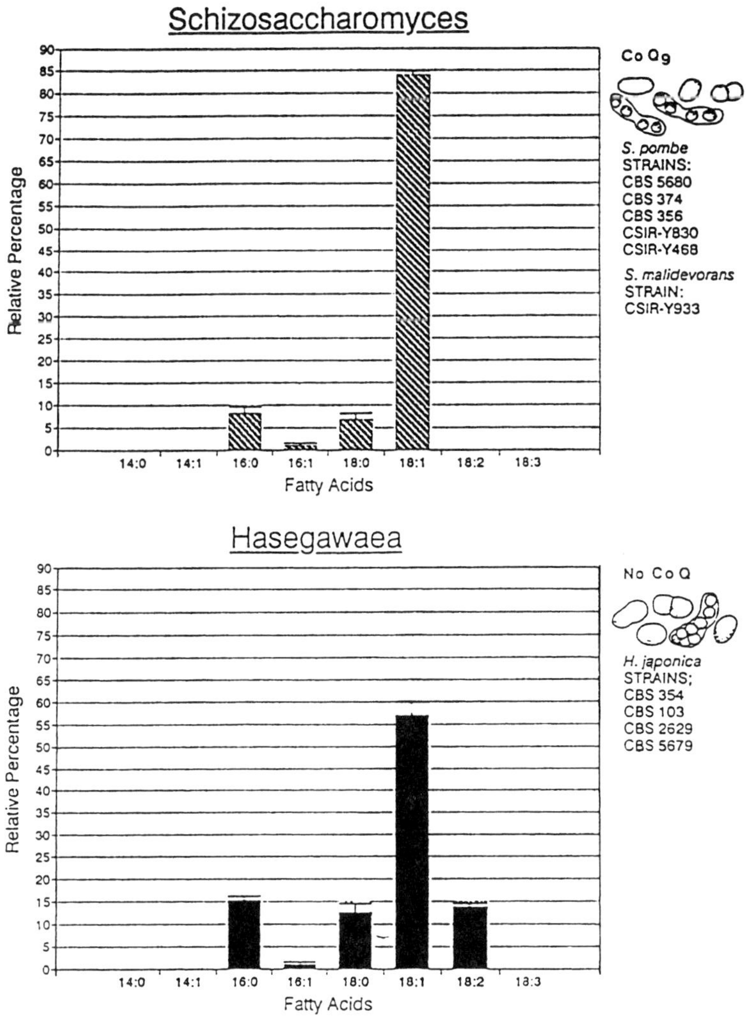

Figure 5 Difference between *Schizosaccharomyces* and *Hasegawaea*. The genus *Schizosaccharomyces* is characterized by a coenzyme Q9 system, rough and warty ascospores and the absence of linoleic acid (18:2). The genus *Hasegawaea* is characterized by the absence of a coenzyme Q system, smooth ascospores and the presence of linoleic acid (18:2). *Source*: Botha and Kock (61).

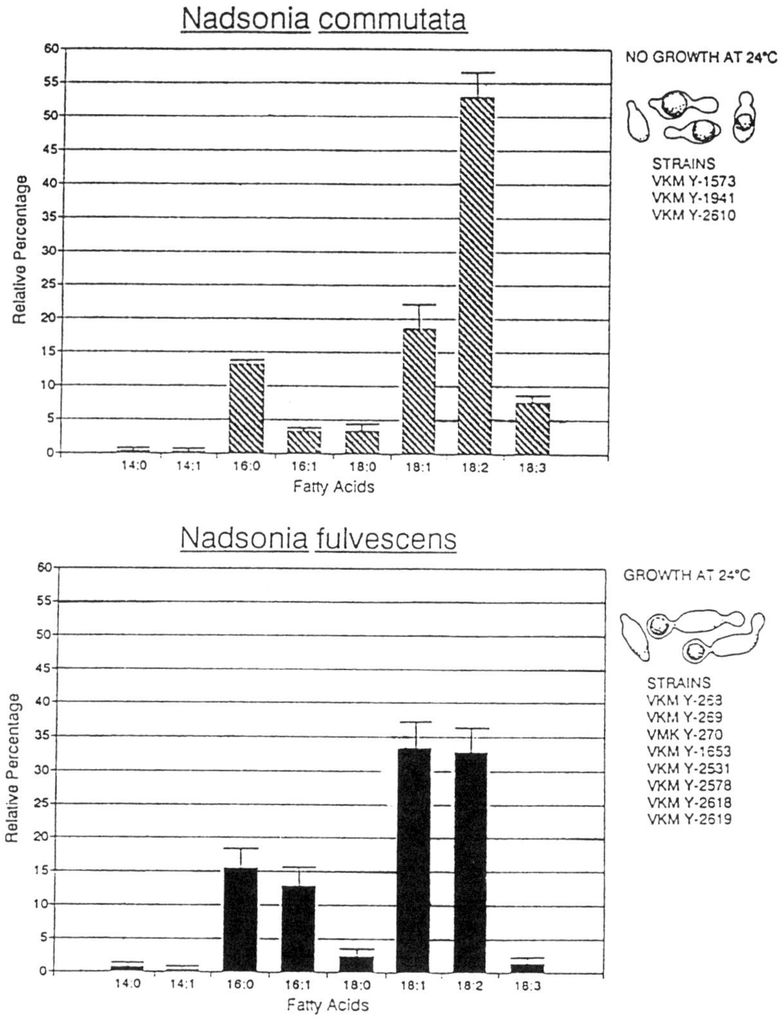

Figure 6 Differences between the FA profiles, maximum growth temperatures, and ascospore formation of *Nadsonia commutata* and *Nadsonia fulvescens*. The species *N. commutata* is characterized by the absence of growth at 24°C, a statistically higher mean percentage linoleic (18:2) and α-linolenic (18:3) acid, and a statistically lower mean percentage palmitoleic (16:1) and oleic (18:1) acid compared to *N. fulvescens*. During ascosporogenesis in *N. commutata* the mother cell becomes the ascus after conjugation with its bud, while in *N. fulvescens* a third meiotic cell develops into an ascus. In contrast to *N. commutata*, *N. fulvescens* is characterized by the ability to grow at 24°C. *Source*: Botha and Kock (61).

immunoassay, blood platelet aggregation studies, and gas-chromatography and mass spectrometry (GCMS), the existence of a certain group of hydroxy FAs, also named eicosanoids (prostaglandins), was indicated in the endomycete yeast *Dipodascopsis uninucleata* (8). Later a novel eicosanoid 3-hydroxy-eicosatetraenoic acid (3-HETE) was also discovered in this species (65). This substance proved to have biological activity on neutrophils, phospholipase A_2, and tumor cells. Furthermore, a chemical synthesis process has now been developed to synthesize 3-HETE (S. Nigam, personal communication). These findings triggered the search for similar eicosanoids in the rest of the Endomycetales.

In 1992, Kock and co-workers (58) continued their research by scanning for the easily detectable precursors of eicosanoids, linoleic and linolenic acid. Two families, both producing these precursors, were selected for further investigation (Lipomycetaceae and Dipodascaceae). Representative strains of the two families were tested for their ability to grow in the presence of aspirin (58), a specific inhibitor of prostaglandin biosynthesis. In contrast to the lipomycetaceous species, the dipodascaceous species were insensitive to this drug. These results were confirmed when representative strains of both families were investigated for their ability to produce eicosanoids from externally fed radiolabeled arachidonic acid along an aspirin sensitive pathway. Thin-layer chromatography of culture extracts, followed by autoradiography, showed that while none of the Dipodascaceae produced aspirin-sensitive arachidonic acid metabolites, the members of the Lipomycetaceae tested positive for these metabolites. The findings correlated with the delimitation of these yeasts in two families (Dipodascaceae and Lipomycetaceae) on the basis of ascospore morphology and the ability to produce extracellular amyloid material (43).

Botha et al. (63) reexamined the relationship between *Dipodascopsis* Batra et Millner, a previous member of the Dipodascaceae and the hyphomycetelike members of the Dipodascaceae by comparing a number of characteristics in a coordinate scheme (48,63,66) (Fig. 7). Using criteria such as the presence of arthroconidia, the presence of 18:3 and 3-HETE, the ability to grow in the presence of aspirin, and the ability to utilize a series of carbohydrate compounds as sole carbon sources, two separate groups could be recognized. Group I, represented by *Dipodascopsis tóthii* and *Dipodascopsis uninucleata*, was characterized by the absence of arthroconidia, the presence of 18:3, the inability to grow in the presence of aspirin and a superior ability to utilize the carbohydrates. Group II contained hyphomycetelike genera of the Dipodascaceae and the anamorphic genus *Geotrichum* and could be distinguished on the basis of the absence of 18:3, the ability to grow in aspirin, and the presence of arthroconidia. Botha et al. (63) pointed out that this grouping agrees with the analyses of aspirin-sensitive arachidonic acid metabolites in some hyphomycetelike genera. The representatives of group I tested positive for the aspirin-sensitive arachidonic acid metabolite 3-HETE, while the representatives of group II tested negative.

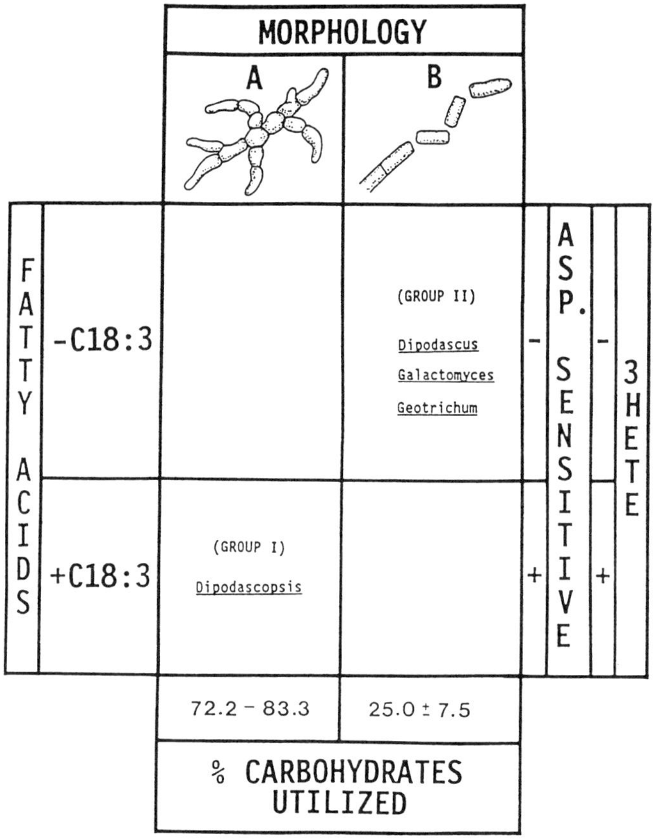

Figure 7 The relationship between the genus *Dipodascopsis* and the hyphomycetelike genera of the Dipodascaceae with their anamorphic genus *Geotrichum*. Criteria used: Morphology (horizontal scale, top). (A) = arthro-conidia absent; B = arthro-conidia present. Fatty acids (vertical scale, left); C18:3 = α-linolenic acid or C18:3 (ω3); + = present; − = absent. Aspirin sensitivity and production of 3-HETE (vertical scale, right). Percentage carbohydrates of a series which can be utilized as sole source of carbon, according to De Hoog et al. (48) and Kreger–van Rij (67) (horizontal scale bottom). This series include D-glucose, D-galactose, L-sorbose, D-ribose, D-xylose, L-arabinose, D-arabinose, L-rhamnose, sucrose, maltose, trehalose, cellobiose, melibiose, lactose, raffinose, melezitose, inulin, and starch. *Source*: Botha et al. (63).

These results were in agreement with studies on septal ultrastructure of *Dipodascopsis uninucleata* and *Dipodascus aggregatus* (66), where electron microscopy of the septa showed plasmodesmata in *D. aggregatus*, and a narrow central pore in *D. uninucleatus*. The results therefore agreed with the transfer of *Dipodascus uninucleatus* to the genus *Dipodascopsis* (67). It also correlated with the delimitation of these fungi in two families (Dipodascaceae and Lipomycetaceae) on the basis of ascospore morphology and ability to produce extracellular amyloid material (43).

C. Industrial Applications

1. Cellular FA Profiles of Saccharomyces *and Related Taxa*

Malfeito-Ferreira and co-workers (68) examined the cellular FA profiles of several strains of yeasts associated with wine spoilage. They used a modification of the protocol described by Kock et al. (42) in which they cultivated pure cultures on solid agar medium before saponification, methylation, and GC analysis of cellular FAs. They unexpectedly found similar results for yeast strains grown on a solid agar medium and the liquid medium described by Kock et al. (42). Utilizing principal-component analysis, Malfeito-Ferreira et al. (68) were subsequently able to rapidly differentiate between *Torulaspora delbreuckii* and *Zygosaccharomyces bailli*.

To obtain a more representative FA profile of *Saccharomyces cerevisiae* and other wine-associated yeast species, Augustyn and his co-workers (52,69–72) undertook an extensive survey of the cellular FA composition of 249 strains representing eight endomycete genera. Using capillary gaschromatography followed by mass spectroscopy, they proved that the minor FAs are useful to statistically differentiate 46 out of 50 strains representing *Saccharomyces cerevisiae* (Fig. 8). They also found that the range of FA profiles within *S. cerevisiae* renders this species indistinguishable from other members of *Saccharomyces* sensu stricto (*Saccharomyces bayanus* and *Saccharomyces pastorianus*). However, they could distinguish *Saccharomyces cerevisiae* from species representing *Saccharomyces* sensu lato (Fig. 9). In addition, they found that with the exception of one strain of *Pachytichospora transvaalensis*, they could distinguish *Saccharomyces* sensu stricto from 105 strains representing *Arxiozyma*, *Pachytichospora*, *Hanseniaspora*, *Saccharomycodes*, *Wickerhamiella*, *Kluyveromyces*, and *Torulaspora*. Augustyn et al. (71) were unable to differentiate among *Hanseniaspora guilliermondii*, *Hanseniaspora occidentalis*, *Hanseniaspora osmophila*, *Hanseniaspora uvarum*, *Hanseniaspora valbyensis*, and *Hanseniaspora viniae*. They also were unable to separate the following species from FA profiles: *Saccharomyces kluyveri*, *Kluyveromyces marxianus* var. *drosophilarum*, *Kluyveromyces waltii*, and *Kluyveromyces marxianus* var. *bulgaricus*.

These findings clearly indicate that cellular long-chain FA profiles used in

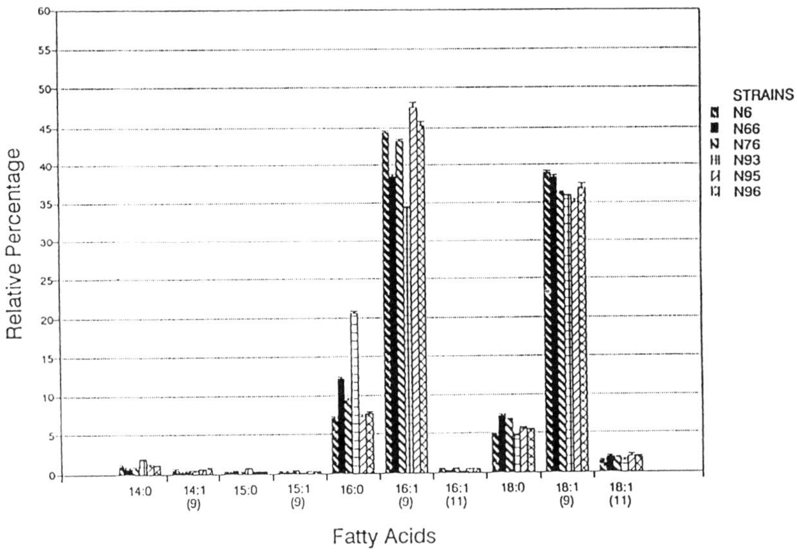

Figure 8 Differences in FA profiles between some strains of *Saccharomyces cerevisiae*. *Source*: Botha and Kock (61).

isolation are not a generally applicable identification technique. However, the technique can be applied to differentiate strains of *S. cerevisiae*. Analysis of FA profiles obtained from capillary gas chromatography is currently being used as a quick, easy, and cheap method to differentiate strains of *Saccharomyces cerevisiae* to help determine the causes of "stuck" fermentations in the South African food and beverage industry (61).

2. Cellular FA Profiles of Yeast Biomass in a Bioprotein Plant

A method has been developed to directly monitor the cellular long-chain FA composition of fungal biomass in a pilot plant aimed at producing bioprotein from *Geotrichum candidum* (61). Within 2 hours of sampling, fungal contaminants in the biomass could be highlighted by comparing the FA profile and cellular morphology of the sample to a standard set by industry. To develop this quality-control process, which utilizes cellular FA composition and morphology as criteria, Botha and Kock (61) first had to determine the following:

1. The potential fungal or yeast contaminants that could grow in the process
2. The cellular FA composition of the above mentioned contaminants when grown in the medium used for biomass production (this was prepared from an industrial effluent devoid of long-chain FAs)
3. The morphology of the contaminants in the medium

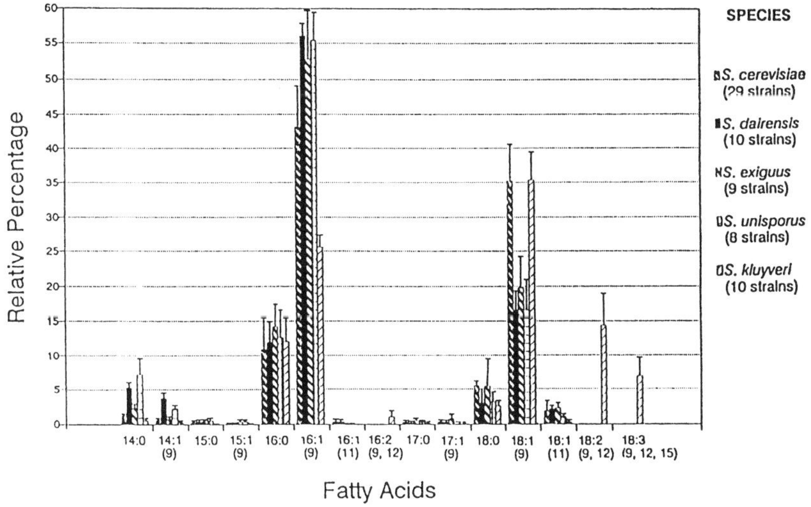

Figure 9 The FA profile of *Saccharomyces cerevisiae* compared to the FA profiles of some species in *Saccharomyces* sensu lato. The species *S. cerevisiae* could be distinguished from *S. dairensis*, *S. exiguus*, and *S. unisporus* on account of its statistically lower mean percentage myristic acid (14:0) and statistically higher mean percentage oleic acid 18:1(9). The species *S. cerevisiae* could also be distinguished from *S. kluyveri* on account of its statistically higher mean percentage palmitoleic acid 16:1(9) and stearic acid (18:0) and the presence of 16:1(11). Unlike *S. cerevisiae*, the species *S. kluyveri* is characterized by the presence of 16:2(9,12), 18:2(9,12), and 18:3(9,12,15). *Source*: Botha and Kock (61).

Two hundred sixty-three strains (representing 44 fungal genera) were examined for growth, cellular FA composition, and morphology in the above-mentioned medium. Botha and Kock (61) were able to differentiate *G. candidum* from potential fungal contaminants. Consequently, a statistical quality-control chart (Fig. 10) utilizing cellular FA composition and morphology as criteria, was created to successfully monitor the fungal contaminants in the bioprotein pilot plant (73).

IV. CONCLUSIONS

The FA composition of lipids in fungi is well established. The presence or absence of the ω3 and ω6 series of FAs and their relative amounts (of C16 and C18 FAs) are important in the identification of these organisms. The presence of the ω3 and ω6 series of FAs in fungi coincides to a large extent with the phylogenetic development of these organisms. The representatives of the ancestral Protoctista (fungi that produce cells or zoospores that swim by means of flagella) are in general characterized by the presence of the ω6 series of PUFAs comprising of

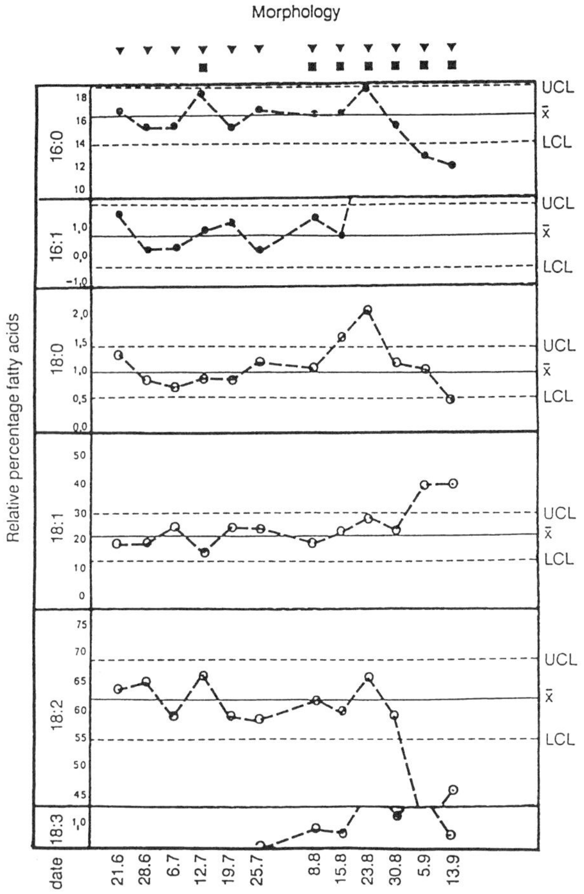

Figure 10 The statistical quality control chart used in the bioprotein plant. Vertical axis, left = relative percentage of different FAs; horizontal axis, bottom = date when the sample was taken; horizontal axis, top = cellular morphology of sample; ▼ = hyphae typical of *Geotrichum candidum*, ■ = budding yeast cells; vertical axis, right = upper control limits (UCL), lower control limits (LCL), and mean (x) for each FA. *Source*: Botha and Kock (61).

C18 and C20 PUFAs. The generally accepted transformation of these primitive fungi to the terrestrial true fungi (coenocytic Zygomycota and septate hyphal and yeastlike Dikaryomycota and associated anamorphs) coincided with a general loss in the C20 polyunsaturated ω6 FAs in the Zygomycota and the loss of C18 polyunsaturated ω6 FAs as well as the emergence of the ω3 (mainly C18 PUFAs) in the Dikaryomycota and associated anamorphs. Furthermore, the more "recently evolved" yeasts have lost the ability to produce dienoic fatty acids or PUFAs.

The desaturation of FAs is generally accepted to occur in the membranes of fungi and to be mediated by different desaturase enzymes; the desaturation is responsible in turn for the membrane structure and to some extent function. Since the desaturation of FAs coincides with the phylogenetic development of the major fungal groups, this function of the membrane seems to be highly conserved. This phenotypic characteristic is also very stable and seems not to change when different cultivation procedures are used which favor *de novo* synthesis of FAs. This generalization does not apply where lipids are incorporated into the growth medium. The presence of ω3 and/or ω6 FAs in the growth medium will influence the FA composition of fungal lipids due to the FA uptake and transesterification into the major lipid classes.

The value of FA profiles (C16 and C18 FAs) in the identification of fungi is at this stage unclear. Many reports have speculated on the value of FA profiles to differentiate between fungi. However, in many of these studies, no effort was made to analyze significant populations of a species represented by strains from various locations. Efforts should be made to accommodate this in the future. FA profiles in conjunction with other phenotypic characters have been used successfully by various yeast taxonomists to differentiate a number of taxa, two examples being *Schizosaccharomyces* and *Nadsonia*, as mentioned earlier.

Since FA profiles proved to be useful in the identification of certain fungal taxa, it is important that in the future more detailed lipid analyses are performed on these microorganisms. Such detailed analyses should include lipid separation into neutral, glycolipid, and phospholipid fractions. Each of these fractions can be further separated and the FA composition of each subfraction determined.

A database concerning the distribution of the ω3 and ω6 series of FAs as well as the relative amounts of FAs in fungi is not only important for identification purposes, but may have far reaching biotechnological applications. At present there are potentially two markets which fungal or single-cell oil (SCO) products can influence. These include markets for cocoa butter containing approx. 30% each of palmitic acid—16:0, stearic acid—18:0, oleic acid—18:1, and γ-linolenic acid (GLA or 18:3 ω6). After several screening attempts for fungi producing high 18:0 contents, it was decided to utilize the yeast *Cryptococcus curvatus* for cocoa butter equivalent production. Equal amounts (30%) each of 16:0, 18:0, and 18:1 were obtained in the triacylglycerol fraction of this organism by partial deletion of the delta 9 desaturase, which catalyzes the formation of 18:1 from 18:0. Consequently, a process for producing a fat equivalent to cocoa butter from yeasts with high 18:0 content is now feasible.

ACKNOWLEDGMENTS

We would like to thank Professor J.P. van der Walt for his advice and encouragement and all the postgraduate students and technical assistants involved in this project. We would also like to thank the Foundation for Research and Development, University of the Orange Free State (in particular, Professors Hans Potgieter and Chris Small, as well as the Sasol Centre for Biotechnology for sponsoring this project.

REFERENCES

1. Ratledge C, Wilkinson SG. An overview of microbial lipids. In: Ratledge C, Wilkinson SG, eds. Microbial Lipids. Vol. 1. London: Academic Press, 1988:3–22.
2. Smit MS. The turnover of exogenous arachidonic acid by the yeast *Dipodascopsis uninucleata*. Ph.D. Thesis. Department of Microbiology and Biochemistry, University of the Orange Free State, Bloemfontein, South Africa, 1993.
3. Van der Westhuizen JPJ. The distribution of the omega 3 and omega 6 series of cellular long-chain fatty acids in fungi associated with disease. Ph.D. Thesis. Department of Microbiology and Biochemistry, University of the Orange Free State, Bloemfontein, South Africa, 1994.
4. Schweizer E. Biosynthesis of fatty acids and related compounds. In: Ratledge C, Wilkinson SG, eds. Microbial Lipids. Vol. 2. San Diego: Academic Press, 1989:3–50.
5. Losel DM. Fungal lipids. In: Ratledge C, Wilkinson SG, eds. Microbial Lipids. Vol. 1. London: Academic Press, 1989:699–806.
6. Rattray JBM. Yeasts. In: Ratledge C, Wilkinson SG, eds. Microbial Lipids. Vol. 1. London: Academic Press, 1988:555–697.
7. Simmons CA, Kerwin JL, Washino RK. Preliminary characterization of lipoxygenase from the entomopathogenic fungus *Lagenidium giganteum*. In: Stumpf PK, Mudd JB, Nes WD, eds. The Metabolism, Structure, and Function of Plant Lipids. New York: Plenum Press, 1987:421.
8. Kock JLF, Coetzee DJ, Van Dyk MS, et al. Evidence for pharmacologically active prostaglandins in yeasts. S Afr J Sci 1991; 87:73–76.
9. Augustyn OPH. Capillary GC-MS fatty acid analysis and yeast identification. Ph.D. Thesis. Department of Microbiology and Biochemistry, University of the Orange Free State, Bloemfontein, South Africa, 1992.
10. Erwin JA. Comparative biochemistry of fatty acids in eukaryotic microorganisms. In: Erwin JA, ed. Lipids and Biomembranes of Eukaryotic Microorganisms. London: Academic Press, 1973:42.
11. Rattray JBM, Schibeci A, Kidby DK. Lipids of yeasts. Bacteriol Rev 1975; 39: 197–231.
12. Alexopoulos CJ. Introductory Mycology. New York: John Wiley and Sons, 1962:5.
13. Ratledge C. Single cell oils—have they a biotechnological future? TIBTECH 1993; 11:278–284.
14. Botha A. Eicosanoid production and nonsteroidal anti-inflammatory drug sensitivity in the yeast genus *Dipodascopsis*. Ph.D. Thesis. Department of Microbiology and Biochemistry, University of the Orange Free State, Bloemfontein, South Africa, 1993.
15. Coetzee DJ. Evidence for and taxonomic value of eicosanoids in the Lipomycetaceae. Ph.D. Thesis. Department of Microbiology and Biochemistry, University of the Orange Free State, Bloemfontein, South Africa, 1992.

16. Cottrell M. The value of cellular long-chain fatty-acid and organic acid composition in the taxonomy of the genera *Kluyveromyces* and *Saccharomyces*. M.Sc. Dissertation. Department of Microbiology and Biochemistry, University of the Orange Free State, Bloemfontein, South Africa, 1986.
17. Cottrell M. Cellular long-chain fatty acid composition, prostaglandins and electrophoretic karyotypes in the yeast family Lipomycetaceae. Ph.D. Thesis. Department of Microbiology and Biochemistry, University of the Orange Free State, Bloemfontein, South Africa, 1989.
18. Jansen van Rensburg EL. The isolation and characterisation of lipomycetaceous yeasts from South African soils. M.Sc. Dissertation. Department of Microbiology and Biochemistry, University of the Orange Free State, Bloemfontein, South Africa, 1992.
19. Miller M. The value of OFAGE, FAME and coenzyme Q in the taxonomy of the genus *Pichia* Hansen *emend.* Kurtzman. M.Sc. Dissertation. Department of Microbiology and Biochemistry, University of the Orange Free State, Bloemfontein, South Africa, 1989.
20. Muller HB. The cellular long-chain fatty acid composition of yeasts representing the genus *Pichia* Hansen *emend.* Kurtzman. M.Sc. Dissertation. Department of Microbiology and Biochemistry, University of the Orange Free State, Bloemfontein, South Africa, 1987.
21. Smit EJ. The value of FAME and PAGE in the taxonomy of *Cryptococcus* and associated yeasts. Ph.D. Thesis. Department of Microbiology and Biochemistry, University of the Orange Free State, Bloemfontein, South Africa, 1991.
22. Tredoux HG. Long-chain fatty acid compositions and volatile metabolite patterns of yeasts associated with wine. M.Sc. Dissertation. Department of Microbiology and Biochemistry, University of the Orange Free State, Bloemfontein, South Africa, 1987.
23. Van der Berg L. Evidence for the production of arachidonic acid metabolites by *Saccharomyces cerevisiae*. M.Sc. Dissertation. Department of Microbiology and Biochemistry, University of the Orange Free State, Bloemfontein, South Africa, 1993.
24. Viljoen BC. Long-chain fatty acid composition and gas pyrolysis of some species belonging to the Endomycetales. M.Sc. Dissertation. Department of Microbiology and Biochemistry, University of the Orange Free State, Bloemfontein, South Africa, 1986.
25. Viljoen BC. The value of cellular long-chain fatty acid composition and electrophoretic karyotypes in the taxonomy of the genus *Candida* Berkhout nom. conserv. Ph.D. Thesis. Department of Microbiology and Biochemistry, University of the Orange Free State, Bloemfontein, South Africa, 1989.
26. Kendrick B. The Fifth Kingdom. Ontario: Mycologue Publications, 1992.
27. Alexopoulos CJ, Mims CW. Introductory Mycology. New York: John Wiley and Sons, 1979.
28. Webster J. Introduction to Fungi. London: Cambridge University Press, 1980.
29. Kock JLF. Chemotaxonomy and yeasts. S Afr J Sci 1988; 84:735–740.
30. Manocha MS, Campbell CD. The effect of growth temperature on the fatty acid composition of *Thamnidium elegans* Link. Can J Microbiol 1978; 24:670.

31. Viljoen BC, Kock JLF, Lategan PM. The influence of culture age on the cellular fatty acid composition of four selected yeasts. J Gen Microbiol 1986; 132:1895–1898.
32. Smit EJ, Van der Westhuizen JPJ, Kock JLF, Lategan PM. A yeast identification method: the influence of culture age on the cellular long-chain fatty acid composition of three selected basidiomycetous yeasts. System Appl Microbiol 1987; 10:38–41.
33. Van der Westhuizen JPJ, Kock JLF, Smit EJ, Lategan PM. The value of long-chain fatty acid composition in the identification of species representing the basidiomycetous genus *Rhodosporidium* Banno. System Appl Microbiol 1987; 10:31–34.
34. Deinema MH. Intra- and extra-cellular lipid production by yeasts. Ph.D. Thesis. Laboratory of Microbiology, Agricultural University, Wageningen, Netherlands, 1961.
35. Ratledge C. Biotechnology of oils and fats. In: Ratledge C, Wilkinson SG, eds. Microbial Lipids. Vol. 2. London: Academic Press, 1989:567–668.
36. Blomquist G, Anderson B, Anderson K, Brondz I. Analysis of fatty acids. A new method for characterization of moulds. J Microbiol Meth 1992; 16:59.
37. Ferreira JHS, Augustyn OPH. Differentiation between *Eutypa lata* and *Cryptovalsa* cf. *ampelina* by means of cellular fatty acid analysis. S Afr J Enol Vit 1989; 10:18–22.
38. Wickerham LJ. Taxonomy of Yeasts. Tech. Bull. No. 1029. Washington, D.C., U.S. Dept. Agric., 1951.
39. Gangopadhyay PK, Thadepalli H, Roy I, Ansari A. Identification of species of *Candida*, *Cryptococcus* and *Torulopsis* by gas-liquid chromatography. J Infect Dis 1979; 40:952–958.
40. Gunasekaran M, Hughes WT. Gas-liquid chromatography: a rapid method for identification of different species of *Candida*. Mycologia 1980; 7:505–511.
41. Moss CW, Shinoda T, Samuels JW. Determination of cellular long-chain fatty acid compositions of various yeasts by gas-liquid chromatography. J Clin Microbiol 1982; 16:1073–1079.
42. Kock JLF, Lategan PM, Botes PJ, Viljoen BC. Developing a rapid statistical identification process for different yeast species. J Microbiol Meth 1985; 4:147–154.
43. Von Arx JA, Van der Walt JP. Ophiostomatales and Endomycetales. In: de Hoog GS, Smith MT, Weijman ACM, eds. The Expanding Realm of Yeast-like Fungi. Amsterdam: Elsevier, 1987:165–176.
44. Barnett JA, Payne RW, Yarrow D. Yeast Characteristics and Identification. 2nd ed., Cambridge: Cambridge University Press, 1990.
45. Kock JLF, Botes PJ, Erasmus SC, Lategan PM. A rapid method to differentiate between four species of the Endomycetaceae. J Gen Microbiol 1985; 131:3393–3396.
46. Kock JLF, Erasmus SC, Lategan PM. A rapid method to identify some species of the Dipodascaceae and Endomycetaceae. J Microbiol Meth 1986; 5:199–203.
47. Miller M, Kock JLF, Botes PJ. The significance of long-chain fatty acid compositions and other phenotypic characteristics in the genus *Pichia* Hansen *emend.* Kurtzman. System Appl Microbiol 1989; 12:70–79.
48. De Hoog GS, Smith MT, Gueho E. A revision of the genus *Geotrichum* and its teleomorphs. Stud Mycol 1986; 29:1–131.
49. Cottrell M, Kock JLF. The yeast family Lipomycetaceae Novak et Zsolt *emend.* Van der Walt et al. and the genus *Myxozyma* Van der Walt et al. 1. A historical account of its delimitation and 2. the taxonomic relevance of cellular long-chain fatty acid

composition and other phenotypic characters. System Appl Microbiol 1989; 12: 291–305.
50. Viljoen BC, Kock JLF, Thoupou K. The significance of cellular long-chain fatty acid compositions and other criteria in the study of the relationship between sporogenous ascomycete species and asporogenous *Candida* species. System Appl Microbiol 1989; 12:80–90.
51. Golubev WI, Smith MT, Poot GA, Kock JLF. Species delineation in the genus *Nadsonia* Sydow. Antonie van Leeuwenhoek 1989; 55:369–382.
52. Augustyn OPH, Kock JLF, Ferreira D. Differentiation between yeast species, and strains within a species, by cellular fatty acid analyses. 3. *Saccharomyces* sensu lato, *Arxiozyma* and *Pachytichospora*. System Appl Microbiol 1990; 13:44–55.
53. Cottrell M, Kock JLF, Lategan PM, Botes PJ, Britz TJ. The long-chain fatty acid compositions of species representing the genus *Kluyveromyces*. FEMS Microbiol Lett 1985; 30:373–376.
54. Cottrell M, Viljoen BC, Kock JLF, Lategan PM. The long-chain fatty acid compositions of species representing the genera *Saccharomyces*, *Schwanniomyces* and *Lipomyces*. J Gen Microbiol 1986; 132:2401–2403.
55. Oosthuizen A, Kock JLF, Botes PJ, Lategan PM. The long-chain fatty acid composition of yeasts used in the brewing industry. Appl Microbiol Biotechnol 1987; 26: 55–60.
56. Viljoen BC, Kock JLF, Lategan PM. Long-chain fatty acid composition of selected genera of yeasts belonging to Endomycetales. Antonie van Leeuwenhoek 1986; 52: 45–51.
57. Muller HB, Kock JLF. *Waltiozyma* gen. nov. (Saccharomycetaceae) a new genus of the Endomycetales. S Afr J Sci 1986; 82:491–492.
58. Botha A, Kock JLF. The distribution and taxonomic value of fatty acids and eicosanoids in the Lipomycetaceae and Dipodascaceae. Antonie van Leeuwenhoek 1993; 63:111–123.
59. Kock JLF, Van der Walt JP. Fatty acid composition of *Schizosaccharomyces* Lindner. System Appl Microbiol 1986; 8:163–165.
60. Von Arx JA. The Genera of Fungi Sporulating in Pure Culture. 3rd ed. Vaduz: J. Cramer, 1981.
61. Botha A, Kock JLF. Application of fatty acid profiles in the identification of yeasts. Int J Food Microbiol 1993; 19:39–51.
62. Tredoux HG, Kock JLF, Lategan PM. The use of cellular long-chain fatty acid composition in the identification of some yeasts associated with the wine industry. System Appl Microbiol 1987; 9:299–303.
63. Botha A, Kock JLF, Coetzee DJ, Van Dyk MS, Van der Berg L, Botes PJ. Yeast eicosanoids 1. The distribution and taxonomic value of cellular fatty acids and arachidonic acid metabolites in the Dipodascaceae and related taxa. System Appl Microbiol 1992; 15:148–154.
64. Yamada Y, Banno I. *Hasegawaea* gen. nov. ascosporogenous yeast genus for the organisms whose asexual reproduction is by fission and whose ascospores have smooth surfaces without papillae and which are characterized by the absence of coenzyme Q and by the presence of linoleic acid in cellular fatty acid composition. J Gen Appl Microbiol 1987; 33:295–298.

65. Van Dyk MS, Kock JLF, Coetzee DJ, Augustyn OPH, Nigam S. Isolation of a novel arachidonic acid metabolite 3-hydroxy 5,8,11,14-eicosatetraenoic acid (3-HETE) from the yeast *Dipodascopsis uninucleata* UOFS-Y128. FEBS Lett 1991; 283: 195–198.
66. Kreger–Van Rij NJW, Veenhuis M. Spores and septa in the genus *Dipodascus*. Can J Bot 1974; 52:1335–1338.
67. Kreger–Van Rij NJW. The species *Dipodascopsis uninucleata* (Biggs) Batra et Millner. Persoonia 1974; 8:105–109.
68. Malfeito-Ferreira M, Aubyn AS, Loureiro V. Long-chain fatty acid composition as a tool for differentiating spoilage wine yeasts. Mycotaxon 1989; 36:35–42.
69. Augustyn OPH, Kock JLF. Differentiation between yeasts species, and strains within a species, by cellular fatty acid analyses. 1. Application of an adapted technique to differentiate between strains of *Saccharomyces cerevisiae*. J Microbiol Meth 1989; 10:9–23.
70. Augustyn OPH. *Saccharomyces cerevisiae*. S Afr J Enol Vit 1989; 10:8–17.
71. Augustyn OPH, Ferreira D, Kock JLF. Differentiation between yeast species, and strains within a species, by cellular fatty acid analyses. *Saccharomyces* sensu stricto, *Hanseniaspora*, *Saccharomycodes*, and *Wickerhamiella*. System Appl Microbiol 1991; 14:324–334.
72. Augustyn OPH, Kock JLF, Ferreira D. Differentiation between yeast species, and strains within a species, by cellular fatty acid analyses. 5. A feasible technique? System Appl Microbiol 1992; 15:105–115.
73. Duncan AJ. Quality Control and Industrial Statistics. Illinois: Richard D. Irwin. 1959.

10

Carbohydrates and Their Impact on Fungal Taxonomy

Gaby E. Pfyffer
Swiss National Center for Mycobacteria, University of Zurich, Zurich, Switzerland

I. INTRODUCTION

When fungi exhibit specific morphological features, there is no doubt that their taxonomy, systematics, and phylogeny may be based on these traits. Where, however, distinctive morphological characteristics are ambiguous or even lacking, the traditional descriptive studies have to be supplemented with additional data derived from ultrastructural, genetic, immunological, and biochemical investigations. With the advent of new and highly sensitive analytical techniques, chemosystematics has evolved within a few decades to become an essential ingredient of taxonomy (1,2). In the recent past, fungi have been screened for distinctive molecular features, and many attempts have been made to introduce new differential chemical characters into fungal taxonomy—not as a merely academic pursuit but in order to overcome the well-known insufficiencies of some traditional schemes. In the quest for such criteria that would be suitable both for identification and for the process of unraveling phylogenetic relationships, many of these efforts have, however, proven to be unsuccessful or, at best, unsatisfactory. Very often, conclusions were drawn from studies encompassing merely a few species. Also, components were found to be restricted to groupings of low taxonomic rank (species) only, or some characteristics studied simply did not display sufficient "taxonomic resolving power."

Bearing in mind that a "good" chemical character, i.e., a diagnostic differential character, has to be distributed widely but not universally (3–5) and must, furthermore, display a high degree of constancy with respect to a variety of other parameters such as the C source of the medium or, in dimorphic fungi, the growth form, the practical value of carbohydrates for fungal taxonomy will be briefly outlined in this chapter. Particular emphasis will be put on compounds which have proven successful for taxonomy at higher ranks of hierarchy—i.e., acyclic polyols and cell wall polysaccharides.

II. MONO- AND DISACCHARIDES

The ability of yeasts to assimilate and ferment a large variety of sugars was recognized very early. Nowadays, the different patterns so obtained constitute an essential part of the standard description of yeasts (6,7). Assimilation, i.e., the ability of a yeast cell to utilize a specific carbohydrate as the sole C source in the presence of oxygen, is the mainstay of yeast identification to the species level. The carbohydrates most frequently tested include glucose, maltose, sucrose, lactose, galactose, melibiose, cellobiose, xylose, raffinose, dulcitol, inositol, and trehalose (8,9). Several ready-to-use kits are commercially available and considerably facilitate yeast identification. Fermentation studies, on the other hand, are rarely needed to identify most of the commonly encountered yeasts if one is familiar with their typical morphology on cornmeal agar.

Although this kind of distinction is applied widely, especially in the taxonomy of medically important yeasts, there are some limitations. Experience has shown that carbohydrates which are utilized by most yeasts, e.g., glycerol, have a low value for the differentiation of taxa. Furthermore, inconsistent results are frequently seen when assimilation or fermentation is weak or slow. Problems may also arise when laboratory strains spontaneously produce mutants which acquire the ability to assimilate certain C sources that are normally not utilized by the parental strain. This was demonstrated for *Candida albicans* in which sorbose- and arabinose-positive strains are associated with distinct chromosomal arrangements (10).

Trehalose, a disaccharide consisting of two α,α'-linked glucose units appears to be a carbohydrate common to all Eumycetes. In these organisms it functions as a reserve compound (11,12) and is involved in developmental processes, such as sporulation (13), establishment/breaking of dormancy, and in the germination of spores (14,15). Since ubiquitous compounds are generally considered valueless as taxonomic characters (3), trehalose is obviously not a suitable candidate.

III. ACYCLIC POLYOLS

The highly water-soluble acyclic polyols (syn. alditols, sugar alcohols) have been isolated from fungal material for nearly two centuries (16). In recent years these compounds have been rapidly analyzed by gas-liquid chromatography (GLC), either as their acetates (17) or as their trimethylsilyl ethers (18,19). Polyols generally contribute significantly to the biomass of fungi (up to 40% in *Agaricus bisporus* [14]), and are therefore considered to be physiologically essential. Currently, the most favored experimental hypotheses concerning their functional role in fungi are the generation of storage pools for carbon and for reducing power, involvement in the control of growth and of water potential, regulation of cytoplasmic pH value, and, where applicable, the maintenance of a proper chemical potential gradient for carbon movement in host-parasite associations (20).

Alditols appear to be among the fungal metabolites studied most intensively inasmuch as there are hundreds of published reports relating to the polyol composition of approximately 500 species (for refs., see [21]) representing some 1.3% of the known lower and 1% of the higher fungi, respectively (based on the estimated total of species given in [22]). Occurrence of at least one type of polyol (Fig. 1) seems to be a feature shared by all fungi except the Oomycetes (Table 1). Mannitol and glycerol are generally considered common fungal constituents, whereas arabitol and xylitol are less frequent. Erythritol, threitol, galactitol, and sorbitol appear to be even more rarely found (for refs., see [21]). The same holds for free ribitol, as the published reports for this sugar alcohol are restricted to *Sclerotinia sclerotiorum* (23), *Puccinia graminis* (24), *Candida albicans*, and some Mucorales (18). Heptitols, e.g., *meso*-glycero-ido- and D-glycero-D-ido-heptitol as well as volemitol, have each been reported in single species, the former two in the yeast *Pichia miso* and the latter to *Lactarius volemus* (21). The presence of further unspecified heptitols in several other yeast species such as *Hansenula*, *Kluyveromyces*, and *Schwanniomyces* has also been reported (25).

Earlier compilations of all available data on the occurrence of free polyols in the Eumycetes has shown that the type of sugar alcohols produced by a given fungus is of taxonomic significance (21). The distribution of these carbohydrates followed three major patterns: P_0 (polyols absent), P_1 (polyols, except mannitol, present), and P_2 (mannitol and perhaps additional polyol[s] present). When fungal species were classified according to the three different stages of the polyol character, chemotaxa were obtained that coincided to the scheme of Ainsworth et al. (26) with the Oomycetes (P_0), the Zygomycetes/Hemiascomycetes (P_1), and the Chytridiomycetes/Ascomycotina (except Hemiascomycetes)/Basidiomycotina/Deuteromycotina (except some imperfect yeasts) group (all P_2).

Any consideration of polyols as taxonomic markers requires their being widespread, which is undoubtedly met by the abundant data. In addition, these compounds must also display distinct and taxonomically restricted patterns which are *qualitatively* unaffected by:

1. Variations of environmental conditions such as the C source of the medium

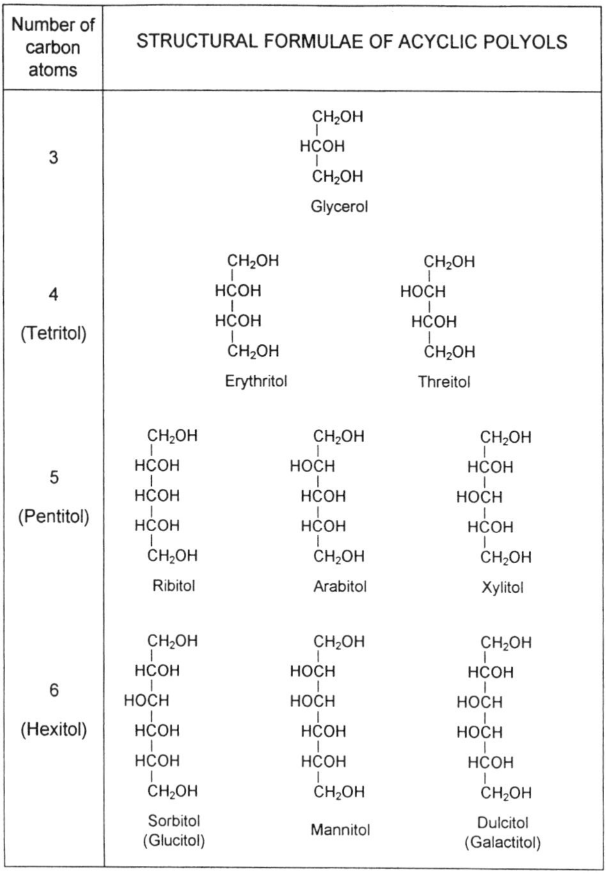

Figure 1 Major acyclic polyols (sugar alcohols) occurring in fungi.

2. Life cycle or phase of the fungus
3. In dimorphic organisms, the growth form.

It has been shown for *Penicillium italicum* that shifts occurred in the relative proportion of the constituents of the polyol fraction on changes in the type of C source supplied in the growth medium. Generally, the synthesis of that alditol was favored where the immediate sugar precursor served as the C source (e.g., in *Claviceps purpurea* [27], *Pyrenochaeta terrestris* [28], and *Sclerotinia sclerotiorum* [23]), whereas the *qualitative* pattern of sugar alcohols remained constant under any of the conditions tested. This has been shown extensively by studies involving *Geotrichum candidum* (29), *Penicillium chrysogenum* (30), *S. scle-*

rotiorum (23), and *Dendryphiella salina* (31) as well as *Allomyces arbuscula, Penicillium italicum*, and *Zygorhynchus moelleri* (32). Investigations into the polyol composition during the different developmental stages of a fungus, i.e., spore, mycelium, and fruiting body, are scant and, from a technical point of view, unsatisfactory, particularly as far as older reports are concerned. Our own experimental studies with *Agaricus bisporus* have demonstrated that mannitol is always present in large quantities. Glycerol, erythritol, and arabitol, on the other hand, although detectable in all developmental stages, are found only in trace amounts in the fruiting body (19). Furthermore, in relation to acyclic polyols and dimorphism, cultures of *C. albicans* have shown that arabitol is the main constituent and that ribitol as well as glycerol are minor components of the polyol fraction of hyphal and yeast cells (18). Finally, it has also been shown that the polyol pattern is retained in phytopathogenic associations (fungus/green plant [33]). In leaves of *Solanum lycopersicum* infected with *Phytophthora infestans*, no sugar alcohols were found (P_0 pattern) while infection of leaves by an ascomycete or deuteromycete (i.e., *Cucumis sativa/Erysiphe cichoracearum, Oryza sativa/Pyricularia oryzae, Gossypium hirsutum/Rhizoctonia solani*) resulted in an accumulation of various polyols with mannitol, mostly, as the major component (P_2 pattern). Taking these facts into account, the polyol character appears to be extremely conservative for a given taxon. Hence, these carbohydrates can successfully be used as markers in the assignment of species of doubtful systematic position.

IV. POLYSACCHARIDES

There are two major ways in which polysaccharides are utilized in fungal taxonomy. Eumycetes are classified either on the basis of the chemical nature of cell wall polysaccharides, or the cleavage products (monomers) of wall polymers obtained by hydrolysis of whole cells.

Fungal cell wall polysaccharides include on the one hand, skeletal polymers, and on the other, amorphous or slightly crystalline matrix substances. Skeletal polysaccharides are generally water-insoluble and highly crystalline (e.g., chitin, β-linked glucans) while matrix polysaccharides are mostly water-soluble. In his pioneering study Bartnicki-Garcia (34) has drawn together the many published data and concluded that both the taxonomic and the phylogenetic potential of the fungal cell wall resides in its carbohydrate variety—more precisely, in the characteristic distribution of chitin, chitosan, and cellulose, as well as of the various types of glucans and mannoproteins. These compounds occur neither universally nor sporadically throughout the entire spectrum of Eumycetes, but appear consistently and characteristically in certain groups of fungi. By selecting dual combinations of principal cell wall polymers, true fungi have been classified into several categories—namely those containing cellulose/β-glucan in their cell walls (Oomycetes), those with cellulose/chitin (Hyphochytridiomy-

Table 1 Composition of the Polyol Fraction in Fungi[a]

Species	Polyols[b]										Polyol character
Oomycetes											P_0
Achlya radiosa										0	
Albugo tragopogonis										0	
Peronospora parasitica										0	
Phytophthora cactorum										0	
Phytophthora infestans										0	
Plasmopara viticola										0	
Chytridiomycetes											P_2
Allomyces arbuscula	G						M				
Blastocladiella emersonii	G						M				
Zygomycetes											P_1
Actinomucor elegans	G										
Circinella mucoroides				R							
Circinella muscae				R							
Mortierella rammaniana	G										
Mucor miehei	G			R							
Mucor rouxii	G			R	A						
Phycomyces blakesleeanus	G			R							
Rhizopus oligosporus	G										
Syncephalastrum racemosum				R							
Zygorhynchus moelleri	G			R							
Hemiascomycetes											P_1
Debaryomyces hansenii	G				A	X					
Hansenula anomala	G				A			D			
Hansenula subpelliculosa	G				A	X					
Pachysolen tannophilus					A	X					
Pichia farinosa	G	E			A	X		D			
Pichia guilliermondii	G			R	A	X					
Pichia membranaefaciens	G				A						
Pichia quercuum					A	X					
Saccharomyces cerevisiae	G				A	X					
Zygosaccharomyces bisporus	G				A						
Zygosaccharomyces fermentati					A						
Zygosaccharomyces rouxii	G				A	X					
Ascomycotina classes											P_2
Chaetomium elatum			T		A		M				
Claviceps purpurea	G	E	T		A	X	M	D			
Dendryphiella salina	G	E			A		M				
Sclerotinia sclerotiorum	G			R	A	X	M	D	S		
Penicillium chrysogenum	G	E			A		M				
Penicillium italicum	G	E			A		M				
Sclerotinia sclerotiorum	G			R	A	X	M	D	S		

Table 1 Continued

Species	Polyols[b]	Polyol character
Basidiomycotina classes		P_2
Agaricus silvaticus	M	
Amanita muscaria	M	
Amanita phalloides	M	
Armillaria mellea	G E T A X M S	
Boletus edulis	M	
Boletus erythropus	G A X M	
Coprinus atramentarius	A M	
Coprinus comatus	M	
Lectarius volemus	M V	
Lentinus edodes	A M	
Lycoperdon pusillum	M	
Pleurotus ostreatus	G E M S	
Puccinia graminis	G E R A X M S	
Schizophyllum commune	G E A X M S	
Ustilago esculenta	E M	
Ustilago maydis	G E T A M	
Ustilago nuda	G E M	

[a]These selected examples represent parts of the compilation list (21).
[b]G, glycerol; E, erythritol; T, threitol; R, ribitol; A, arabitol; X, xylitol; M, mannitol; D, dulcitol; S, sorbitol; V, volemitol; 0, no polyols

cetes), those with chitin/chitosan (Zygomycetes), those with mannan/β-glucan (Hemiascomycetes), those with chitin/β-glucan (Chytridiomycetes, Euascomycetes, Homobasidiomycetes, Deuteromycetes), and those with mannan/chitin (Heterobasidiomycetes). The close correlation between overall cell wall composition and higher taxonomic position (22,26) indicates the highly conservative nature of this particular chemical trait.

The Oomycetes are traditionally regarded as cellulosic fungi, although cellulose (β-1,4-linked glucose) appears to be a minor component of hyphal cell walls, ranging from 4% (*Apodachlya* [35] to up to 20% in *Pythium* [36]). The major cell wall component in this group of fungi is a β-glucan consisting primarily of β-1,3- and β-1,6-linked glycosyl units. Chitin (β-1,4-linked *N*-acetyl-D-glucosamine) in combination with β-1,3-glucan appears to be quite universal in filamentous fungi inasmuch as it is only absent in Oomycetes and Hemiascomycetes. The β-1,4-linked polymer of glucosamine, on the other hand, is an exclusive feature of Zygomycetes, whereas a large variety of different mannoproteins only occur in Hemiascomycetes.

Even with the growth of knowledge in chemotaxonomy, the correlation of these dual combinations of major structural cell wall polysaccharides (chitin, β-glucans, cellulose, chitosan, α- and β-mannans) with taxonomic groupings has still held true, apart from minor adjustments (37). For instance, Hoddinott and Olsen (38) demonstrated that the chitosan/chitin wall type characteristic of most Zygomycetes does not extend to the Entomophthorales, which exhibit the chitin/β-glucan wall type instead. Furthermore, Bulone et al. (39) have found very small amounts of chitin occurring as small globular particles in the cell wall of *Saprolegnia monoica*. This has been characterized by x-ray as well as electron diffraction and infrared spectroscopy. However, the amount of chitin found in this fungus was very low and comparable to the amounts determined in the Leptomitales (40). These recent findings in Oomycetes, however, do not change the fact that these fungi have still the typical components of an oomycetous wall, cellulose/β-1,3-1,6-glucan/cellulose.

Cell wall polymers can also be used as taxonomic characters at the genus level, as documented by numerous studies on filamentous fungi. The major components of the water-soluble fraction of purified cell walls have been determined by various analytical techniques such as GLC, GLC/mass spectrometry, and nuclear magnetic resonance (NMR) spectrometry. Different galactofuranans have been found in species of *Eupenicillium*, *Penicillium*, and *Aspergillus* and its teleomorphs (41); a complex glucomannogalactan has been found in most *Talaromyces* species (42); and α-1,2-1,6-mannans have been found in members of the *Aphanoascus* genus (43). A hyphal nigeran, a linear α-1,3-1,4-glucan, has been proposed as a potential phylogenetic marker for *Aspergillus* and *Penicillium* species (44). Although water-soluble polysaccharides appear to be universal components of any fungal cell wall, the distribution of the many different types has been found to follow a distinct pattern, making these components suitable for the delimitation of genera or lower taxa.

Cleavage products of wall polymers, on the other hand, have been used extensively for taxonomic studies of yeasts, again, mainly on the genus level. Comparative studies of the carbohydrate composition of whole yeast cells reveal two major groups of sugars. The first group comprises glucose, mannose, *N*-acetylglucosamine, and usually galactose, which are of cell wall (glucans, mannans, chitin, galactomannans) or of intracellular origin (glycogen). The second group includes xylose, fucose, and rhamnose and originates from capsules that are present in whole-cell hydrolysates of basidiomycetous yeasts and are generally not found in yeasts of ascomycetous affinity (45). Occurrence of xylose can, for instance, be regarded as a reliable criterion of some basidiomycetous yeasts, such as *Bullera*, *Cryptococcus*, *Filobasidiella*, and other fungi now classified in the Filobasidiaceae, whereas in the Sporobolomycetaceae this particular pentose is absent (46). Brondz and Olsen (47) have shown that the spectrum of monoses obtained after whole-cell hydrolysis does not, however, always reveal sufficient taxonomic resolving power. Although *C. albicans* and *Torulopsis glabrata* can be distinguished from *Saccharomyces cerevisiae* on the basis of the quantitative distribution of mannose, glucose, and galactose, no distinction is possible between *C. albicans* and *T. glabrata*. Clear differences appear only among the three species when their total fingerprints including cellular fatty acids are compared. Among the yeasts and yeastlike fungi different monose patterns may occur within a single genus (48–50), illustrating the well-known heterogeneous nature of these taxa.

V. EVALUATION OF CARBOHYDRATES AS DIFFERENTIAL CHARACTERS FOR CLASSIFYING FUNGI

Despite their ubiquity, some selected carbohydrates can be utilized as excellent differential characters for taxonomic purposes. This holds in particular, for components that are present in many but not all taxa. Some of these, e.g., certain monoses obtained after hydrolysis of whole cells or distinct polysaccharides of the water-soluble cell wall fraction, are helpful criteria in the delimitation of taxa at the species or genus level. Others, such as acyclic polyols (P [19,21]) and the dual combination of the major cell wall polymers (W [34,51]) appear to be suitable for fungal taxonomy at higher ranks. At these levels only two other chemical criteria can be successfully applied. These concern the intermediates of lysine biosynthesis (L; via diaminopimelic acid or via α-amino adipic acid [52]) and the specific characteristics of enzymes involved in tryptophan biosynthesis (T [53]). These enzymes, anthranilate synthetase, phosphoribosyl transferase, *N*-(5′-phosphoribosyl)-anthranilate isomerase, and indole-3-glycerophosphate synthetase, differ in certain chemical properties such as stability requirements, differential precipitation with ammonium sulfate, and, in particular, their sedimentation behavior after zonal centrifugation.

In contrast to the extensive investigations of polyols, the studies focusing on the W and T features have been based on a limited number of fungal organisms (some 20 and 100 species, respectively). Nevertheless, as outlined above, each of these features meets the requirements for providing a good diagnostic character and thus yields an objective criterion for the classification of fungi at higher ranks (e.g., classes). Integrating these characters into a common scheme (Fig. 2) can give chemotaxa which agree with the higher categories of the mainly morphologically based, conventional, classification systems:

1. Oomycetes are singled out from all other fungi by their great diversity in P, W, L, and T. A number of authors have outlined earlier the very special position of the Oomycetes with respect to all other classes of fungi (54,55). These include several biochemical characteristics, viz., the presence of significant amounts of hydroxyproline in cell wall protein (56), a higher molecular weight of the 25S ribosomal RNA (Oomycetes $1.4–1.43 \times 10^6$; all other fungi $1.3–1.36 \times 10^6$ [57]), the occurrence of desmosterol, 24-methylenecholesterol, and fucosterol

TAXON (classes)	CHEMICAL CHARACTER
Oomycetes	P_0 W_1 L_1 T_1
Zygomycetes	P_1 W_2 L_2 T_2
Hemiascomycetes	P_1 W_3 L_2 T_3
Blastomycetes	$P_{1,2}$ $W_{3,4}$ L_2 T_2
Teliomycetes	P_2 W_4 L_2 T_2
Chytridiomycetes	P_2 W_5 L_2 T_4
Ascomycotina classes	P_2 W_5 L_2 T_4
Hypho-/Coelomycetes	P_2 W_5 L_2 T_4
Hymeno-Gasteromycetes	P_2 W_5 L_2 T_4

Figure 2 Correlation between major higher groups of the Eumycetes as given in Ainsworth et al. (26) and chemotaxa obtained by using as differential characters features of the type of polyol (P_0, P_1, P_2 [18,21]), the dual combinations of major cell wall polymers (W_1–W_5 [34,51]), the lysine biosynthetic pathway (L_1, L_2 [52]), and the enzymes involved in tryptophan biosynthesis (T_1–T_4 [53]).

in the absence of ergosterol (58), and the obvious inability to accumulate polyphosphates (59).

2. The P_1 group is subdivided into two major taxa on the basis of their W and T characteristics, i.e., into W_2/T_2 and W_3/T_3 (where W_2/W_3 means a chitosan/chitin vs. a mannan/glucan type of cell wall). The phenetic groups obtained include the Zygomycetes in one group and the Hemiascomycetes in the other. Both taxa can be further delineated by using the P character as ribitol is prevalent in the former and arabitol/xylitol in the latter (18,21).

3. The Chytridiomycetes are unique among the lower fungi in possessing the $P_2/W_5/T_4$ combination, and belong chemosystematically to the same group as the Ascomycotina classes (except Hemiascomycetes) and the Basidiomycotina classes (Hymeno-/Gasteromycetes).

4. The Blastomycetes are well known as an artificial group of fungi growing as a yeast form, and generally having unknown teleomorphs, if at all. Heterogeneity within a single genus is therefore considered a common feature in the

taxonomy of yeasts. Not only is this well reflected in the P and W characters, but also, in particular, in the monoses of whole-cell hydrolysates, as mentioned above. Whereas the large group of species constituting the Hyphomycetes in the 1971 dictionary (26) is perfectly homogeneous with respect to the polyol criterion P_2, the merging (in the 1983 dictionary [22]) of Blastomycetes with Hyphomycetes has created heterogeneity within this taxon. If polyol characters are given a high weighting within established "natural" higher taxa, the heterogeneous assembly of imperfect yeasts should, therefore, be kept as a separate entity, class Blastomycetes, as before (26).

As stated more than half a century ago (5), it is unlikely that any diagnostic differential character will ever be used with ultimate certainty because all of them lack an absolute value. Comparative studies with taxonomic characters other than P, W, and T are necessary to prove the taxonomic potential of carbohydrates. In particular, such characters would include conserved ribosomal nucleotide sequences which have proven to be of considerable taxonomic significance, for instance, the 5S rRNA sequences (60) or large parts of the 18S (61) and 28S rRNA (62) from which far more information on the phylogeny of organisms can be gained from than from the 5S rRNA.

VI. CONCLUSIONS

At first sight, due to their wide distribution and fundamental importance, carbohydrates may not appear promising differential characters for fungal chemotaxonomy. However, some of these compounds have provided very significant taxonomic information. Among these are those carbohydrates that are widely distributed (but not universally), and which display a high degree of constancy with respect to a variety of other parameters such as, e.g., the C source of the medium. While monoses are useful compounds for the taxonomy of yeasts at the species or genus level, acyclic polyols (glycerol, arabitol, mannitol, etc.) and polysaccharides are excellent criteria for the taxonomy of Eumycetes at higher taxonomic levels. Fungi can be divided into three groups on the basis of their polyol composition: organisms that lack polyols (P_0); fungi that contain polyols except mannitol (P_1); and species that contain mannitol and perhaps additional polyols (P_2). Delimitation of groups based on P_0, P_1, and P_2 yields chemotaxa which equate to the Oomycetes (P_0), the Zygo- and Hemiascomycetes (P_1), and the Chytridiomycetes, Ascomycotina (except Hemiascomycetes), Basidiomycotina, and Deuteromycotina (except some imperfect yeasts; P_2]). Similarly, the occurrence of dual combinations of major structural cell wall polysaccharides (chitin, β-glucans, cellulose, chitosan, α- and β-mannans) constitutes an essential element of fungal taxonomy. The complete spectrum of fungi may be subdivided into various categories according to the chemical nature of their cell walls. These groups closely parallel conventional taxonomic boundaries: cellulose/β-glucan, for instance, is a characteristic of Oomycetes; chitin/β-glucan of Chytridio-, Euasco-, Homobasidio-, and Deuteromycetes; chitin/chitosan of Zygomycetes; and mannan/β-glucan of Hemiascomycetes. Hence, both acyclic polyols and cell wall polysaccharides appear to be extremely conservative and may therefore be considered reliable markers for fungal taxonomy.

REFERENCES

1. Harborne JB. Chemical data in practical taxonomy. In: Heywood VH, Moore DM, eds. Current Concepts in Plant Taxonomy. London: Academic Press, 1984:237–243.
2. Heywood VH. The current scene in plant taxonomy. In: Heywood VH, Moore DM, eds. Current Concepts in Plant Taxonomy. London: Academic Press, 1984:3–21.
3. Herout V. A chemical compound as taxonomic character. In: Bendz G, Santesson J, eds. Chemistry in Botanical Classification. Nobel Symposium 25. Uppsala: Almquist-Wiksell, 1973:55–62.
4. Heywood VH. Phytochemistry and taxonomy. In: Swain T, ed. Comparative Phytochemistry. London: Academic Press, 1966:1–20.
5. Just T. The relative value of taxonomic characters. Am Midl Nat 1946; 36:291–297.
6. Kreger-van Rij NJW. General classification of the yeasts. In: Kreger-van Rij NJW, ed. The Yeasts—A Taxonomic Study. Amsterdam: Elsevier Science Publishers, 1984: 1–44.
7. Kreger-van Rij NJW. Classification of yeasts. In: Rose AH, Harrison JS, eds. The Yeasts. 2nd ed. Vol. 1. London: Academic Press, 1987:5–61.
8. D'Amato RF, Bottone EJ, Amsterdam D. Substrate profile systems for the identification of bacteria and yeasts by rapid and automated approaches. In: Balows A, Hausler WJ, Herrmann KL, Isenberg HD, Shadomy HJ, eds. Manual of Clinical Microbiology. 5th ed. Washington, D.C.: American Society for Microbiology, 1991:128–136.
9. Warren NG, Hazen KC. *Candida*, *Cryptococcus*, and other yeasts of medical importance. In: Murray PR, Baron EJ, Pfaller MA, Tenover FC, Yolken RH, eds. Manual of Clinical Microbiology. 6th ed. Washington, D.C.: American Society for Microbiology Press, 1995:723–737.
10. Rustchenko EP, Howard DH, Sherman F. Chromosomal alterations of *Candida albicans* are associated with the gain and loss of assimilating functions. J Bacteriol 1994; 176:3231–3241.
11. Elbein AD. The metabolism of α,α-trehalose. In: Tipson RS, Horton D, eds. Advances in Carbohydrate Chemistry. Vol. 30., New York: Academic Press, 1974: 227–256.
12. Thevelein JM. Regulation of trehalose mobilization in fungi. Microbiol Rev 1984; 48:42–59.
13. Hanks DL, Sussman RS. The relationships of trehalose and its metabolism to conidiation in *Neurospora*. Neurospora News Lett 1967; 11:9–11.
14. Rast DM. Zur stoffwechselphysiologischen Bedeutung von Mannit und Trehalose in *Agaricus bisporus*. Planta 1965; 64:81–93.
15. van Assche JA, Carlier AR, Deekersmaeker HI. Trehalose activation in dormant and activated spores of *Phycomyces blakesleeanus*. Planta 1972; 103:327–333.

16. Braconnot H. Recherches analytiques sur la nature des champignons. Ann Chim (Paris) 1811; 79:265–304.
17. Pfyffer GE, Rast DM. The polyol pattern of some fungi not hitherto investigated for sugar alcohols. Exp Mycol 1980; 4:160–170.
18. Pfyffer GE, Rast DM. Accumulation of acyclic polyols and trehalose as related to growth form and carbohydrate source in the dimorphic fungi *Mucor rouxii* and *Candida albicans*. Mycopathologia 1989; 105:25–33.
19. Rast DM, Pfyffer GE. Acyclic polyols and higher taxa of fungi. Bot J Linn Soc 1989; 99:39–57.
20. Jennings DH. Polyol metabolism in fungi. Adv Microbial Physiol 1984; 25:149–193.
21. Pfyffer GE, Pfyffer BU, Rast DM. The polyol pattern, chemotaxonomy, and phylogeny of the fungi. Sydowia 1986; 39:160–201.
22. Hawksworth DL, Sutton BC, Ainsworth GC. Ainsworth and Bisby's Dictionary of the Fungi. 7th ed. Kew, Surrey: Commonwealth Mycological Institute, 1983.
23. Wang S-YC, Le Tourneau D. Carbon sources, growth, sclerotium formation and carbohydrate composition of *Sclerotinia sclerotiorum*. Arch Microbiol 1971; 80: 219–233.
24. Maclean DJ, Scott KJ. Identification of glucitol (sorbitol) and ribitol in a rust fungus, *Puccinia graminis* f. sp. *tritici*. J Gen Microbiol 1976:97:83–89.
25. Stankovic L, Kovacovska R. Production of alditols from D-xylose by yeasts. Folia Microbiol 1991; 36:542–548.
26. Ainsworth GC, James PW, Hawksworth DL. Ainsworth and Bisby's Dictionary of the Fungi. 6th ed. Kew, Surrey: Commonwealth Mycological Institute, 1971.
27. Vining LC, Taber WA. Analysis of the endogenous sugars and polyols of *Claviceps purpurea* (Fr.) Tul. by chromatography on ion exchange resins. Can J Microbiol 1964; 10:647–657.
28. Wright JR, Le Tourneau D. Utilization and production of carbohydrates by *Pyrenochaeta terrestris*. Physiologia 1965; 18:1044–1053.
29. Chang S-C, Li K-H. Identification of D-mannitol in *Geotrichum candidum* Link and the mechanism of its formation. Sci Sin 1964; 13:621–630.
30. Ballio A, Di Vittorio V, Russi S. The isolation of trehalose and polyols from the conidia of *Penicillium chrysogenum* Thom. Arch Biochem Biophys 1964; 107:177–183.
31. Holligan PM, Jennings DH. Carbohydrate metabolism in the fungus *Dendryphiella salina*. II. The influence of different carbon and nitrogen sources on the accumulation of mannitol and arabitol. New Phytol 1972; 71:583–594.
32. Pfyffer GE, Rast DM. The polyol pattern of fungi as influenced by the carbohydrate nutrient source. New Phytol 1988; 109:121–126.
33. Pfyffer GE, Boraschi-Gaia C, Weber B, Hoesch L, Orpin CG, Rast DM. A further report on the occurrence of acyclic sugar alcohols in fungi. Mycol Res 1990; 94: 219–222.
34. Bartnicki-Garcia S. Cell wall chemistry, morphogenesis, and taxonomy of fungi. Annu Rev Microbiol 1968; 22:87–108.
35. Sietsma JH. Protoplast formation and cell wall composition of some Oomycetes species. Ph.D. Thesis, University of Amsterdam, Netherlands, 1969.
36. Novaes-Ledieu M, Jimenez-Martinez A, Villanueva JR. Chemical composition of hyphal wall of Phycomycetes. J Gen Microbiol 1967 47:237–245.

37. Bartnicki-Garcia S. The cell wall: a crucial structure in fungal evolution. In: Rayner ADM, Brasier CM, Moore D, eds. Evolutionary Biology of the Fungi. Cambridge: Cambridge University Press, 1987:389–402.
38. Hoddinott J, Olsen OA. Ultrastructure of freeze-substituted hyphae of the basidiomycete *Laettisaria arvalis*. Protoplasma 1980; 103:281–297.
39. Bulone V, Chanzy H, Gay L, Girard V, Fèvre M. Characterization of chitin and chitin synthase from the cellulosic cell wall fungus *Saprolegnia monoica*. Exp Mycol 1992; 16:8–21.
40. Bertke CC, Aronson JM. Hyphal wall composition of *Mindeniella spinospora* and *Araiospora* sp. Am J Bot 1985; 72:467–471.
41. Leal JA, Guerrero C, Gomez-Miranda B, Prieto A, Bernabé M. Chemical and structural similarities in wall polysaccharides of some *Penicillium*, *Eupenicillium* and *Aspergillus* species. FEMS Microbiol Lett 1991; 90:165–168.
42. Parra E, Jimenez-Barbero J, Bernabé M, Leal JA, Prieto A, Gomez-Miranda B. Structural studies of fungal cell-wall polysaccharides from two strains of *Talaromyces flavus*. Carbohydr Res 1994; 251:315–325.
43. Leal JA, Gomez-Miranda B, Bernabé M, Cano J, Guarro J. The chemical composition of the wall of six species of *Aphanoascus*: the taxonomic significance of the presence of α-(1-2)(1-6) mannan and α-(1-4) glucan. Mycol Res 1992; 96:363–368.
44. Bobbitt F, Nordin JH. Hyphal nigeran as a potential phylogenetic marker for *Aspergillus* and *Penicillium* species. Mycologia 1978; 70:1201–1211.
45. Boekhout T. Classification of heterobasidiomycetous yeasts: characteristics and affiliation of genera to higher taxa of Heterobasidiomycetes. Can J Microbiol 1993; 39: 276–290.
46. Weijman ACM, Rodrigues de Miranda L. Xylose distribution within and taxonomy of the genera *Bullera* and *Sporobolomyces*. Antonie van Leeuwenhoek 1983; 49: 559–562.
47. Brondz I, Olsen I. Multivariate analyses of cellular carbohydrates and fatty acids of *Candida albicans*, *Torulopsis glabrata*, and *Saccharomyces cerevisiae*. J Clin Microbiol 1990; 28:1854–1857.
48. Weijman ACM. Carbohydrate composition and taxonomy of the genus *Dipodascus*. Antonie van Leeuwenhoek 1977; 43:323–331.
49. Weijman ACM, Golubev WI. Carbohydrate patterns and taxonomy of yeasts and yeast-like fungi. In: de Hoog GS, Smith MT, Weijman ACM, eds. The Expanding Realm of Yeast-like Fungi. Studies in Mycology 30. Amsterdam: Elsevier Science Publishers, 1987:361–371.
50. Weijman ACM, Rodrigues de Miranda L. Carbohydrate patterns of *Candida*, *Cryptococcus* and *Rhodotorula* species. Antonie van Leeuwenhoek 1988; 54:535–543.
51. Bartnicki-Garcia S. Cell wall composition and other biochemical markers in fungal phylogeny. In: Harborne JB, ed. Phytochemical Phylogeny. London: Academic Press, 1970:81–103.
52. Vogel HJ. Lysine biosynthesis and evolution. In: Bryson V, Vogel HJ, eds. Evolving Genes and Proteins. New York: Academic Press, 1965:25–40.
53. Hütter R, DeMoss JA. Organisation of the tryptophan pathway: a phylogenetic study of the fungi. J Bacteriol 1967; 94:1896–1907.
54. Deacon JW. Introduction to Modern Mycology. Oxford: Blackwell Scientific Publications, 1980.

55. Ragan MA, Chapman DJ. A Biochemical Phylogeny of the Protists. New York: Academic Press, 1978:211–232.
56. Bartnicki-Garcia S, Lippman E. Fungal cell wall composition. In: Laskin AI, Lechevalier HA, eds. CRC Handbook of Microbiology. 2nd ed. Vol. 4. Boca Raton: CRC Press, 1982:229–252.
57. Lovett JS, Haselby JA. Molecular weights of the ribosomal ribonucleic acid of fungi. Arch Microbiol 1971; 80:191–204.
58. Weete JD. Lipid Biochemistry of Fungi and Other Organisms. New York: Plenum Press, 1980.
59. Chilvers GA, Lapeyrie FF, Douglas PA. A contrast between Oomycetes and other taxa of mycelial fungi in metachromatic granule formation. New Phytol 1985; 99: 203–210.
60. Blanz PA, Gottschalk M. Systematic position of *Septobasidium*, *Graphiola* and other Basidomycetes as deduced on the basis of their ribosomal RNA nucleotide sequences. System Appl Microbiol 1986; 8:121–127.
61. Nishida H, Blanz PA, Sugiyama J. The higher fungus *Protomyces inouyei* has two group I introns in the 18S rRNA gene. J Mol Evol 1993; 37:25–28.
62. Blanz PA, Unseld MG, Rauh I. Group-specific differences in the secondary structure of the ribosomal RNA of yeasts. Yeast 1989; 5:S399–S404.

11

Volatiles in Fungal Taxonomy

Thomas Ostenfeld Larsen
Technical University of Denmark, Lyngby, Denmark

I. INTRODUCTION

Fungi are known to biosynthesize a variety of metabolic products, including volatile metabolites which can be products of both primary and secondary metabolism (1,2). When liberated to the surroundings in a gaseous form, many volatile metabolites contribute to the intense and characteristic odors of fungi. Characteristics such as fruity, floral, moldy, or earthy have been used to describe odors produced by fungal species cultivated under defined conditions, as a supplement to traditional taxonomic descriptions based primarily on morphological characters (3–6). The development of modern analytical techniques such as gas chromatography and mass spectrometry has facilitated the identification of such volatiles, even though they occur in small quantities (7,8). However, despite this advance in technology and the fact that volatiles, and especially terpenes, have been used successfully in the chemosystematics of plants, flowers, and lichens (9–13), only a few reports on fungal chemotaxonomic studies based on the production of volatile metabolites are found in the literature. The major efforts dealing with fungal volatiles have been undertaken in the area of food and feed research, in biotechnology, and in studies on the possible role of fungal volatiles in chemical interactions between microorganisms. The present chapter will deal with fungal characterization based on volatiles and will also cover the biosynthesis of common fungal volatiles and methods for their analysis.

II. PRIMARY AND SECONDARY METABOLISM OF FUNGAL VOLATILES

Primary metabolites are characterized by their broad distribution in all living things. Primary metabolism might be viewed as all integrated metabolic processes involved in the essential growth processes of an organism. A characteristic of primary metabolism is that it often proceeds in cycles—such as the citric acid cycle (14,15). Secondary metabolites are more restricted in their distribution and are often characteristic of individual genera, species, or even strains (14,16). In contrast to primary metabolism, secondary metabolism might be regarded as nonessential to the growth of the producing organism (17). Secondary metabolites are typically biosynthesized from a few key intermediates of primary metabolism (18,19). This has led to theories such as secondary metabolites being simply waste products produced in order to keep primary metabolism going at unfavorable conditions, when some metabolites may accumulate in the cell (20).

There is, however, a growing understanding of secondary metabolites being of importance to the producing organisms due to the increasing evidence of specialized ecological functions of specific secondary metabolites (21–24). The capability of fungi to interact with their surroundings through the gas phase seems very relevant, since the majority of fungi develop part of their mycelium and their entire reproductive structure in air, above the usually solid substrate that they grow on. Thus, if the transfer can be mediated through the gas phase, by for example volatile compounds, the chance of reaching a target organism seems much higher (25).

Williams et al. (22) proposed that all secondary metabolites serve the producing organisms by improving their survival fitness—"by acting at specific receptors in competing organisms." Christophersen (26) extended this hypothesis suggesting that "all metabolites, even minor ones, are expressed as a result of stimuli and are directed against or support actions on receptor systems," thus including both primary and secondary metabolites in one general biological role. From a chemotaxonomic perspective, secondary metabolism might be viewed as chemical differentiation, analogous to the acquisition of morphological differences (21), in agreement with Frisvad (24), who concluded that secondary metabolites are differentiating characteristics of high taxonomic information value.

A. Biosynthesis of Volatile Metabolites

The most important building block (precursor) in biosynthesis of volatile fungal metabolites is acetate, present as acetyl-coenzyme A in cells (19). Acetyl-CoA is derived via pyruvate generated by glycolysis (15). Acetyl-CoA is the main precursor of fatty acids (FAs) and mevalonate, an important intermediate in the secondary metabolism of terpenes. FAs are further metabolized to other primary metabolites. In the following pages the biosynthesis of the major acetate derived types of volatile compounds are presented together with the biosynthesis of some important amino acids derived compounds.

1. Fatty Acid-Derived Volatiles

The biosynthesis of FAs is well described (15). FAs are synthesized in a multienzyme complex by condensation of acetyl (or acyl) and malonyl enzyme bound units, leading to straight-chain FAs (14). Unsaturated FAs can be produced by elimination of a pair of neighboring hydrogen atoms in the corresponding saturated FA without the involvement of oxygenated intermediates (2). Both saturated and unsaturated FAs are precursors for a number of different types of compounds. Two major pathways leading to important compounds are β-oxidation of saturated FAs, and oxidation of unsaturated FAs such as linoleic acid by lipoxygenases (27) (Fig. 1). β-oxidation of free FAs results in the formation of β-keto acids. Methyl ketones and secondary alcohols are biosynthesized from these β-keto acids, by either decarboxylation or decarboxylation and subsequent reduction of the β-keto acids. The methyl ketones and secondary alcohols produced thus contain one carbon less than the precursors (27–29) (Fig. 1).

Edible mushrooms and filamentous fungi are known to produce a number of aliphatic eight-carbon compounds (30–33). 1-Octene-3-ol, or the "mushroom alcohol" which is prevalent in fresh edible fungi, originates from linoleic acid (C18:2) and is biosynthesized through two enzyme-catalyzed reactions (34,35). Firstly, linoleic acid is oxidized by a lipoxygenase in the presence of atmospheric oxygen giving 10-hydroxyperoxide (10HPOP). The 10HPOP is then cleaved by a lyase giving eight- and 10-carbon fragments (Fig. 1).

Esters are synthesized from alcohols and FAs are bound as acyl-CoA. Ester formation has been reported to be suppressed by high concentrations of unsaturated fatty acids and oxygen since these compounds stimulate growth; consequently, FAs are required for synthesis of essential cellular lipids (36). Lactones are derived from mainly γ- and δ-hydroxy acids originating from the breakdown of lipids.

2. Terpenes

All terpenes originate from isopentenyl diphosphate and the isomeric dimethylallyl diphosphate, which are formed from acetyl CoA. An important step is the conversion into mevalonic acid, since it is irreversible. Mevalonic acid has no known metabolic function except in the formation of terpenes and steroids (and C_5 alcohols) (14). The above-mentioned diphosphates are activated molecules, which easily undergo nucleophilic substitution at the α-carbon position or electrophilic substitution at the double bond, explaning their condensation into geranyl diphosphate (Fig. 2). Geranyl diphosphate is the major precursor of *monoterpenes* (C_{10}). Condensation between geranyl diphosphate and another molecule of isopentenyl diphosphate leads to farnesyl diphosphate, which is a major precursor of *ses-*

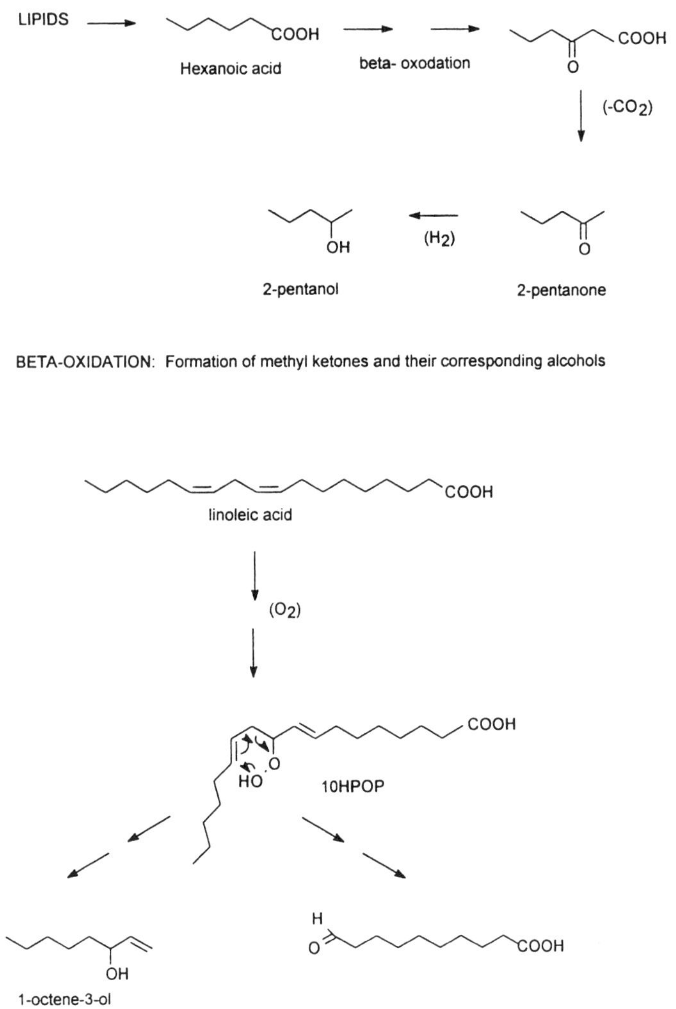

Figure 1 Examples of β-oxidation (top) and lipoxygenation (bottom) of fatty acids.

quiterpenes (C_{15}) (Fig. 2). Similarly repeated condensations lead to diterpenes (C_{20}), triterpenes (C_{30}), etc. (2,14).

Sesquiterpenes form the largest group of terpenes, and more than 100 sesquiterpene skeletals are known (37). Reports on sesquiterpenes produced by ascomycetes like penicillia and aspergilli are scarce, and identifications are often tentative (8,38–40). By contrast, numerous sesquiterpenes have been reported

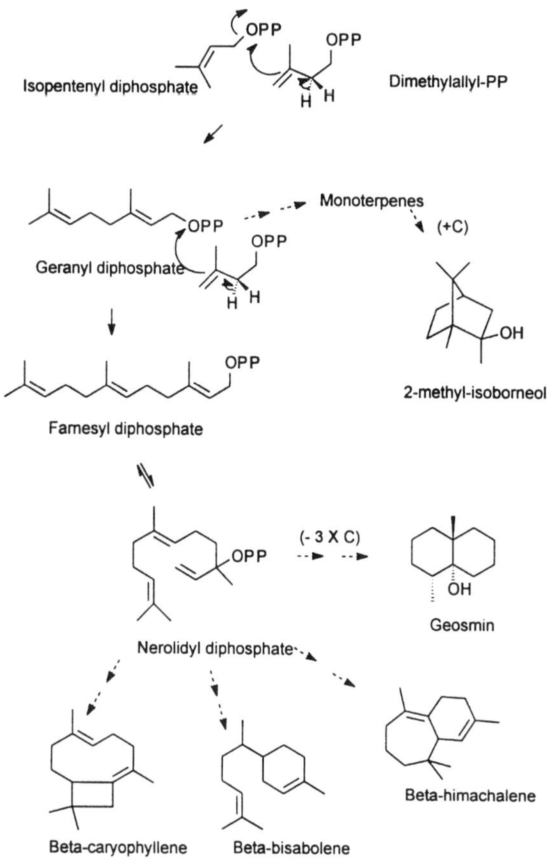

Figure 2 Formation of mono- and sesquiterpenes produced by among others penicillia (8).

from ophiostomatoid fungi (7) and basidiomycetes (41). However, most of the compounds produced by the basidiomycetes contain several oxygen atoms and are not likely to be volatile compounds. Interestingly, many of the tentatively identified terpenes from fungi are also plant products (8) and are likely to be produced by the same biosynthetic pathways. An example of this is some monoterpenes (42). The two moldy and earthy odorous fungal metabolites, 2-methyl-isoborneol and geosmin (Fig. 2), have both been suggested to originate from terpenes, even though they are C_{11} and C_{12} carbon compounds (43). Thus, 2-methyl-isoborneol is probably synthesized by methylation of isoborneol, whereas geosmin is likely to originate from a sesquiterpene precursor by loss of three carbon atoms.

3. *Compounds Derived from Amino Acids*

Several different volatile compounds are derived from amino acids, which themselves may originate from the enzymatic degradation of proteins or simply from the growth medium. Some widely occurring fungal alcohols like isobutanol (2-methyl-propanol), isopentanol (3-methyl-1-butanol), 2-methyl-1-butanol, and 2-phenylethanol can be formed from valine, leucine, isoleucine, and phenylalanine, respectively. They are formed via the analogue α-keto acids, which can be decarboxylated to give aldehydes. The aldehydes can subsequently be either reduced to alcohols or oxidized to fatty acids (Fig. 3). In yeast fermentations alcohols are formed at stages when there is nitrogen limitation, leading to the accumulation of α-keto acid, as the yeast cells appear unable to turn off the amino acid biosynthetic pathways (36). Isopentanol can also be biosynthesized by hydrogenation of the unsaturated alcohols isopentenol and 3,3-dimethylallyl alcohol, in turn derived from mevalonic acid (2). Primary alcohols can also be formed from fatty-acid CoA esters by two reductive steps (2). Thus, ethanol can be formed from acetaldehyde. In yeast fermentations this process takes place under anaerobic conditions (44). *Pyrazines*, known as strong flavorings, are also produced from amino acids. Thus, 2-methoxy-3-isopropylpyrazine reported from among others *P. camemberti* and *P. vulpinum* (45,46) is probably produced from valine, glyoxylic acid, and ammonia, as suggested by Leete et al. (47).

B. Influence of Environmental Factors on Production of Volatiles

The metabolic pathways described above can be influenced by a number of different factors, and environmental factors and substrate compositions can have a great influence on both the qualitative and quantitative production of volatile

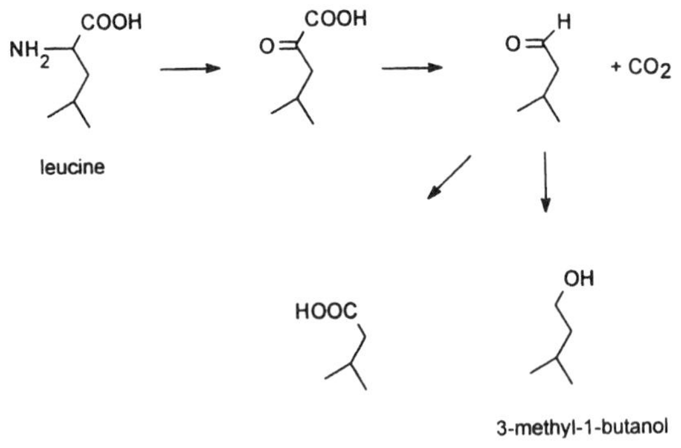

Figure 3 Formation of 3-methyl-1-butanol from leucine.

fungal metabolites. Important environmental factors that have been reported to influence volatile metabolism are water activity, pH, atmospheric composition (including carbon dioxide), agitation, and temperature (25,48–53). In general, it might be concluded that conditions favoring growth also favor the production of volatile metabolites (54). Likewise, the composition of volatiles produced can vary significantly depending on the types of carbon and nitrogen sources used (55). Thus, FA-derived volatiles are more dominant when fungi are grown on high-lipid-containing media, such as cheese (50,56,57). Trace metals have also been demonstrated to be important in the production of volatiles (58), probably because trace metals affect sporulation (59). As mentioned above, limitation in nitrogen might lead to increased formation of alcohols instead of amino acids; however, this has only been demonstrated for yeast fermentations.

III. TECHNIQUES FOR SAMPLING, ANALYSIS, AND CHARACTERIZATION OF VOLATILES

In general, the analysis of volatile compounds comprises three steps: sample preparation, separation methods, and identification techniques; these are often followed by a statistical evaluation of the results (also see Chapter 2).

A. Sample Preparation

Methods for collection of fungal volatiles can be roughly divided into two groups. There are different *headspace* (HS) methods using the direct collection of volatiles released to the surroundings and which are therefore present in the gas phase. Volatiles are collected either by active or passive sampling onto an adsorbent such as activated carbon, or a synthetic polymer such as Tenax TA (46,60,61). A second group of methods includes methods for *extraction* of volatiles present in the fungal biomass or growth medium (e.g., cheese). These methods nearly always involve some kind of extraction or distillation step, or their combination, such as simultaneous steam distillation and extraction (SDE). These methods may be performed under vacuum or with supercritical fluid extraction (SFE), to avoid high temperatures and oxidative conditions (38,54,62–64).

Charpentier et al. (65) compared the effect of different extraction methods on the volatile flavor composition of the edible mushroom *Lentinus edodes* and found advantages and disadvantages of HS, SDE, and SFE methods. HS was emphasized as a rapid technique representative of the true aroma of the sample; however, high-boiling-point components of significant importance to the aroma of a sample were poorly collected. The high-boiling-point components were more effectively collected by SDE and SFE. A general advantage with SDE is that it is possible to concentrate volatile compounds from a dilute mixture within a few hours; however, thermal degradation can be a major problem, especially when

SDE is carried out at atmospheric pressure. SFE with CO_2 was found to be an excellent method for extraction of hydrocarbons and other lipophilic compounds, whereas SFE tended to discriminate against more polar compounds like alcohols and acids. A major advantage of SFE is that a very gentle extraction can be performed at the critical conditions of CO_2, so thermal degradation is rarely a problem; the easy removal of CO_2 from extracts also makes SFE an attractive method.

Larsen and Frisvad (51) compared HS to SDE for samples produced during active fungal growth and found large differences in the composition of volatile fungal metabolites collected. Some of the major compounds released to the surroundings during growth could not be detected in SDE samples, while typical lipid degradation products were present. This study showed that one should not compare production of fungal volatile metabolites from different studies, when the same methods (and media) have not been used.

A new method for the collection of volatiles is headspace solid-phase microextraction (HS-SPME) (66). With HS-SPME, volatiles are extracted from the headspace onto, e.g., a fused silica fiber coated with a polymeric organic liquid. The volatiles are then concentrated in the coating until an equilibration between the gas phase and solid phase is reached. Once the sampling is completed, the silica fiber (housed inside a syringe) can be directly transferred to a GC injector for thermal desorption and analysis (67). This method has been applied to the analysis of fungal volatiles (68). HS-SPME was found useful in the direct recording of sesquiterpenes from penicillia with a sampling time of 30 min. Being a very simple and practically noninterfering technique, HS-SPME has good potential for studying microbial biosynthesis of volatile metabolites with time (68).

B. Separation Methods

The most important separation technique for mixtures of volatile compounds is gas chromatography (GC) or gas-liquid chromatography (GLC). Capillary columns, which give high resolution and low retention times have, for example, been widely used in analysis of mono- and sesquiterpenes (69). Different stationary phases can be chosen depending on the polarity of the compounds to be separated, and the film thickness can be varied depending on the concentration and volatility of the components to be analyzed; therefore run conditions have to be chosen to suit the specific analytical problem (70). An important step involved in the separation process is the injection method for introduction of the aroma sample into the capillary column. Some of the most commonly used techniques are split, splitless, or on-column injection (71,72).

When components in a sample are especially difficult to separate, multidimensional gas chromatography (MDGC) can be applied. With MDGC two different types of columns are connected in series (in two different GC ovens). The relevant fraction of the effluent from the precolumn is transferred into the main column for a subsequent separation, occasionally including intermediate "trapping" (71). In recent years special columns for chiral separation of enantiomers have been introduced and are often used successfully in MDGC (73,74). The latter authors separated the two enantiomers of *cis*-6-γ-dodecenolactone from *Fusarium poae* on a fused silica capillary column coated with a modified α-cyclodextrin or β-cyclodextrin phase.

C. Detection Techniques

Flame ionization detection (FID) is probably the most commonly used detection method for GC. In FID carbon atoms coupled to hydrogen atoms are detected; however, this gives no structural information for the compounds detected. Similarity in retention times of an unknown and a standard may be used as indirect identification, when used together with authentic standard compounds and analyzed on at least two different columns of substantially different polarity (75).

The most important detection method for identification purposes in modern flavor research is *mass spectrometry* (MS), which can give the molecular mass of an unknown compound and its typical fragmentation pattern (the mass spectrum). A limitation of MS in the identification of unknowns can be that insufficient information is present in the the mass spectrum to determine stereoisomers and positional isomers in aromatic systems (76). Mass spectrometry has potential in scanning for a selected number of characteristic ions—*selected ion recording* (SIR). SIR can be applied to the analysis of compounds with one or few characteristic ions in the mass spectrum and can improve sensitivity from detecting at the ng level to pg levels (71). An example of this is the mass spectrum of geosmin, which has a very characteristic ion at 112 m/z, which can be used for SIR (Fig. 4).

Another specific detection method is *infrared spectroscopy* (IR), which can also be performed on compounds in the gas phase, using Fourier transformation methods for spectrum accumulation and processing (GC-FTIR) (69). Most compounds possess a unique fingerprint absorption, and IR can deliver an unambiguous identification, if a comparison can be made between the spectrum of an unknown and the spectrum of an authentic standard (77). The method is nondestructive, as the effluent from the column is only passed through a so-called light-pipe for IR detection, and consequently it can be used in combination with either FID or MS. IR is especially useful in the detection of functional groups such as carbonyl groups (C=O). It is often possible to discriminate between structural isomers and even stereoisomers with GC-FTIR, making the method a very good supplement to GC-MS (78). Two major drawbacks with GC-FTIR are the lower sensitivity compared to FID and MS, and the lack of adequate vapor phase IR spectral data banks (71).

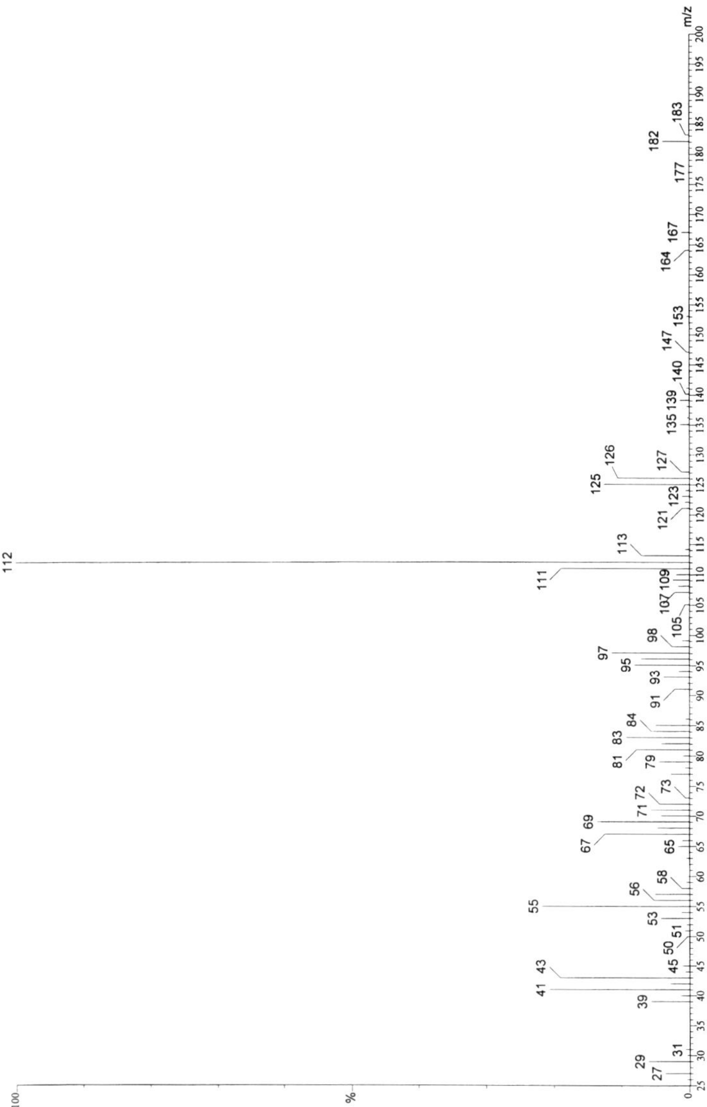

Figure 4 Mass spectrum of geosmin.

Nuclear magnetic resonance (NMR) spectroscopy is normally considered the most powerful tool available for the elucidation of structures of unknown compounds (69). However, since the technique cannot be coupled directly to gas chromatography, unknown compounds have to be separated and enriched before NMR analysis. The major limitation of the use of NMR in research of volatiles is that many important compounds occur in extremely low amounts, which makes preparative sampling very difficult (71).

Finally, it should be mentioned that research is in progress to develop an "artificial nose" for use in quality control. This research involves different kinds of gas-sensitive sensors in combination with pattern recognition techniques and neural networks, in order to characterize gas mixtures (79,80).

IV. FUNGAL CHARACTERIZATION BASED ON VOLATILES

Detection of volatile fungal metabolites by nose has probably been used to recognize fungi present in nature as long as mankind has existed. However, fungi are estimated to have existed and possibly coevolved with insects some 100 million years longer than they have with mammals (23). From an evolutionary point of view, it is therefore possible that a major part of the biological active metabolites, like mycotoxins and volatiles produced by fungi, are part of a fungal chemical defense system directed toward insects (21). Fungi are generally more nutritious than plant tissue, due to higher levels of proteins, making them potentially desirable sources of foods for predatory insects (23).

Karahadian et al. (33) reported that enhanced production of 1-octene-3-ol and the analog 1,5-octadiene-3-ol occurred on crushing the mycelium of *Penicillium camemberti*. Both these compounds have been demonstrated as attractants of insects (81–83). Based on these observations it might be suggested that the two compounds, which are widespread in fungi (31), have a broad ecological function in attracting insects at times when fungi are being destroyed. The insects, often adapted to the toxic metabolites produced by the fungi, may serve as vectors for fungal spores (21). Reports on biological effects of fungal volatiles toward other fungi are numerous (84,85), as are reports of bacterial volatiles affecting or inhibiting fungal growth and sporulation (86–88): thus in competitive environments, fungal and bacterial volatiles may take part in chemical interactions between microorganisms. Fungal volatiles may have a self-inhibitory role, for example, on germination as shown for *Geotricum candidum* (84). Carbon dioxide is probably the metabolite most often reported as affecting growth and metabolism of microorganisms, probably due to its almost ubiquitous formation (25).

A. Volatiles as Indicators of Fungal Growth

Fungal growth on stored food or feedstuffs is a worldwide problem, partly because the food will be deteriorated, but also because fungi are potential mycotoxin producers. A major problem in discovering fungal growth in huge stocks is the often very local nature of the infection, making representative sampling difficult. However, analysis of volatile metabolites from the enclosed atmosphere above the stored product might be a way of producing results representative of the entire product. A major research area concerning the problems of fungi in stored feedstuffs has been the quality control of stored cereal grains (89). In several studies volatile production from selected grain deteriorating species of penicillia and aspergilli have been studied in laboratory scale, to identify compounds useful as indicators for growth (61,63,90–94). In situ detection of volatiles in different granaries, with special emphasis on the influence of moisture content of the grains and the degree of ventilation on volatile production, has also been studied (60,95–97). The same few compounds including 3-methyl-1-butanol, 3-octanone, and 1-octene-3-ol were always detected together when grains were infected, showing that the detection of fungal growth based on their volatile metabolites is at present rather unspecific.

One example of a possible volatile-specific chemical marker was reported by Mattheis and Roberts (98). They suggested that geosmin might be a specific indicator for detection of *Penicillium expansum*, infecting postharvest-stored fruits like apples, pears, and cherries. However, as mentioned by the authors, other fruit-associated species like *Botrytis cinerea* need to be investigated before this compound can be assigned as a specific marker for *P. expansum* in fruit storage environments.

B. Taxonomy

As stated in the introduction, traditional mycologists have used characteristic odors in their morphological description of fungi like penicillia. An example might be the description by Raper and Thom (3) of *Penicillium commune* as having a "fairly moldy" odor when grown on synthetic media like Czapek agar. This description agrees with chemical investigations performed by Larsen and Frisvad (8,99), who demonstrated that *P. commune* among others produces the moldy-smelling compound 2-methyl-isoborneol. One might describe this rather subjective discipline as "sniffing taxonomy." However, due to several suggestions that airborne mycotoxins and particular volatile metabolites from growing fungal cultures may be important factors in symptoms such as headache, eye, nose, and throat irritations and fatigue (100–103), sniffing must be considered a dangerous and not recommended practice. During recent years several authors have suggested the use of fungal volatiles in "chemotaxonomy" (8,38,40,46,90, 104–108). Volatile metabolites, especially terpenes, have been used in chemical classification of plants, flowers, and lichens (9–13), indicating that this may be true for fungi as well. In the following pages the literature supporting the use of volatiles in fungal classification will be presented.

1. Studies Only Including Few Isolates of the Same Species

Pacioni et al. (105) studied the *Tuber melanosporum* complex. Although they found small qualitative differences in the composition of volatiles from four different taxa, they suggested that the relative amounts of some of the major substances were significant for chemical taxonomy. Likewise, they concluded that such quantitative differences for the major compounds resulted in different overall odors from the four taxa investigated. This is a very important point when fungal odors are used in characterization of fungi. A similar conclusion was made by Gallois et al. (109), who studied the influence of culture conditions on the production of flavor compounds by 29 ligninolytic basidiomycetes. They studied the flavor quality of their fungi on six different media either with or without agitation. Some of the fungi developed the same odor on nearly all media, whereas others produced a specific odor on each medium. However, they concluded that variations of odor assessments have to be considered carefully since the perception of different odors produced by the same basidiomycete isolate on different media did not result in different qualitative compositions of volatiles in their gas chromatographic analysis.

Another important point concerning the use of fungal volatiles in chemosystematics was also pointed out by Pacioni et al. (105). When they compared their findings on *Tuber melanosporum* volatiles with other studies, they found large qualitative differences in the total composition of volatile metabolites. They concluded that the use of different methods of analysis were probably the reason for the differences observed. A recent study by Larsen and Frisvad (51) supports these conclusions, since very different compositions of volatile metabolites were collected from *Penicillium vulpinum*, when the collection of volatiles by steam distillation and extraction (SDE) was compared to headspace sampling methods. In studies where some kind of distillation step has been used, aldehydes are often described as fungal volatiles, whereas aldehydes are seldom found when headspace sampling techniques have been used (8). Likewise, lipid degradation products such as 1-octene-3-ol are often reported as fungal volatiles when SDE has been the sampling method (51). Larsen and Frisvad (51) rejected SDE for ecological studies since the major compounds liberated to the surroundings during active fungal growth ("the true volatiles") could not found by SDE, in agreement with the concept of odor stated by Pacioni et al. (105).

Berger et al. (104) studied volatile production of one isolate of each of the basidiomycetes *Bjerkandera adusta*, *Poria aurea*, and *Tyromyces sambuceus*. The three isolates all produced a number of lactones, aromatics, and fusel alcohols, but nonetheless it was found that each species had characteristic chromatographic patterns and odor profiles. In another screening of basidiomycetes Gross et al. (110) studied three isolates of *Phlebia radiata*. One of the isolates was a very poor producer of volatiles on all media investigated; however, the two others were found to have very similar profiles, consisting of mainly a large number of alcohols.

In addition to the above-mentioned studies on volatile production from truffles and basidiomycetes, there are several other papers which likewise only include one or very few isolates of each species investigated. These studies can therefore hardly be regarded as truly chemotaxonomic, although they all involve the use of volatiles in fungal taxonomy to some degree (8,38,46,90,110). If some of the few isolates used were found to be misidentified, then false taxonomic conclusions might be made. Thus Börjesson (54) concluded that volatile metabolites could not be used in the classification of penicillia since significant differences could be observed between isolates of the same species. However, one of his only two isolates of *Penicillium aurantiogriseum* has later been reidentified as *P. crustosom* (8). Likewise, Frisvad (111) demonstrated that several misidentifications can be found in the literature.

2. *Studies Including Several Isolates of the Same Species*

When several isolates of the same species have been investigated for the production of volatiles, large differences in the quantitative production of the individual compounds can often be seen in combination with some qualitative differences (7,55). Such observations in studies on isolates of *Ophiostoma* and *Ceratocystis* led Hanssen (7) to conclude that formation of volatiles is frequently a strain-dependent feature. In a study on penicillia, Jollivet and Belin (106) found that the major volatile metabolites produced by 10 strains of *P. camemberti* were produced consistently; however, qualitative differences were observed for compounds produced in relatively minor amounts. They therefore grouped the 10 isolates into six aromatic strain groups. The differences in aromatic profiles may be used by the dairy industry in the rapid selection of strains for cheese making.

Zerinque et al. (40) found very similar profiles of sesquiterpenes (e.g., α-gurjunene, *trans*-caryophyllene, and cadinene) from four wild-type aflatoxin-producing strains of *Aspergillus flavus*. Interestingly, four nonaflatoxigenic isolates also included in the study did not produce any sesquiterpenes. That study indicated that there may be a correlation between the release of volatile compounds and the initiation of aflatoxin biosynthesis; similarly, there was also a correlation between the decline of aflatoxin synthesis and the disappearance of volatile sesquiterpenes. Since aflatoxins are biosynthesized via the polyketide pathway and not by the terpenoid pathway, Zerinque et al. suggested that these two different families of secondary metabolites are regulated at a common precursor step. Similar correlations between production of volatile sesquiterpenes and nonvolatile trichothecene mycotoxins of *Fusarium sambucinum* have been demonstrated by Jelén et al. (112). They studied 25 strains grown on wheat kernels. Ten of the strains, all producing toxic trichothecenes, all produced a very similar profile of volatile sesquiterpenes including trichodiene. Trichodiene is the first unique sesquiterpene intermediate in the formation of the trichothecene skeleton. The nontoxigenic strains all exhibited similar patterns of volatile sesquiterpene production, some of the major compounds being longifolene and β-farnesene. In this case true chemotaxonomists might suggest that two different species in *Fusarium sambucinum* could be described based on the findings of Jelén et al. (112). Similar correlations in the biosynthesis of volatile and nonvolatile secondary metabolites probably exist for many other mycotoxin-producing fungi such as the penicillia. This hypothesis is strongly supported by the successful use of both nonvolatile (113–116) and volatile secondary metabolites in the chemosystematics of penicillia (8,108).

Larsen and Frisvad (108) studied volatile production from 132 isolates of 25 different terverticillate *Penicillium* taxa. Isolates of the same species were found to produce similar profiles of volatile metabolites when grown on a sucrose- and yeast extract-based medium (Fig. 5). Likewise, profiles from different species such as *P. roqueforti* and *P. commune* could easily be distinguished (Fig. 5). The latter species is the most frequently occurring contaminant on cheeses including Blue cheese (117). Figure 5 gives an example of how these two important *Penicillium* species can be identified based solely on their production of volatile secondary metabolites. For most taxa four or five isolates were investigated; however, for *P. commune* 18 isolates were included in the chemotaxonomic study. All *P. commune* isolates were found to cluster, but the isolates separated in two groups (Fig. 6). All isolates had a similar production of low-boiling volatiles like isopentanol, 3-heptanone, styrene, 1,3-octadiene, 3-octanol, 1-octene-3-ol, 3-octanone, and 2-methyl-isoborneol; however, only the 10 isolates at top in the dendrogram (Fig. 6) produced some unique sesquiterpenes. Interestingly, Lund (118) included the eight nonsesquiterpene isolates of *P. commune*, investigated by Larsen and Frisvad, in the species *P. palitans*. *P. palitans* is closely related to, but different from, *P. commune* when morphological criteria are considered in combination with a chemotaxonomy based on nonvolatile secondary metabolites. The classification based on volatile metabolites for all other *Penicillium* taxa (Fig. 6) were in good agreement with a previous classification based on production of nonvolatile secondary metabolites (116).

V. CONCLUSIONS

Fungi often develop characteristic odors due to the production of unique combinations of volatile metabolites like alcohols, ketones, esters, terpenes, and other hydrocarbons. The production of volatiles is very dependent on the environmental conditions such as carbon and nitrogen sources, trace metals, pH, temperature, water activity, etc. In natural environments volatile metabolites are likely to be of ecological significance in chemical interactions with other organisms.

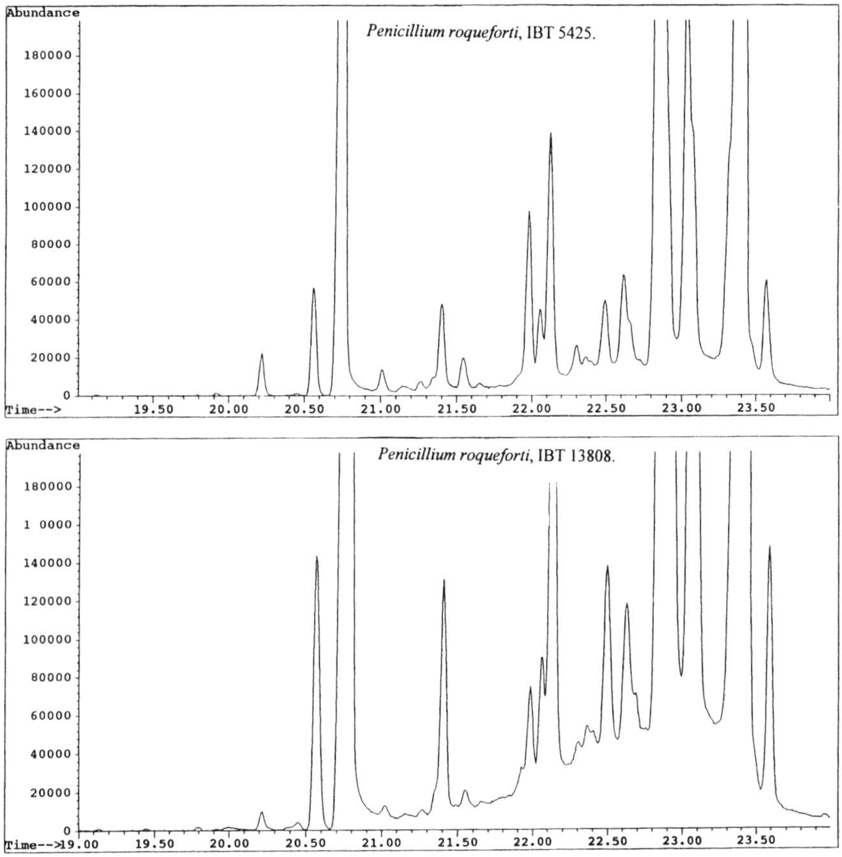

Figure 5 Profiles of volatile sesquiterpenes collected by solid-phase microextraction (68) from two isolates of *Penicillium roqueforti* (left) and two isolates of *Penicillium commune* (right).

The ability of fungi to produce a specific combination of volatile metabolites and thus aroma is widely used for industrial production of edible fungi and fermented food products. On the other hand, volatile metabolites have only scarcely been used for chemotaxonomic purposes. A major reason why some literature reports question the use of fungal volatiles in taxonomy is probably due to the use of only one or very few isolates of the species studied. Furthermore, some of these few isolates might be incorrectly identified. However, recent studies based on several isolates have shown that toxic and nontoxic isolates of both

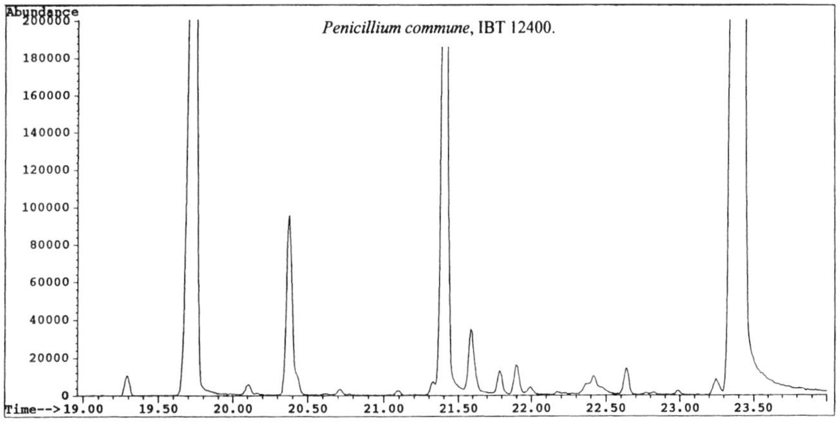

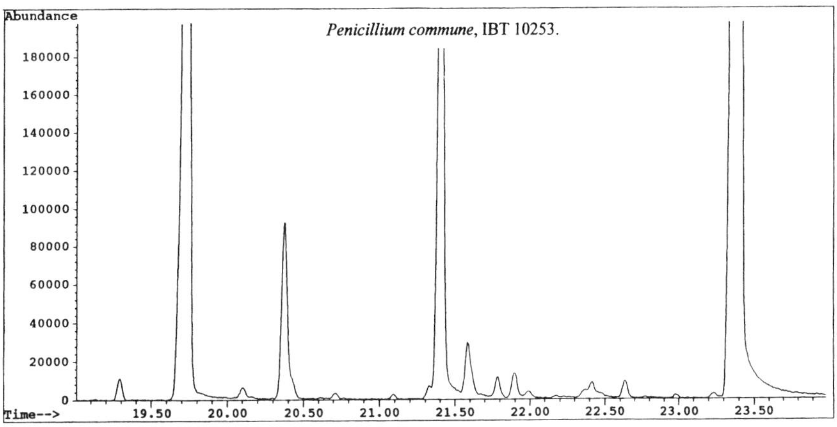

Figure 5 Continued

Aspergillus and *Fusarium* species can be distinguished from each other based on their production of sesquiterpenes. Similarly it has been demonstrated that a large number of related species in genus *Penicillium* could be classified based solely on their profiles of volatile metabolites, a finding that may be true for other genera. The scope for the use of fungal volatiles in the detection and classification of fungi is likely to be in the rapid detection of unwanted fungal growth and in the separation of closely related species that are difficult to distinguish by other methods.

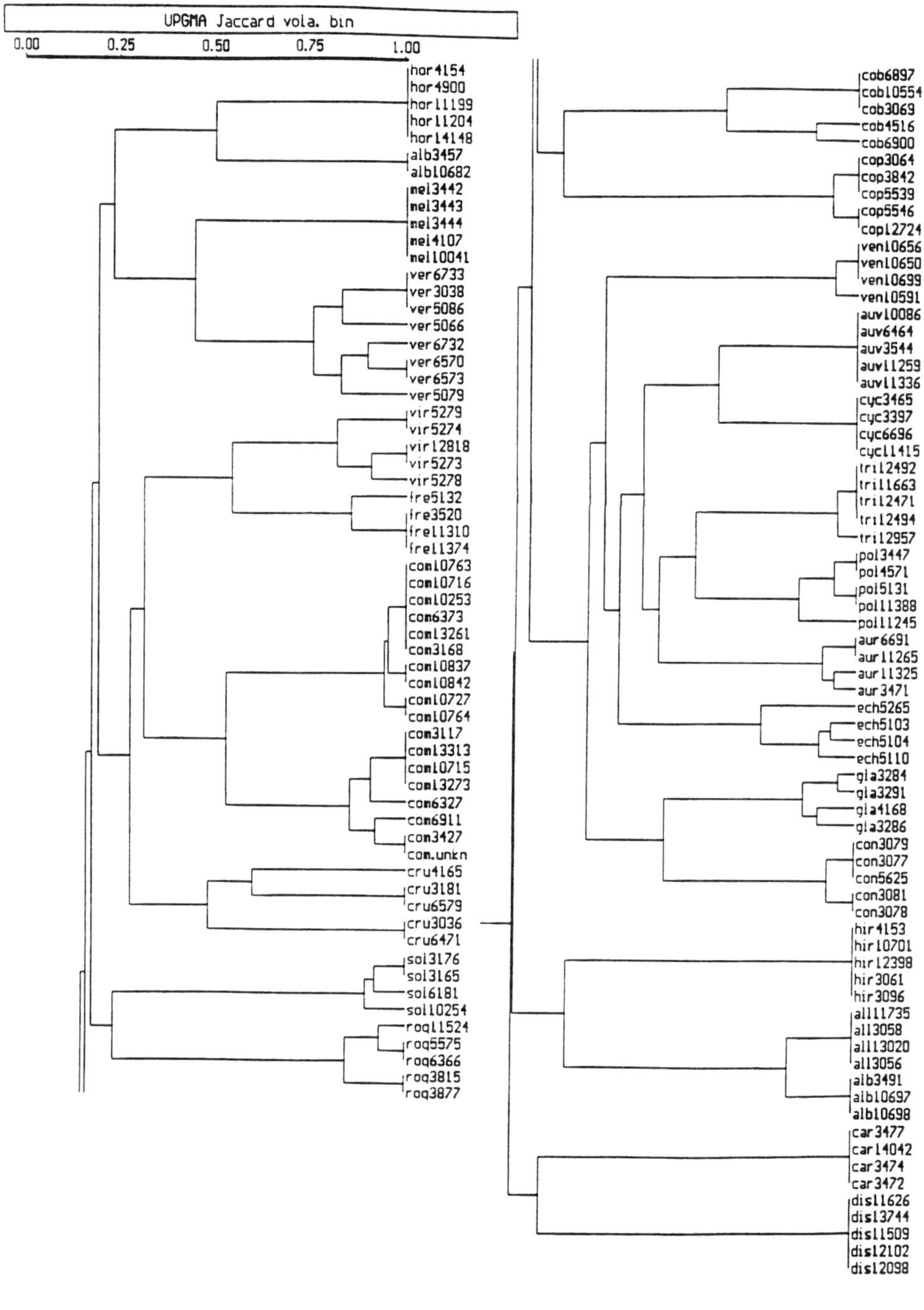
UPGMA Jaccard vola. bin
0.00
0.25
0.50
0.75
1.00
hor4154
hor4900
hor11199
hor11204
hor14148
alb3457
alb10682
mel3442
mel3443
mel3444
mel4107
mel10041
ver6733
ver3038
ver5086
ver5066
ver6732
ver6570
ver6573
ver5079
vir5279
vir5274
vir12818
vir5273
vir5278
fre5132
fre3520
fre11310
fre11374
com10763
com10716
com10253
com6373
com13261
com3168
com10837
com10842
com10727
com10764
com3117
com13313
com10715
com13273
com6327
com6911
com3427
com.unkn
cru4165
cru3181
cru6579
cru3036
cru6471
sol3176
sol3165
sol6181
sol10254
roq11524
roq5575
roq6366
roq3815
roq3877
cob6897
cob10554
cob3069
cob4516
cob6900
cop3064
cop3842
cop5539
cop5546
cop12724
ven10656
ven10650
ven10699
ven10591
auv10086
auv6464
auv3544
auv11259
auv11336
cyc3465
cyc3397
cyc6696
cyc11415
tri12492
tri11663
tri12471
tri12494
tri12957
pol3447
pol4571
pol5131
pol11388
pol11245
aur6691
aur11265
aur11325
aur3471
ech5265
ech5103
ech5104
ech5110
gla3284
gla3291
gla4168
gla3286
con3079
con3077
con5625
con3081
con3078
hir4153
hir10701
hir12398
hir3061
hir3096
all11735
all3058
all13020
all3056
alb3491
alb10697
alb10698
car3477
car14042
car3474
car3472
dis11626
dis13744
dis11509
dis12102
dis12098

Figure 6 UPGMA dendrogram of 132 *Penicillium* isolates using the Jaccard distance coefficient on a binary data matrix. The variables are the 131 different volatile fungal metabolites detected. Abbreviations of fungal names are the three first characters in the species names in combination with the culture collection number from the Fungal Culture Collection (IBT) at the Technical University of Denmark (e.g., *P. hordei*, IBT 4154 = hor4154, except for *P. coprobium* and *P. aurantiovirens*, which have the abbreviations cop and auv). Taken from Larsen and Frisvad (108).

ACKNOWLEDGMENTS

I thank Mycological Research for permission to include already published material in this chapter. I thank scientists and technical staff of the Mycology Group at Department of Biotechnology, Technical University of Denmark, for their invaluable help.

REFERENCES

1. Turner WB, Aldridge DC. Fungal Metabolites. Vol 2. London: Academic Press, 1983.
2. Luckner M. Secondary Metabolism in Microorganisms, Plants and Animals. Berlin: Springer-Verlag, 1990.
3. Raper KB, Thom C. A Manual of the Penicillia. Baltimore: Williams & Wilkins, 1949.
4. Hunt J. Taxonomy of the genus *Ceratocystis*. Lloydia 1956; 19:1–58.
5. Raper KB, Fennell DI. The Genus *Aspergillus*. Baltimore: Williams & Wilkins, 1965.
6. Ramírez C. Manual and Atlas of the Penicillia. Amsterdam: Elsevier Biomedical Press, 1982.
7. Hanssen H-P. Volatile meabolites produced by species of *Ophiostoma* and *Ceratocystis*. In: Wingfield MJ, Seifert KA, Webber JF, eds. Ceratocystic and Ophiostoma. Taxonomy, Ecology and Pathogenicity. St. Paul, Minn.: APS Press, American Phytopathological Society, 1995:117–125.
8. Larsen TO, Frisvad JC. Characterization of volatile metabolites from 47 taxa in genus *Penicillium*. Mycol Res 1995; 99:1153–1166.
9. Von Rudloff E. Scope and limitations of gas chromatography of terpenes in chemosystematic studies. Rec Adv Phytochem 1969; 2:127–162.
10. Harborne JB, Turner BL. Plant Chemosystematics. London: Academic Press, 1984.
11. Barberio J, Twibell J. Chemotaxonomy of plant species using headspace sampling, thermal desorption and capillary GC. J High Resolution Chromatogr 1991; 14: 637–639.
12. Adams RP. Geographic variation in the volatile terpenoids of *Juniperus monosperma* and *J. osteosperma*. Biochem Systemat Ecol 1994; 22:65–71.
13. Tollsten L, Knudsen J, Bergström LG. Floral scent in generalistic *Angelica* (Apiaceae)—an adaptive character? Biochem Systemat Ecol 1994; 22:161–169.

14. Herbert RB. The Biosynthesis of Secondary Metabolites. London: Chapman and Hall, 1981.
15. Stryer L. Biochemistry. New York: W.H. Freeman, 1981.
16. Campbell IM. Secondary metabolism and microbial physiology. Adv Microb Physiol 1984; 25:1–60.
17. Bennett JW, Bentley R. What's in a name? Microbial secondary metabolism. Adv Appl Microbiol 1989; 34:1–28.
18. Bennett JW. Differentiation and secondary metabolism in mycelial fungi. In: Bennett JW, Ciegler A, eds. Secondary Metabolism and Differentiation in Fungi. New York: Marcel Dekker, 1983:1–32.
19. Mantle PG. Secondary Metabolites of *Penicillium* and *Acremonium*. In: Peberdy JF, ed. *Penicillium* and *Acremonium*. Biotechnology Handbooks. Vol. 1. New York: Plenum Press, 1987:161–243.
20. Bu'lock JD. Mycotoxins as secondary metabolites. In: Steyn PS, ed. The Biosynthesis of Mycotoxins. New York: Academic Press, 1980:1–16.
21. Wicklow DT. Metabolites in the coevolution of fungal defence systems. In: Pirozynski KA, Hawksworth DL, eds. Coevolution of Fungi with Plants and Animals. London: Academic Press, 1988:173–201.
22. Williams DH, Stone MJ, Hauck PR, Rahman SK. Why are secondary metabolites (natural products) biosynthesized? J Nat Prod 1989; 52:1189–1208.
23. Dowd PF. Insect interactions with mycotoxin-producing fungi and their hosts. In: Bhatnagar D, Lillehøj EB, Arora DK, eds. Handbook of Applied Mycology. Vol. 5. Mycotoxins in Ecological Systems. New York: Marcel Dekker, 1992:137–155.
24. Frisvad JC. Classification of organisms by secondary metabolites. In: Hawkworth DL, ed. The Identification and Characterization of Pest Organisms. Wallingford: CAB International, 1994:303–321.
25. Hutchinson SA. Biological activities of volatile fungal metabolites. Annu Rev Phytopathol 1973; 11:223–246.
26. Christophersen C. Evolution in molecular structure and adaptive variance in metabolism. Comp Biochem Physiol 1991; 98B:427–432.
27. Kinderlerer J. Volatile metabolites of filamentous fungi and their role in food flavour. J Appl Bacteriol Symp Suppl 1989:133S–144S.
28. Hawke JC. Reviews of the progress of dairy science. Section D. Dairy chemistry. The formation and metabolism of methyl ketones and related compounds. J Dairy Res 1966; 33:225–243.
29. Kinsella JE, Hwang DH. Enzymes of *Penicillium roqueforti* involved in the biosynthesis of cheese flavour. Crit Rev Food Sci Nutr 1976; 8:191–228.
30. Pyysalo H. Identification of volatile compounds in seven edible fresh mushrooms. Acta Chem Scand 1976; B30:235–244.
31. Maga JA. Mushroom flavour. J Agric Food Chem 1981; 29:1–4.
32. Tressl R, Bahri D, Engel KH. Formation of eight-carbon and ten-carbon components in mushrooms (*Agaricus campestris*). J Agric Food Chem 1982; 10:89–93.
33. Karahadian C, Josephson DB, Lindsay RC. Contribution of *Penicillium* sp. to the flavor of Brie and Camembert cheese. J Dairy Sci 1985; 68:1865–1877.
34. Wurzenberger M, Grosch W. The enzymatic oxidative breakdown of linoleic acid in mushrooms (*Psalliota bispora*). Z Lebensmittel Untersuchung Forschung 1982; 175:186–190.

35. Wurzenberger M, Grosch W. Origin of the oxygen in the products of the enzymatic cleavage reaction of linoleic acid to 1-octen-3-ol and 10-oxo-*trans*-8-decenoic acid in mushrooms (*Psalliota bispora*). Biochim Biophys Acta 1984; 794:18–24.
36. Berry DR. Products of primary metabolic pathways. In: Berry DR, ed. Physiology of Industrial Fungi. Oxford: Blackwell Scientific Publications, 1988:130–160.
37. Fraga BM. Sesquiterpenoids. In: Charlwood BV, Banthorpe DV, eds. Methods in Plant Biochemistry. Vol. 7. London: Academic Press, 1991:145–185.
38. Seifert RM, King DA Jr. Identification of some volatile constituents of *Aspergillus clavatus*. J Agric Food Chem 1982; 30:786–790.
39. Chalier P, Crouzet J. Production of volatile components by *Penicillium roqueforti* cultivated in the presence of soya bean oil. Flavour Fragrance J 1993; 8:43–49.
40. Zeringue HJ, Bhatnagar D, Cleveland E. $C_{15}H_{24}$ volatile compounds unique to aflatoxigenic strains of *Aspergillus flavus*. Appl Environ Microbiol 1993; 59:2264–2270.
41. Ayer WA, Browne LM. Terpenoid metabolites of mushrooms and related basidiomycetes. Tetrahedron 1981; 37:2199–2248.
42. Lanza E, Palmer JK. Biosynthesis of monoterpenes by *Ceratocystis moniliformis*. Phytochemistry 1977; 16:1555–1560.
43. Bently R, Meganathan R. Geosmin and methylisoborneol biosynthesis in streptomycetes. FEBS Lett 1981; 125:220–222.
44. Gadd GM. Carbon nutrition and metabolism. In: Berry DR, ed. Physiology of Industrial Fungi. Oxford: Blackwell Scientific Publications, 1988:21–57.
45. Karahadian C, Josephson DB, Lindsay RC. Volatile compounds from *Penicillium* sp. contributing musty-earthy notes to Brie and Camambert cheese flavors. J Agric Food Chem 1985; 33:339–343.
46. Larsen TO, Frisvad JC. A simple method for collection of volatile metabolites from fungi based on diffusive sampling from Petri dishes. J Microbiol Methods 1994; 19: 297–305.
47. Leete E, Bjorklund JA, Reineccius GA, Cheng TB. Biosynthesis of 3-isopropyl-2-methoxypyrazine and other alkylpyrazines: widely distributed flavour compounds. In: Patterson RLS, Charlwood BV, MacLeod G, Williams AA, eds. Bioformation of Flavours. Cambridge: Royal Society of Chemistry, 1992:75–95.
48. Gervais P. Water activity: a fundamental parameter of aroma production by microorganisms. Appl Microbiol Biotechnol 1990; 33:72–75.
49. Chen C-C, Chen S-D, Chen J-J, Wu C-M. Effects of pH value on the formation of volatiles of shiitake (*Lentinus edodes*) an edible mushroom. J Agric Food Chem 1984; 32:999–1001.
50. Gallois A, Langlois D. New results in the volatile odorous compounds of French cheeses. Lait 1990; 70:89–106.
51. Larsen TO, Frisvad JC. Comparison of different methods for collection of volatile chemical markers from fungi. J Microbiol Methods 1995; 24:135–144.
52. Richard-Molard D, Cahagnier B, Poisson J, Drapon R. Evolutions comparées des constituants volatils et de la microflore de maïs stockés sous différentes conditions de température et d'humidité. Ann Technol Agric 1976; 25:29–44.
53. Dionigi C, Ingram DA. Effects of temperature and oxygen concentration on geosmin production by *Streptomyces tendae* and *Penicillium expansum*. J Agric Food Chem 1994; 42:143–145.

54. Börjesson T. Volatile fungal metabolites as indicators of mould growth in stored cereals. Ph.D. thesis. Swedish University of Agricultural Sciences. Uppsala. ISBN 91-576-4706-2, 1993.
55. Sprecher E, Hanssen H-P. Influence of strain specificity and culture conditions on terpene production by fungi. J Med Plant Res 1982; 44:41–43.
56. Yagi T, Kawaguchi M, Hatano T, Fukui F, Fukui S. Screening of methylalkylketone-accumulating fungi from type culture strains. J Ferment Bioeng 1990; 70:2, 94–99.
57. Yong L. The effect of carbon and nitrogen sources on the growth and aroma production of *Penicillium italicum*. In: Charalambous G, ed. Food Science and Human Nutrition. Amsterdam: Elsevier Science, 1992:115–122.
58. Bjurman J, Kristensson J. Production of volatile metabolites by the soft rot fungus *Chaetomium globosum* on building materials and defined media. Microbios 1992; 72:47–54.
59. Smith G. The effects of adding trace metals to Czapek-Dox culture medium. Trans Br Mycol Soc 1949; 32:280–283.
60. Abramson D, Sinha RN, Mills JT. Mycotoxin and odor formation in moist cereal grain during granary storage. Cereal Chem 1980; 57:346–351.
61. Börjesson T, Stöllman UM, Adamek P, Kaspersson A. Analysis of volatile compounds for detection of molds in stored cereals. Cereal Chem 1989; 66:300–304.
62. Halim AF, Narciso JA, Collins RP. Odorous constituents of *Penicillium decumbens*. Mycologia 1975; 67:1158–1165.
63. Kaminski E, Libbey LM. Stawicki S, Wasowicz E. Identification of the predominant volatile compounds produced by *Aspergillus flavus*. Appl Microbiol 1972; 24:721–726.
64. Vanhaelen M, Vanhaelen-Fastré R, Geeraerts J. Volatile constituents of *Trichothecium roseum*. Sabouraudia 1978, 16:141–150.
65. Charpentier BA, Sevenants MA, Sanders RA. Comparison of the effect of extraction methods on the flavor volatile composition of Shiitake mushrooms (*Lentinus edodes*) via GC/MS and GC/FTIR. In: Charalambous G, ed. The Shelf Life of Foods and Beverages. Proceedings of the 4th International Flavor Conference, Rhodes, Greece, 23–26 July 1985. Amsterdam: Elsevier, 1986:413–433.
66. Zhang Z, Pawliszyn J. Headspace solid-phase microextraction. Anal Chem 1993; 65:1843–1852.
67. Zhang Z, Pawliszyn J. Analysis of organic compounds in environmental samples by headspace solid phase microextraction. J High Resolution Chromatogr 1993; 16: 689–692.
68. Nilsson T, Larsen TO, Montanarella L, Madsen JØ. Application of solid-phase microextraction for analysis of volatile fungal metabolites. J Microbiol Methods 1996; 25:245–255.
69. Banthorpe DV. Classification of terpenoids and general procedures for their characterisation. In: Charlwood BV, Banthorpe DV, eds. Methods in Plant Biochemistry. Vol. 7. London: Academic Press, 1991:1–41.
70. Murphy RE, Heath RR, Dorsey JG. The optimization of capacity and efficiency when coupling fused silica open tubular columns in gas chromatography. Chromatographia 1993; 37:65–72.
71. Schreier P, Idstein H. Advances in the instrumental analysis of food flavours. Z Lebensmittel Untersuchung Forschung 1985; 180:1–14.

72. Grob K. Split and Splitless Injection in Capillary GC. Heidelberg: Hüthig Buch Verlag, 1993.
73. Full G, Wintherhalter P, Schmidt G, Herion P, Schreier P. MDGC-MS: a powerful tool for enantioselective flavor analysis. J High Resolution Chromatogr 1993; 16: 642–644.
74. Guichard E, Mosandl A, Hollnagel A, Latrasse A, Henry R. Chiral γ-lactones from *Fusarium poae*. Enantiomeric ratios and sensory differentiation of *cis*-6-γ-dodecenolactone enantiomers. Z Lebensmittel Untersuchung Forschung 1991; 193:26–31.
75. Davies NW. Gas chromatographic retention indices of monoterpenes and sesquiterpenes on methyl silicone and Carbowax 20M phases. J Chromatogr 1990; 503: 1–24.
76. Ramaswami SK, Briscese P, Gargiullo J, Geldern T. Sesquiterpene hydrocarbons: from mass confusion to orderly line-up. In: Lawrence BM, Mookherjee BD, Willis BJ, eds. Flavors and Fragrances: A World Perspective. Proceedings of the 10th International Congress of Essential Oils, Fragrances and Flavors, Washington, D.C. 16–20 November 1986. Amsterdam: Elsevier Science, 1988:951–980.
77. Herres W. HRGC-FTIR: Capillary Gas Chromatography—Fourier Transform Infrared Spectroscopy (Theory and Application). New York: Dr. Alfred Hüthig Verlag, 1987.
78. Kubezcka K-H, Schultze W, Formacek V, Herres W. New developments in essential oils analysis by Fourier-transform spectroscopy. In: Lawrence BM, Mookherjee BD, Willis BJ, eds. Flavors and Fragrances: A World Perspective. Proceedings of the 10th International Congress of Essential Oils, Fragrances and Flavors, Washington, D.C. 16–20 November 1986. Amsterdam: Elsevier Science, 1986:931–950.
79. Aishima T. Discrimination of headspace flavor volatiles using semiconductor gas sensor array and pattern recognition techniques. In: Bessiére Y, Thomas AF, eds. Flavour Science and Technology. New York: John Wiley, 1990:199–202.
80. Schnürer J, Essen C, Winquist F, et al. An electronic nose for the detection of mould growth in cereals. Abstract at ICMF-3 workshop. Copenhagen, June 26–30, 1994.
81. Pierce AM, Pierce HD, Oehlschlager AC, Borden JH. 1-octen-3-ol, attractive semiochemical for foreign grain beetle, *Ahasverus advena* (Waltl) (Coleoptera: Cucujidae). J Chem Ecol 1991.
82. Pierce AM, Pierce HD, Borden JH, Oehlschlager AC. Fungal volatiles: semiochemicals for stored-product beetles (Coleoptera: Cucujidae). J Chem Ecol 1991.
83. Vanhaelen M, Vanhaelen-Fastré R, Geeraerts J, Wirthlin T. *Cis*- and *trans*-octa-1,5-dien-3-ol, new attractants to the cheese mite *Tyrophagus putrescentiae* (Schrank) (Acarina, Acaridae) identified in *Trichotecium roseum* (fungi imperfecti). Microbios 1979; 23:199–212.
84. Tariq V-N, Campbell VM. Influence of volatile metabolites from *Geoticum candidum* on other fungi. Mycol Res 1991; 95:891–893.
85. Lanciotti R, Guerzoni ME. Competitive inhibition of *Aspergillus flavus* by volatile metabolites of *Rhizopus arrhizus*. Food Microbiol 1993; 10:367–377.
86. Moore-Landecker E, Stotzky G. Inhibition of fungal growth and sporulation by volatile metabolites from bacteria. Can J Microbiol 1972; 18:957–962.
87. Barr JG. Effects of volatile bacterial metabolites on the growth, sporulation and mycotoxin production of fungi. J Sci Food Agric 1976; 27:324–330.

88. Beattie SE, Torrey GS. Toxicity of methanethiol produced by *Brevibacterium linens* toward *Penicillium expansum*. J Agric Food Sci 1986; 34:102–104.
89. Kaminski E, Wasowicz E. The usage of volatile compounds as indicators of grain deterioration. In: Chelkowski J, ed. Cereal Grain. Mycotoxins, Fungi and Quality in Drying and Storage. New York: Elsevier, 1991:229–258.
90. Wilkins CK, Scholl S. Volatile metabolites of some barley storage molds. Int J Food Microbiol 1989; 8:11–17.
91. Kaminski E, Stawicki S, Wasowicz E. Volatile flavor compounds produced by molds of *Aspergillus*, *Penicillium*, and fungi imperfecti. Appl Microbiol 1974; 27:1001–1004.
92. Wasowicz E, Kaminski E, Kollmannsberger H, Nitz S, Berger G, Drawert F. Volatile components of sound and musty wheat grains. Chem Mikrobiol Technol Lebensmittel 1988; 11:161–168.
93. Börjesson T, Stöllman UM, Schnürer J. Volatile metabolites and other indicators of *Penicillium aurantiogriseum* growth on different substrates. Appl Environ Microbiol 1990; 56:3705–3710.
94. Börjesson T, Stöllman UM, Scnhnürer J. Volatile metabolites and other indicators of *Penicillium aurantiogriseum* growth on different substrates. Appl Environ Microbiol 1992; 58:2599–2605.
95. Abramson D, Sinha RN, Mills JT. Mycotoxin and odor formation in barley stored at 16 and 20% moisture in Manitoba. Cereal Chem 1983; 60:350–355.
96. Sinha RN, Tuma D, Abramson D, Muir WE. Fungal volatiles associated with moldy grain in ventilated and non-ventilated bin-stored wheat. Mycopathologia 1988; 101: 53–60.
97. Tuma D, Sinha RN, Muir WE, Abramson D. Odor volatiles associated with microflora in damp ventilated and non-ventilated bin stored bulk wheat. Int J Food Microbiol 1989; 8:103–119.
98. Mattheis JP, Roberts RG. Identification of geosmin as a volatile metabolite of *Penicillium expansum*. Appl Environ Microbiol 1992; 58:3170–3172.
99. Larsen TO, Frisvad JC. Production of volatiles and presense of mycotoxins in conidia of common indoor *Penicillium* and *Aspergillus* species. In: Samson RA et al., eds. Health Implications of Fungi in Indoor Environments. Amsterdam: Elsevier Science, 1994:251–279.
100. Samson RA. Occurrence of moulds in modern living and working environments. Eur J Epidemiol 1985; 1:54–61.
101. Flannigan B, McCabe EM, McGarry F. Allergenic and toxigenic microorganisms in houses. J Appl Bacteriol Symp Suppl 1991; 70:61S–73S.
102. Miller JD, Laflamme AM, Sobol Y, Lafontaine P, Greenhalgh R. Fungi and fungal products in some Canadian houses. Int Biodeterioration 1988; 24:103–120.
103. Sorenson W. Health impact of mycotoxins in the home and workplace—an overview. In: O'Rear CE, Llewellyn GC, eds. Biodeterioration Research. 2nd ed. New York: Plenum Press, 1989:201–215.
104. Berger RG, Neuhäuser K, Drawert F. Characterization of odour principles of some basidiomycetes: *Bjerkandera adusta*, *Poria aurea*, *Tyromyces sambuceus*. Flavour Fragrance J 1986; 1:181–185.
105. Pacioni G, Bellina-Agostinone C, D'Antonio M. Odor composition of the *Tuber melanosporum* complex. Mycol Res 1990; 94:201–204.

106. Jollivet N, Belin J-M. Comparison of volatile flavor compounds produced by ten strains of *Penicillium camemberti* Thom. J Dairy Sci 1993; 76:1837–1844.
107. Emokhonow VN, Groznov IN, Monastirskii OA, Permogorov A. Detection of grain infection with specific toxigenenous fungal species. In: Fenselau C, ed. Mass Spectrometry for the Characterization of Microorganisms. Washington, DC: American Chemical Society, 1994:85–90.
108. Larsen TO, Frisvad JC. Chemosystematics of species in genus *Penicillium* based on profiles of volatile metabolites. Mycol Res 1995; 99:1167–1174.
109. Gallois A, Gross B, Langlois D, Spinnler HE, Brunerie P. Influence of culture conditions on production of flavor compounds by 29 lignolytic *Basidiomycetes*. Mycol Res 1990; 94:494–504.
110. Gross B, Gallois A, Spinnler H-E, Langlois D. Volatile compounds produced by the lignolytic fungus *Phlebia radiata* Fr. (Basidiomycetes) and influence of the strain specificity on the odorous profile. J Biotechnol 1989; 10:303–308.
111. Frisvad JC. The connection between the penicillia and aspergilli and mycotoxins with special emphasis on misidentified isolates. Arch Environ Contam Toxicol 1989; 18:452–467.
112. Jelén HH, Mirocha CJ, Wasowicz E, Kaminski E. Production of volatile sesquiterpenes by *Fusarium sambucinum* strains with different abilities to synthesize trichothecenes. Appl Environ Microbiol 1995; 61:3815–3820.
113. Frisvad JC, Filtenborg O. Classification of terverticillate penicillia based on profiles of mycotoxins and other secondary metabolites. Appl Environ Microbiol 1983; 46: 1301–1309.
114. Frisvad JC, Filtenborg O. Terverticillate penicillia: chemotaxonomy and mycotoxin production. Mycologia 1989; 81:837–861.
115. Lund F, Frisvad JC. Chemotaxonomy of *Penicillium aurantiogriseum* and related species. Mycol Res 1994; 98:481–492.
116. Svendsen A, Frisvad JC. A chemotaxonomic study of the terverticillate penicillia based on high performance liquid chromatography of secondary metabolites. Mycol Res 1994; 98:1317–1328.
117. Lund F, Frisvad JC, Filtenborg O. Associated mycoflora on cheese. Food Microbiol 1995; 12:173–180.
118. Lund F. Diagnostic characterization of *Penicillium palitans*, *P. commune* and *P. solitum*. Lett Appl Microbiol 1995; 21:60–64.

12

Role and Use of Secondary Metabolites in Fungal Taxonomy

Jens C. Frisvad, Ulf Thrane, and Ole Filtenborg
Technical University of Denmark, Lyngby, Denmark

I. INTRODUCTION

Secondary metabolites are products of normal cellular metabolism that are more restricted in their distribution, being found in less than every species in a single fungal family (1). They can also be regarded as outward-directed (extrovert) differentiation products that function as chemical signals between organisms or species (2–4). Fungal species have traditionally been characterized by similar morphological differentiation in representative isolates, and consequently chemical differentiation products could also be expected to be species-specific. For example, the basidiomycete *Chalciporus piperatus* (formerly *Suillus piperatus*) can be considered to be characterized by its chemical features (the peppery taste, the red gills, and the yellow basidiocarp stipe) rather than morphological ones. One chemotaxonomic approach to the characterization of that species could thus be to analyze chemically or elucidate the structures of the secondary metabolites responsible for the specific taste, smell, and colors of representative isolates in that particular taxon and compare the results with other species of the same genus. For instance, the peppery taste in *C. piperatus* is caused by chalciporone, chalciporonyl propionate, isochalciporone, and dehydrochalciporone, which until now have been found only in that taxon (5). Such comparative studies are rather rare in mycology, with the exception of the extensive use of secondary metabolite data in the taxonomy of lichenized fungi.

Secondary metabolites have not been used extensively in fungal taxonomy. Early reviews of fungal chemotaxonomy were descriptive and rarely comparative for the simple fact that few chemical data were available (6–9). The compounds were ordered artificially, for example, as acetylenic compounds, pigments, tetronic acids, etc. (8,9). Turner (10) and Turner and Aldridge (11) suggested a subdivision of fungal secondary metabolites according to biosynthetic origin—e.g., compounds originating from shikimic acid, polyketides, compounds from the tricarboxylic acid cycle, terpenes, compounds from amino acids and mixed compounds—but treated the compounds from a biochemical rather than a taxonomic point of view. Mantle (12) reviewed *Penicillium* secondary metabolites from the same biosynthetic point of view and admitted that the renaming of original isolates according to new taxonomic systems may lead to errors, especially when those isolates are no longer available to the scientific community. A large number of fungi which produce secondary metabolites have been misidentified in *Penicillium*, and this situation may also apply to other genera (13,14). For this reason a large number of isolates in each species need to be examined.

Other reviews have also been written from the point of view of organic chemistry or biological activity rather than chemotaxonomy (15). The former approach has been especially successful when taxonomists have been working together with organic chemists (6,19), while the latter method needs taxonomists with a knowledge of basic chemical separation and detection methods.

Chemotaxonomic studies of a large number of isolates and taxa in the xylariaceous fungi and *Penicillium* have shown, however, that secondary metabolites have a potential for the characterization of species and for phylogenetic relationships (2,6). Thus, secondary metabolites complement morphological data to give a fuller description of an important part of the phenotype that may be perceived by other organisms. Chemical analysis of secondary metabolites will provide more objective and comparable results than traditional descriptions of color and odor (22). In several genera of filamentous fungi, morphological data are limited or very variable, and in these cases secondary metabolites have helped in resolving taxonomic problems. The taxonomy of species in the genus *Penicillium* is a good example. The foodborne terverticillate penicillia have often been characterized as difficult to classify by traditional characters (24–28). Chemotaxonomic studies of many of these species have shown, however, that secondary metabolites alone can give very clear classifications for some terverticillate penicillia (29–31). The relative merit of taxonomies based on morphology, secondary metabolites, physiology, or molecular characters will not be evaluated here, but it is the aim of this chapter to show that secondary metabolites have been valuable, occasionally even indispensable, in classifications that are expected to be unequivocal and stable. The role and applications of secondary metabolites in the taxonomy of filamentous fungi are reviewed.

II. ANALYTICAL METHODS FOR SEPARATION AND DETECTION OF SECONDARY METABOLITES

A. Thin-Layer Chromatography

Gross separations were originally introduced using paper chromatography or chemical paper tests, but these have not been used recently in chemotaxonomy except in a few simple diagnostic tests (9,32–35). Thin-layer chromatography (TLC) has been the method of choice for many years in chemotaxonomic studies concerning nonvolatile secondary metabolites (36–42), but this can now be replaced or rather supplemented by the more efficient HPLC methods described below. Nearly all known secondary metabolites can be detected by TLC, but several systems have to be used in order to separate all fungal secondary metabolites (22,37,43).

B. Gas Chromatography

Gas chromatography (GC) has mostly been used for volatile chemical compounds (See Chapter 11, by Larsen, in this volume). However, in some cases GC has also been used for less volatile compounds without derivatization, especially in connection with mass spectrometric detection (44). Some nonvolatile secondary metabolites can be separated without thermal degradation by GC, for instance, certain xanthones, anthraquinones, and usnic acid (45,46), but most other examples involve chemotaxonomic studies based on GC-MS or mycotoxin analysis using silylation (44,47–56). Whereas most chemotaxonomic studies which include *Fusarium* secondary metabolites are now based on HPLC analysis (57), GC-MS has the advantage of confirming identifications using MS data. Most identifications have been based on comparison with authentic standards and mass spectra. "Lichen mass spectrometry" was suggested by Santesson in 1970 (46) and involves introducing a small sample of lichen directly into the inlet system of the MS apparatus. With this method some of the xanthones gave recognizable molecular ions with some low mass decomposition products, but usnic acid and pulvinic acid derivatives gave complex mixtures of mass peaks.

C. High-Performance Liquid Chromatography

High-performance liquid chromatography (HPLC) has become the method of choice for most nonvolatile secondary metabolites (37,41,57). HPLC using gradient elution on reversed-phase material allows a good separation of a large range of secondary metabolites with different polarities, and UV and fluorescence detectors can be used to detect small amounts of most compounds with a chromophore. Identification or partial characterization of the compounds can be achieved with a diode array detector, and many secondary metabolites have characteristic UV spectra (10,11,41,57–59). Correct identification of secondary metabolites requires that analytical standards be available. Retention indices, such as those based on a series of alkylphenones, can be used to minimize the effect of different HPLC conditions on the retention time, but should be used with caution between laboratories (57). There have been few HPLC-MS applications, and these have not been used in purely chemotaxonomic secondary metabolite studies.

D. Micellar Capillary Electrophoresis

Only a few papers have been published on the sue of micellar capillary electrophoresis (MCE) of fungal secondary metabolites, which was developed for chemotaxonomic applications in mycology by Nielsen et al. (66). The large number of secondary metabolites produced by some isolates of fungi can only be separated effectively using MCE with its large number of theoretical plates; however, the less effective separation in HPLC can be circumvented by using diode array detection. Such a three-dimensional dataset can be treated by chemometric methods to obtain pure spectra and good estimations of the amounts of all secondary metabolites (61).

E. Flow Injection Electrospray Mass Spectrometry

This new method is based on the limited fragmentation pattern and high sensitivity of electrospray mass spectrometry (ESMS) (62,63). ESMS has been used mostly for the accurate mass determination of proteins and other macromolecules, but it can also be used for smaller molecules (64–67). Injecting a small volume of a fungal extract can give a "mass profile" of the fungus, consisting of peaks for the protonated secondary metabolites and isotopes (68). This profile can be used for classification, identification, or verification of the presence of known secondary metabolites in a mixture (31,69). Chlorinated compounds such as griseofulvin are easily recognized by their characteristic isotope pattern. The combination of data from TLC, HPLC, and ESMS is particularly useful in verifying the presence of secondary metabolites in complex mixtures, especially if analytical standards are also available. Thus the identity of a secondary metabolite could be verified by comparison with a standard concerning, first, similar retardation factors in two eluents using normal-phase TLC systems and similar color reactions after chemical spray reagent treatments; second, similar retention times or indices and UV spectra in reversed-phase HPLC using diode array detection (42,43,70); and, third, by the presence of the mass ion (+1) in the mass profile and occasionally supplemented with comparison of isotope pattern to the standard (68).

F. Ultraviolet and Diode Array Detection

Diode array detection is particularly relevant in connection with HPLC but can also be used with MCE. The ultraviolet (UV) spectra of secondary metabolites can be used to identify families of compounds with similar chromophores. Often these chromophore families are the same as biosynthetic families, but there are examples where the chromophore changes considerably during the biosynthesis of a secondary metabolite due to molecular rearrangements and addition or disappearence of conjugated double bonds. Thus, methoxylations, the addition of a dimethylallyl group, and several other common changes during the biosynthesis of secondary metabolites will only change the chromophore slightly, whereas extra conjugated double bonds will change a chromophore considerably. It is thus valuable to know the biosynthetic route of the secondary metabolites from a group of fungi that is being examined in a chemotaxonomy.

Reflectance UV spectra can be used in connection with the scanning of TLC plates, but has so far only been used to a very limited degree in fungal chemotaxonomy (71,72). A good separation is needed to obtain valuable reflectance UV spectra, and often two-dimensional high-performance TLC is necessary to achieve such a separation (42). Furthermore, there are few data on reflectance spectra on silica gel for most secondary metabolites, so the use of pure standards is very important.

G. Fluorescence

Certain secondary metabolites fluoresce strongly and so can be detected in very low amounts with a fluorescence detector. These detectors can often be placed after a diode array detector in HPLC analyses and thus result in better sensitivity for some secondary metabolites. Such analyses may also help in confirming the identity of any fluorescing compounds. An obvious application of fluorescence measurements would be to analyze fungal extracts directly using perhaps four different excitation wavelengths and then comparing the emission spectra chemometrically.

H. Nuclear Magnetic Resonance Detection

Nuclear magnetic resonance (NMR) has been suggested for use in identification of human pathogenic bacteria, provided a test set of many bacteria has been developed (73). NMR or HPLC-NMR has also been used for direct quantification and identification of amino acids and other chemical compounds at very low levels of detection (74,75). These methods appear to be potentially useful in fungal chemical taxonomy; however, the use of this method is still to be exploited for taxonomic purposes.

I. Chemometric Methods

Many secondary metabolites are nonreducible unit characters, in contrast to some morphological features and can therefore be treated as binary characters in numerical analyses (30,76,77). Results based on these binary characters can be analyzed by cluster or correspondence analysis.

Though secondary metabolites are only a part of the fungal biomass, they have properties that may dominate when using certain methods. For example, the color of organic solvent extracts of fungi are mostly caused by secondary metabolites and, in rare cases, their polymerization products. As an alternative to the partial or full identification of all secondary metabolites present in a fungus, unpurified extracts may be characterized by multivariate chromatographic or spectrometric methods, such as UV-VIS spectrometry, near infrared spectrometry, mass spectrometry, pyrolysis GC, or pyrolysis MS (78,79). These methods can also be used for detection of cell wall and cytoplasmic components, although these may be the same in closely related fungi. Thus chemometric methods such as SIMCA Disjoint Principal Component Analysis or Discriminant Analysis may be used for the detection of differences between species (Bridge and Saddler, Chap. 2, this volume).

III. DIRECT AND INDIRECT USE OF SECONDARY METABOLITES IN TAXONOMY

Secondary metabolites have been a significant part of fungal descriptions and identifications, but have usually been included as morphological rather than chemical features. Important features, used in diagnoses, descriptions, and keys include color and odor of basidiocarp and ascocarp, and color of meiospore, mycelium, exudate, and soluble pigment. Mitospore (conidium) color is a special problem as it will often include both color from melanin-protein complexes and mixtures of secondary metabolites that exhibit visible colors. Whereas the brown or green colors caused by melanins can be useful at the genus or subgenus level, some color differences of specific significance are probably caused by a mixture of the amount and quality of protein linked to melanin and secondary metabolites. Colorless secondary metabolite mixtures can be visualized by chemical spot tests or by the use of UV illumination, approaches that have been used in both ascomycete and basidiomycete taxonomy (33,34). Some of the chemical spot tests may also indicate enzyme activity, complex formation, and pH-mediated changes in color. All these characters may be of value in routine identification work, although they are often variable and can only be subjectively recorded. Qualitatively, their value is limited, as, for example, a red color in one fungus may be due to a completely different mixture of secondary metabolites in another fungus with approximately the "same" red color, and clearly there is a problem of character homology in such comparisons. Quantification of such characters can also present problems. These types of problems have been widely discussed with the use of unordered multistate characters, such as, for example, flower petals red, green, blue, or yellow (80). One solution is to avoid the use of such characters in systematics and perhaps reserve them for diagnostic purposes.

There are different opinions on the use of secondary metabolites in taxonomy. Some workers claim that secondary metabolites are strain-specific (81,82), whereas others believe that the secondary metabolites present or produced in culture are very sensitive to environmental factors (83–86) in contrast to molecular genetic data. These critisisms have, however, been rejected in many major studies involving several strains of each taxon (6,16–18,29–31,40,71,87–92). The problem of the influence of the environment on phenetic features, including micromorphology and secondary metabolites, can be solved by careful attention to optimization and standardization of culture conditions and to work with "normkultur" (93): "one in which all the forms characteristic of a fungus are present and of good development" (94). In the case of secondary metabolites, this would involve giving the fungi under consideration the best conditions suitable for the expression of all differentiation products. That in turn may require the combination of conditions and/or extensive testing or sampling and will therefore present several practical problems.

This testing will not be possible for fungal isolates that are difficult to cultivate, for example, mycorrhizal and lichenized fungi, but extensive sampling and observation and a knowledge of the ecology of the organisms involved are of paramount importance in these cases if the full potential for production of secondary metabolites is to be realized.

Primary metabolites have also been used for classification purposes, but as they are present in nearly all fungi, they have to be quantified. Profiles of amino acids were used to discriminate between orders and families in basidiomycetous fungi (95,96), but a report by Krzeczkowska et al. (97) indicated that they were less suitable for species differentiation. Similarly polyols are of major value for taxonomic levels higher than species level (Pfyffer, Chap. 10, this volume), but may be less informative in species differentiation (77). Profiles of free fatty acids, which have been of value in chemotaxonomy at the species level in bacteria and yeasts, have later been shown to be somewhat variable and may present problem in species recognition in yeasts (98–100). On the other hand, the nature of lipids in procaryotic organisms is extremely complex, and some of the cell wall and structural lipids in fungi may also reflect differentiation and secondary rather than primary metabolism (101–103).

IV. SECONDARY METABOLITES AS EXPRESSIONS OF DIFFERENTIATION

A major critisism raised against using secondary metabolites is that in many cases the same secondary metabolite can be produced by unrelated species, even between different fungal orders or kingdoms (Table 1). In some cases secondary

Table 1 Examples of Secondary Metabolites Produced by Phylogenetically Unrelated Species Within Fungi, Plants, Bacteria, and Animals

Metabolite family	Examples		Kingdom	Species
Polyketides	Emodin	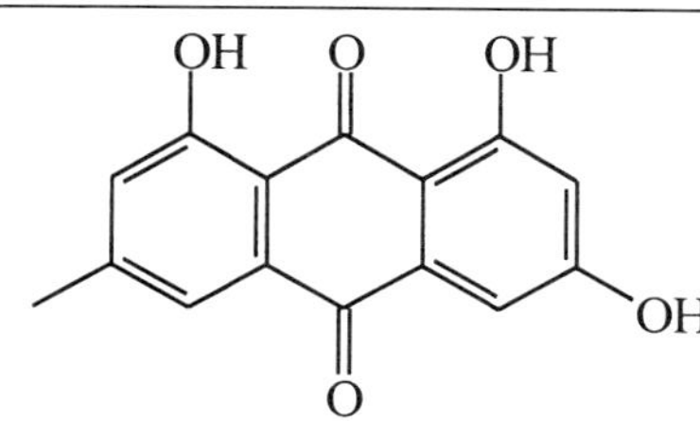	Plant	*Rheum* sp. (Rhubarb) (10)
			Fungi[a]	*Penicillium brunneum* (148)
				Penicillium tardum[b] (150)
				Hamigera avellanea (140)
				Eurotium cristatum[b] (110
				Aspergillus wentii (154)
				Penicilliopsis clavariaeformis (149)
			Fungi[c]	*Cladosporium fulvum* (108)
			Fungi[d]	*Nephroma laevigata* (139)
				Dermocybe sanguinea[b] (136
	Griseofulvin	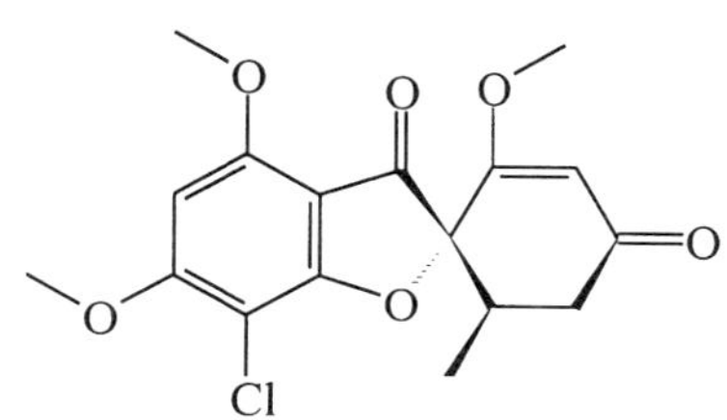	Fungi[a]	*Thamnomyces chordalis* (6)
				Xylaria psamathos (6)
				Penicillium griseofulvum (143)
				Penicillium janzewskii[b] (114)
				Aspergillus cf. *versicolor* (135)
				Khuskia oryzae (115)
				Khuskia sacchari (115)
				Nigrospora oryzae (124)
				Memnoniella echinata (134)

	Citrinin	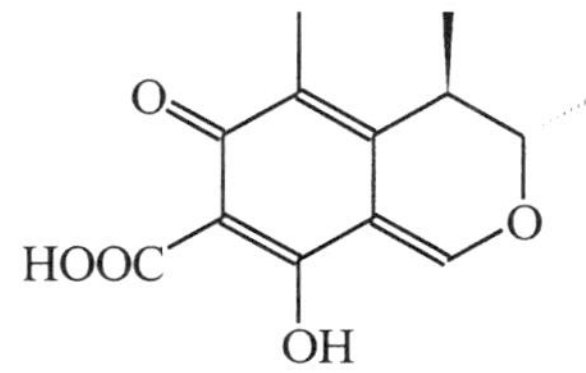	Fungi[a]	*Penicillium citrinum* (131)
				Penicillium verrucosum (16)
				Aspergillus carneus (121)
				Aspergillus terreus (144)
				Clavariopsis aquatica (115)
				Blennoria sp. (115)
			Fungi[e]	*Pythium ultimum* (126)
			Plant	*Crotolaria crispata* (127)
	Sterigmatocystin	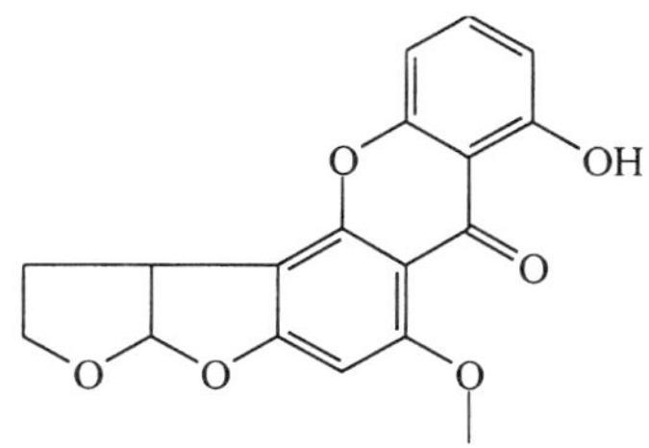	Fungi[a]	*Monocillium nordinii* (112)
				Chaetomium udagawae (152)
				Chaetomium thielavioideum (151)
				Farrowia sp. (151)
				Aspergillus versicolor (123)
				Aspergillus flavus[b] (146)
				Emericella nidulans[b] (133)
				Bipolaris sorokiniana (133)
Terpenes	Limonene	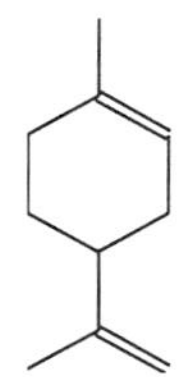	Fungi[a]	*Penicillium brasilianum* (92)
				Penicillium vulpinum[b] (92)
			Plants	*Citrus limon* (125)
				Thymus moroderi (118)
				Eucalyptus spp. (138)
			Animals	*Oxycarenus hyalinipennis* (142)
				Hotea gambiae (130)

Table 1 Continued

Metabolite family	Examples		Kingdom	Species
	Taxol	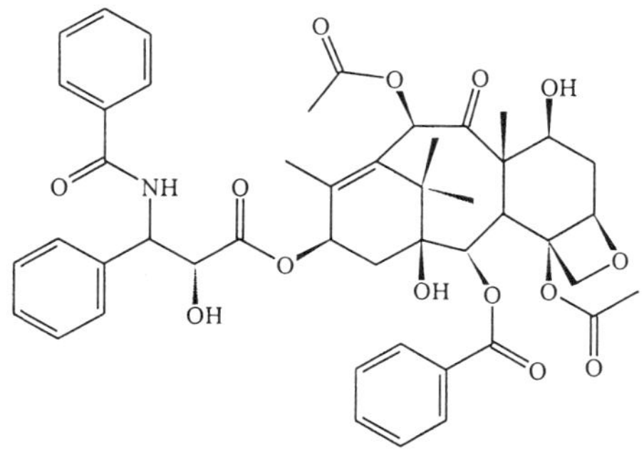	Fungi[a] Plant	*Taxomyces andreanea* (147) *Taxus brevifolia* (147)
Amino acid derived	3-nitropropionic acid	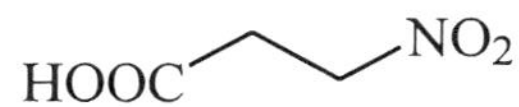	Fungi[a] Bacteria Plants 	*Aspergillus flavus* (117) *Pencillium atrovenetum* (145) *Arthrinium sacchari* (137) *Arthrinium saccharicola* (137) *Streptomyces* sp. (111) *Hiptage madoblota* (129) *Indigophera endecaphylla* (122) *Corynecarpus laevigata* (119)
	Chrysogine	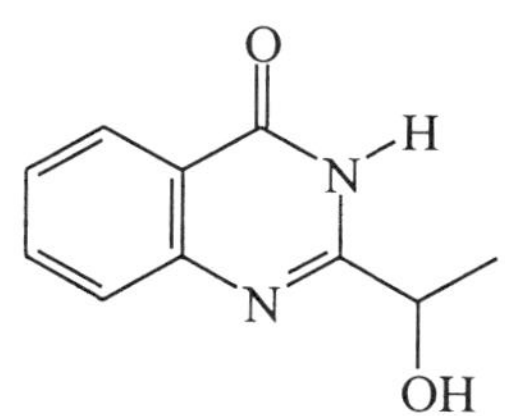	Fungi[a]	*Alternaria citri* (120) *Fusarium sambucinum* (141) *Fusarium culmorum* (113) *Penicillium chrysogenum*[b] (132)

Amino acid derived (cont.)	Chaetoglobosin C	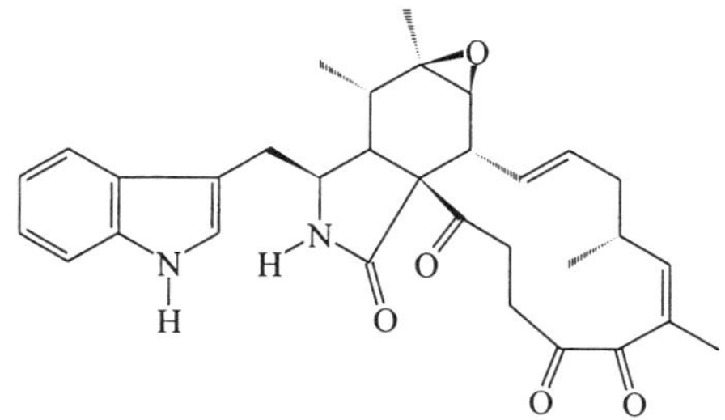	Fungi[a]	*Chaetomium globosum* (151) *Chaetomium cochlioides*[b] (151) *Penicillium expansum*[b] (16)
	Cytochalasin E	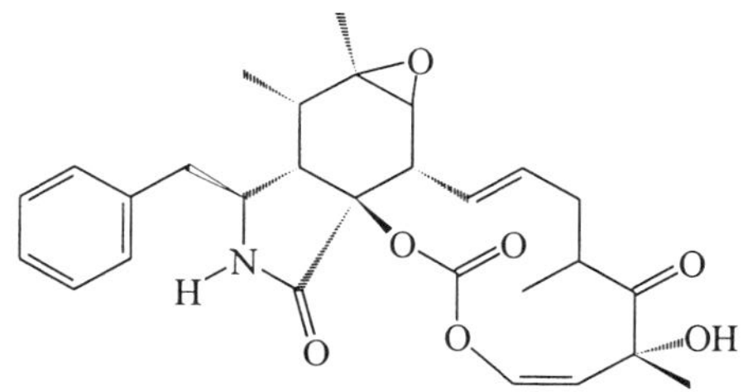	Fungi[a]	*Rosellinia necatrix* (109) *Hypoxylon terricola* (6) *Aspergillus clavatus* (124)

[a]Ascomycetes.
[b]And several other species in the same genus.
[c]Lichenized Ascomycetes.
[d]Basidiomycetes.
[e]Oomycetes.

metabolites are produced by very closely related species only and probably indicate close phylogenetic relationship. For instance, the amino acid- and terpene-containing secondary metabolite roquefortine C has only been found in several closely related species of *Penicillium* (16). Another example is the zaragozic acids, which are only produced by some species in the Pleosporales and Onygenales, but not in any basidiomycetous or zygomycetous fungi or procaryotes (107).

Secondary metabolites produced by organisms belonging to different kingdoms appear to be derived from completely different biosynthetic routes. For example, 3-nitropropionic acid is produced from L-aspartate in fungi and from malonate in the plant *Indigofera spincata* (154). Similarly, emodin is also produced from different pathways in fungi and in plants (155). Emodin and related compounds are produced by both basidiomycetous and ascomycetous fungi and from quite unrelated groups within each phylum (Table 1) (11,21). It therefore appears that the biosynthetic routes to secondary metabolites such as emodin have evolved more than once, and thus homoplasies in unrelated groups of fungi may be expected. The transfer of certain species from one genus to another based on single metabolites should therefore not be undertaken without other corroborating evidence (6,21,156). Crowfords (157) remarks, "given similar selection pressures, the same compounds may arise independently in distantly related plants. Clearly, this has occurred with floral anthocyanins, where the distribution of particular compounds is related to pollinators"; it could also be expected that fungal secondary metabolites could have arisen more than once. Thus secondary metabolites are extremely important for the ecotype, but obviously less important regarding phylogenetic interpretations, at least in distantly related taxa.

It is important to note that while individual secondary metabolites are of value in chemotaxonomy, it is the profile of secondary metabolites that is valuable in characterizing fungal species. The secondary metabolite profile (SMP) introduced by Frisvad and Filtenborg (40) in *Penicillium* taxonomy has later been modified to include other expressions of differentiation, but also to include considerations of biosynthetic families (16,158). The advantage in emphasizing biosynthetic families rather than individual compounds, including early precursors, is that chemosyndromic variation (87) is less likely to distort the chemotaxonomic interpretation of the results. In many chemosystematic studies only one biosynthetic family has been considered (158), but it is advantageous to consider several biosynthetic families (systems) simultaneously.

Primary metabolic products, such as most amino acids, are often widespread and can be valuable at higher taxonomic levels, while expressions of differentiation are of value at the species or genus level (2,76). Secondary metabolites that are of value in protecting against physical stress such as radiation may be present in all species in a genus and may therefore be genus- or species-specific. Examples of this are the anthraquinones physcion and erythroglaucin, which are present in all species of the genus *Eurotium* (110,159).

More and more secondary metabolites have been shown to have possible ecological functions, and probably all of them have functions (2,15,160–169). This will not be explored in this chapter, but if all secondary metabolites have ecological functions, the significance of secondary metabolites in taxonomy must be very high.

V. APPLICATIONS OF SECONDARY METABOLITES IN FUNGAL TAXONOMY

A. Zygomycota and Chytridiomycota

Secondary metabolites have not yet been used in chemotaxonomy of any zygomycetous genera or species. Species of *Rhizopus*, such as *R. stolonifer*, have been claimed to produce ergot alkaloids (170) and other secondary metabolites (171,172). Rhizonin is a cyclic heptapetide produced by *Rhizopus microsporus* (171). These peptides, which do not have strong chromophores, may be difficult to detect using traditional chemotaxonomic methods. Secondary metabolites could be important in zygomycetous classification if the analytical methods are modified to make use of the chemical nature of most secondary metabolites. Though fungi belonging to Chytridiomycota produce many signaling metabolites, very little is known concerning their significance in chemotaxonomy (173).

B. Ascomycota

Many more secondary metabolite based chemotaxonomic studies have been made on ascomycetous genera and their anamorphic states than any other fungal phylum. Nevertheless, Mantle (174) proposed that secondary metabolism is incompatible with sex. This is clearly not the case, however, as a very large number of secondary metabolites have been isolated from sclerotia and ascomata of a large number of ascomycete species (6,15,17,18,37,162,175–178). It is clear, however, that ascomycetous yeasts do not produce many secondary metabolites (10,11). Five orders of Ascomycota—Diaporthales, Eurotiales, Hypocreales, Sordariales, and Xylariales—have been examined extensively for secondary metabolites.

Few genera of the Diaporthales have been examined in comparative chemotaxonomic studies, but the two anamorphic species, related to *Lewia*, *Alternaria infectoria*, and *A. alternata*, can be separated by their different profiles of secondary metabolites (179,180). In another study of *Pleospora* and *Stemphylium*, Andersen et al. (181) discriminated between eight species based on known and unknown secondary metabolites. Even though a large number of secondary metabolites have been characterized from these genera, the complex taxonomy and difficulty in obtaining a sufficient number of correctly identified strains from different geographic regions and habitats have precluded larger studies. The reviews written by Kachlicki (182), Sivanesan (183), and Montemurro and Visconti (184) contain compilations of mycotoxins produced by *Alternaria*, *Stemphylium*,

Ulocladium, Drechslera, Embellisia, Curvularia, Exserophilum, Helmintosporium, Cochliobolus, and associated teleomorphs without any chemotaxonomic considerations.

Some genera in the Eurotiales have been studied chemotaxonomically, particularily the anamorph genera *Penicillium* and *Aspergillus*. The genus *Talaromyces* was studied by Frisvad et al. (17), and the secondary metabolites detected were very useful in separating taxa and in showing the connections to the anamorphic state, *Penicillium* subgenus *Biverticillium*. Secondary metabolites such as mitorubrins, certain bisanthraquinones (rugulosin, skyrin, etc.), vermicellin, vermistatin, vermiculine, duclauxin, and glauconic acid were detected in *Talaromyces* species and subgenus *Biverticillium*. Interestingly these compounds are never found in the genus *Eupenicillium* and its associated anamorphs (17,18). A connection between teleomorphs and anamorphs was also clear in the genus *Neosartorya* and the anamorphic section *Fumigati* in *Aspergillus*. Secondary metabolites often found in some species of both the teleomorphs and anamorphs included tryptoquivalins, fumagillin, fumitremorgins, gliotoxin, and other secondary metabolites (176,185).

Several chemotaxonomic studies have shown that closely related species of *Penicillium* (the terverticillate penicillia) can be separated using either volatile or other secondary metabolites (16,29–31,40,68,71,91,92). These penicillia have always been regarded as difficult to classify and identify (24–26), but both HPLC with diode array detection or flow injection analysis electrospray mass spectrometry can be used for that purpose. An example of the consistent profiles of secondary metabolites of two strains are shown in Figures 1 and 2. This consistency is typical for isolates of terverticillate penicillia both qualitatively and quantitatively, although isolates that have been kept in culture collections for many years occasionally lose their ability to produce one or two secondary metabolite families (71,76). Cluster analysis or correspondence analysis based on these results shows that the profiles of secondary metabolites are species-specific and very different from species to species, even in closely related taxa (2,30,31,68,76, 77). It is an advantage to identify the secondary metabolites produced by comparison to authentic metabolite standards, but the chromatographic or mass spectrometric traces can also often be used for classification or identification without any further characterization (68,69).

In the Hypocreales, *Fusarium*, the polyphyletic anamorph of some *Nectria* and *Gibberella* species, have been extensively studied because of the many potent mycotoxins produced. Many *Fusarium* species can be distinguished by their unique profiles of secondary metabolites (89,90), but some species—e.g., *F. graminearum, F. culmorum,* and *F. cerealis* (= *F. crookwellense*)—appeared to differ only quantitatively (88). New and more sensitive methods may show whether there are qualitative differences between these closely related species. Leslie et al. (186) and Moretti et al. (187) also reported on secondary metabolites

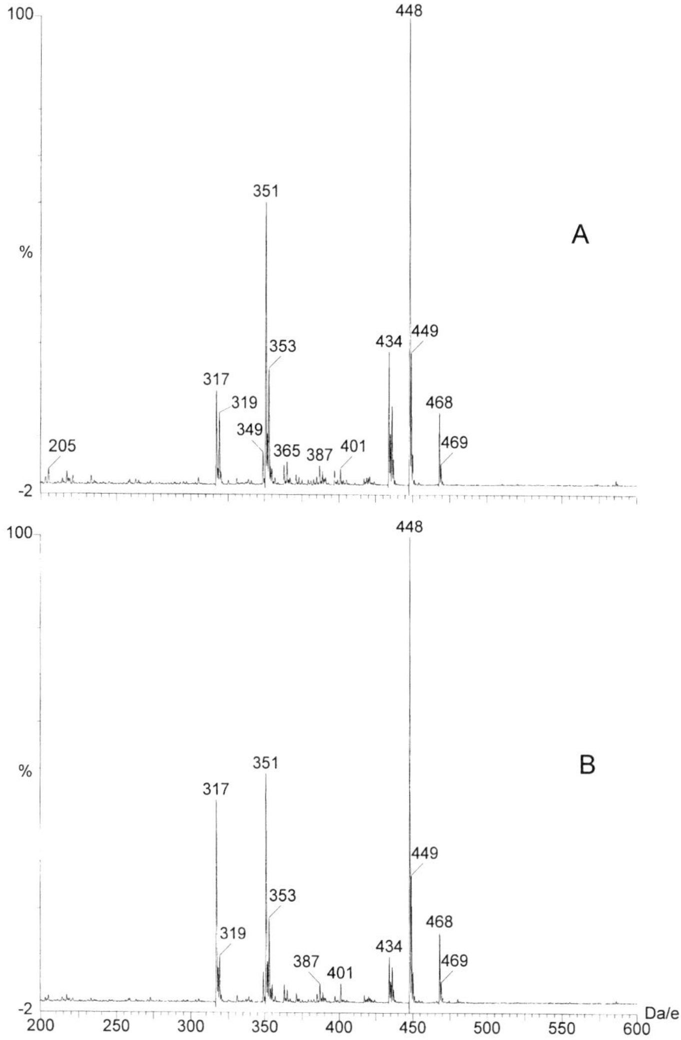

Figure 1 Flow injection electrospray mass spectrometry profile of the same two extracts as in Figure 1. The peak at 317 represents dechlorogriseofulvin; the peak at 351 represents griseofulvin: the peaks at 434 and 448 represent meleagrin and oxaline, respectively.

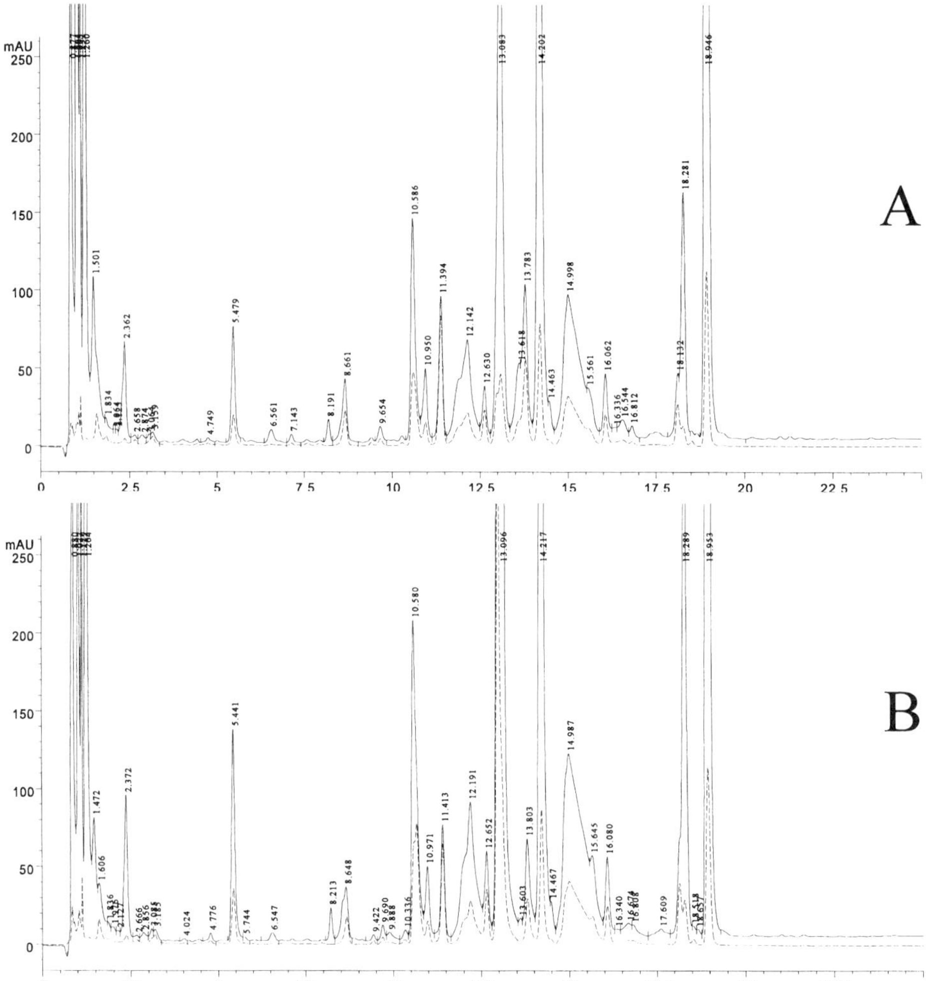

Figure 2 Comparison of chromatographic traces after HPLC analysis of two isolates of *Penicillium coprophilum* showing the similarity between extracts of two strains A (IBT 5551) and B (IBT 12992) of the same species. The broad peaks at retention time 12.142 (12.191) and 14.998 (14.987) are meleagrin and oxaline, respectively. The peak at 12.630 (12.652) is griseofulvin. The secondary metabolite extract (from the fungus grown on yeast extract sucrose agar) was analyzed using a water acetonitrile gradient with trifluoracetic acid in both eluents.

that could be used for differentiating mating populations in *Gibberella fujikuroi* sensu lato. There is still some debate on whether some groups in certain *Fusarium* species should be regarded as populations, subspecies, or species (187), so it is difficult to state how significant secondary metabolites are compared to other characters in *Fusarium* taxonomy.

In the Sordariales, the secondary metabolites of the genus *Chaetomium* have been particularly well studied by Udagawa and co-workers (151,152,189,190), and these seem to be of chemotaxonomic value. Few strains of each taxon have as yet been analyzed, however, and the data have not been synthesized into a chemotaxonomic overview.

The Xylariales also contain several genera which have been extensively studied, principally by Whalley and Edwards (6,19). These authors show that certain families of secondary metabolites, especially dihydroisocoumarins, succinic acid derivatives, butyrolactones, cytochalasins, punctaporonins, naphthalenes, mitorubrins, and griseofulvins, have phylogenetic significance among *Hypoxylon*, *Biscogniauxia*, *Camillea*, *Ustulina*, *Xylaria*, and *Rhopalostroma*. Other secondary metabolites are only produced in small quantities in the ascostroma as found in nature. These compounds will be difficult to detect, but they may be useful additions to the full picture of the secondary metabolite profiles of these often colorful species. Whalley and Edwards (6) stress that the secondary metabolites should be looked at in concert and that they may be especially valuable in separating closely related species and to predict intra- and intergeneric associations.

C. Basidiomycota

The different kinds of pigments, often directly observed in many basidiomycetous fungi, have been the basis of many isolations and the elucidation of structures of secondary metabolites of chemotaxonomic interest (5,9–11,21,191). As with many lichen chemotaxonomic studies, usually one biosynthetic family of metabolites has been examined at a time. For example, Høiland (192) studied only styrylpyrones in the genus *Gymnopilus* and suggested that all other (unidentified) secondary metabolites detected by TLC were related compounds. The styrylpyrones are also produced by species in the genera *Pholiota*, *Cortinarius*, and *Hypholoma* and in the unrelated polypores *Phaeolus* and *Phellinus* (9,21,189–191). In his study Høiland (192) found the styrylpyrones in all six accepted species of *Gymnopilus*. However, other families of secondary metabolites are present in species of *Gymnopilus*, including psilocybin in *G. purpuratus* (194,195), although the psilocybin reported in other *Gymnopilus* species may have been misidentified (196). As in the Ascomycota there are several other examples of the same secondary metabolite being produced by quite unrelated species—i.e., a betalain pigment such as muscaflavin is produced by both *Hygrocybe* and *Amanita* species (197–199). Often the discovery of the same secondary metabolites in both closely and less closely related species and genera can give rise to rethinking of the morphologically based taxonomy (21,200–206). Bresinsky and Besl (202) recommended that a larger number of specimens in each taxon be examined to evaluate the relationships between genera and species in the Boletales. Their results indicated that secondary metabolites such as variegatic acid and tetronic acid derivatives were widespread in the Boletales, including species in the Paxillaceae, Gomphridiaceae, Boletaceae, Strobilomycetaceae, and the gastroid Boletales.

As in other chemotaxonomic studies, a large number of correctly identified isolates from different geographic regions and a broad analytical chemical screening techniques are needed to evaluate the value of profiles of secondary metabolites in fungal systematics. In a chemotaxonomic study of some *Amanita* species, Beutler and der Marderosian (207) examined specimens for three families of toxic secondary metabolites, tryptamines, cyclopeptides, and isoxazoles. Tryptamines were present consistently in all isolates tested of *Amanita porphyria* (two isolates) and *A. citrina* (17 isolates), while *A. ocreata*, *A. phalloides*, *A. virosa*, and *A. bisporigera* specimens contained amanitins and phalloidin. Phallacidin and phallisin were detected only in *A. ocreata* and *A. phalloides*, but not in *A. virosa* and *A. bisporigera*. Finally, only one or two isoxazoles were found in specimens of *A. gemmata* (two of 12 isolates analyzed) and *A. muscaria* (204). Several other secondary metabolites are known from *Amanita*, so it would be interesting to compare *Amanita* species on a even broader basis in the future. One problem with some basidiomycetous taxa is the difficulty in growing all life-cycle stages in culture, so it may be necessary to sample several growth stages to be sure whether the secondary metabolites are produced or not.

Antibiotically active, antifeedant, or other biologically active compounds are also known from several basidiomycetous taxa (169,208). *Lactarius* species produce characteristic and species-specific combinations of sesquiterpenes. The compounds are present as relatively inactive esters that are enzymatically converted to the biologically active compounds when the fungus is injured (169).

The considerable data that are available on isolates and taxa of the Basidiomycota have rarely been synthesized and treated by multivariate statistical or cladistic methods, but there are indications that profiles of secondary metabolites would also be of great value in basidiomycete taxonomy. The secondary metabolites detected and listed in different studies have until now mostly been used as additional information for drawing taxonomic conclusions and, rarely, as a part of the primary features used in the classification or in identification keys.

VI. CONCLUSIONS

Secondary metabolites have been shown to be very reliable and highly diagnostic taxonomic characters in some of the few genera of filamentous fungi, where they have been applied on a large scale. Combined morphological and secondary metabolite data have yielded clearly circumscribed species in cases where a large number of isolates from different geographic regions and habitats have been examined. In *Penicillium*, *Aspergillus*, and *Fusarium*, which are regarded as especially difficult to classify and identify, secondary metabolites have been particularly effective. Secondary metabolites can be expected to be important aspects of descriptions of new species in the future, as has been the case for lichenized fungi in the past decades. It has often been implied that only highly specialized taxonomists are able to correctly identify species of *Penicillium*, *Fusarium*, and many other fungal genera. Secondary metabolites may be characterized objectively by standardized chemical methods available in many laboratories and provide data for reliable identifications. It is expected that secondary metabolites will play a major role in future revisions of many fungal genera, especially as they provide more accurate data than those used in the rather subjective descriptions of composite features such as color and odor. It is expected that data on profiles of secondary metabolites will be provided by analyses using advanced chromatographic and spectrometric methods for taxonomic purposes. It will be the responsibility of the taxonomist using these advanced analytical chemical methods to develop diagnostic and more easily recordable chemical characters for identifications purposes. This simplification step is necessary if chemical methods are to be used in traditional mycological laboratories in the future. While the use of secondary metabolites in fungal taxonomy in general may or may not increase, molecular methods will certainly be applied on many fungal species. The combination of morphological and secondary metabolite data with molecular data will undoubtedly be particularily effective in taxonomic revisions. Other features of filamentous fungi such as physiological responses to substrate and physical or chemical environmental factors may also be included. These physiological data are of great value in predicting the growth of fungi on different substrates, although they do add to the overall taxonomic circumscription of the species. Other chemotaxonomic methods based on cell wall constituents, membrane lipids, isozymes and other proteins, polysaccharides, etc., may also add significantly to the overall characterization of filamentous fungi.

There are some problems in using secondary metabolites in the taxonomy of filamentous fungi. Most fungal species have the ability to produce secondary metabolites, but in some cases a particular taxon or isolate needs specific stimuli to initiate the accumulation of some of these metabolites. Other species may not even by culturable and can only be collected in situ and so are very dependent on environmental conditions. Even when they are culturable, few general statements can be provided regarding medium constituents and laboratory conditions optimal for the production of all potential secondary matabolites. There are only speculations on the average number of families of secondary metabolites produced by fungal species, and many secondary metabolites are discovered only because of biological activity in large-scale screenings by major pharmaceutical companies. Certain types of secondary metabolites may also remain unnoticed because of inefficient extraction procedures or low analytical sensitivity. Reproducibility in metabolite production between different laboratories may also be a problem, especially concerning the use of secondary metabolites for identification purposes, as media composition may vary. Peptones and yeast, vegetable, and malt extracts can differ significantly, and this may affect secondary metabolite production in some cases. This is of course also a problem for the reproducible recording of other phenotypic features. Despite these occasional difficulties, more research on the chemical structure, physical properties, biosynthesis, and genetic regulation of secondary metabolites may help in devising efficient and improved methods for their detection. Furthermore, determination of the function of secondary metabolites in nature and the molecular basis for their presence and production may be important research areas in the future, that also bear on their taxonomic and phylogenetic significance.

ACKNOWLEDGMENTS

We thank Jørn Smedsgaard for help with preparation of the table and figures, and Birgitte Andersen and an anonymous referee for reading the paper.

REFERENCES

1. Campbell IM. Secondary metabolism and microbial physiology. Adv Microb Physiol 1984; 25:1–60.
2. Frisvad JC. Classification of organisms by secondary metabolites. In: Hawksworth DL, ed. The Identification and Characterization of Pest Organisms. Wallingford: CAB International, 1994:303–320.
3. Chadwick DJ, Whelan J, eds. Secondary Metabolites: Their Function and Evolution. Ciba Foundation Symposium 171. Chichester: John Wiley & Sons, 1992.
4. Christophersen C. Theory of the origin, function, and evolution of secondary metabolites. In: Atta-ur-Rahman, ed. Studies in Natural Products Chemistry 18. Stereoselective Synthesis (part K). Amsterdam: Elsevier, 1996:677–737.
5. Gill M. Pigments of fungi (Macromycetes). Nat Prod Rep 1996; 13:513–528.
6. Whalley AJS, Edwards RL. Secondary metabolites and systematic arrangement within the Xylariaceae. Can J Bot 1995; 73:S802–S810.
7. Hawksworth DL. Lichen chemotaxonomy. In: Brown DH, Hawksworth DL, Bailey RH, eds. Lichenology: Progress and Problems. London: Academic Press, 1976: 139–184.
8. Tyrrell D. Biochemical systematics and fungi. Bot Rev 1969; 35:305–316.
9. Benedict RG. Chemotaxonomic relationships among basidiomycetes. Adv Appl Microbiol 1970; 13:1–23.
10. Turner WB. Fungal Metabolites. London: Academic Press, 1971.
11. Turner WB, Aldridge DC. Fungal Metabolites II. London: Academic Press, 1983.
12. Mantle PG. Secondary metabolites of *Penicillium* and *Acremonium*. In: Peberdy JF, ed. *Penicillium* and *Acremonium*. New York: Plenum Press, 1987:161–243.

13. Frisvad JC. The connection between the penicillia and aspergilli and mycotoxins with special emphasis on misidentified isolates. Arch Environ Contam Toxicol 1989; 18:452–467.
14. Kozakiewicz Z. *Aspergillus* toxins and taxonomy. In: Powell KA, Renwick A, Peberdy JF, eds. The Genus *Aspergillus* From Taxonomy and Genetics to Industrial Application. New York: Plenum Press, 1994:303–311.
15. Gloer J. The chemistry of fungal antagonism and defence. Can J Bot 1995; 73: S1265–S1274.
16. Frisvad JC, Filtenborg O. Terverticillate penicillia: chemotaxonomy and mycotoxin production. Mycologia 1989; 81:837–861.
17. Frisvad JC, Filtenborg O, Samson RA, Stolk AC. Chemotaxonomy of the genus *Talaromyces*. Antonie Leeuwenhoek 1990; 57:179–189.
18. Frisvad JC, Samson RA, Stolk AC. Chemotaxonomy of *Eupenicillium javanicum* and related species. In: Samson RA, Pitt JI, eds. Modern Concepts in *Penicillium* and *Aspergillus* Classification. New York: Plenum Press, 1990:445–453.
19. Whalley AJS, Edwards RL. Xylariaceous fungi: use of secondary metabolites. In: Rayner ADM, Brasier CM, Moore D, eds. The Evolutionary Biology of Fungi. Cambridge: Cambridge University Press, 1987:423–434.
20. Hawksworth DL. Problems and perspectives in the systematics of the Ascomycotina. Proc Indian Acad Sci (Plant Sci) 1985; 94:319–339.
21. Moser M. The relevance of chemical characters for the taxonomy of the Agaricales. Proc Indian Acad Sci (Plant Sci) 1985; 94:381–386.
22. Huneck S, Yoshimura I. Identification of Lichen-Substances. Berlin: Springer, 1996.
23. Kozakiewicz Z. *Aspergillus* species on stored cereals. Mycol Pap 1989; 161:1–188.
24. Ciegler A, Fennell DI, Sansing GA, Detroy RW, Bennett GA. Mycotoxin-producing strains of *Penicillium viridicatum*: classification into subgroups. Appl Microbiol 1973; 26:271–278.
25. Onions AHS, Bridge PD, Paterson RRM. Problems and prospects for the taxonomy of *Penicillium*. Microbiol Sci 1984; 1:185–189.
26. Samson RA, Gams W. The taxonomic situation in the hyphomycete genera *Penicillium*, *Aspergillus* and *Fusarium*. Antonie Leeuwenhoek 1984; 50:815–824.
27. Pitt JI, Samson RA. Approaches to *Penicillium* and *Aspergillus* systematics. Stud Mycol (Baarn) 1990; 32:77–90.
28. Seiler H, Zillinger C, Hoffmann P. Identification of moulds with microtitration plates. Milschwiss 1994; 49:248–252.
29. Larsen TO, Frisvad JC. Chemosystematics of *Penicillium* based on profiles of volatile metabolites. Mycol Res 1995; 99:1167–1174.
30. Svendsen A, Frisvad JC. A chemotaxonomic study of the terverticillate penicillia based on high performance liquid chromatography of secondary metabolites. Mycol Res 1994; 98:1317–1328.
31. Smedsgaard J, Frisvad JC. Terverticillate penicillia studies by direct electrospray mass spectrometric profiling of crude extracts. I. Chemosystematics. Biochem Syst Ecol 1997; 25:51–64.
32. Harborne JB, Turner BL. Plant Chemosystematics. London: Academic Press, 1984.
33. Frank HM. Makrochemische Farbreaktionen bei Grosspilzen. I. Voraussetzungen für eine systematische Untersuchung. Z Mykol 1987; 53:93–98.

34. Tan CS, Hoekstra ES, Samson RA. Fungi that Cause Superficial Mycoses. Baarn: Centraalbureau voor Schimmelcultures, 1994.
35. Lund F. Differentiating *Penicillium* species by detection of indole metabolites using a filter paper method. Lett Appl Microbiol 1995; 20:228–231.
36. Culberson CF, Kristinsson H. A standardized method for the identification of lichen substances. J Chromatogr 1970; 46:85–93.
37. Culberson CF, Elix J. Lichen substances. Meth Plant Biochem 1989; 1:509–535.
38. Filtenborg O, Frisvad JC. A simple screening method for toxigenic fungi in pure cultures. Lebensm Wiss-Technol 1980; 13:128–130.
39. Filtenborg O, Frisvad JC, Svendsen JA. Simple screening method for moulds producing intracellular mycotoxins in pure cultures. Appl Environ Microbiol 1983; 45:581–585.
40. Frisvad JC, Filtenborg O. Classification of terverticillate penicillia based on profiles of secondary metabolites. Appl Environ Microbiol 1983; 46:1301–1310.
41. Frisvad JC, Thrane U. Standardized high-performance liquid chromatography of 182 mycotoxins and other fungal metabolites based on alkylphenone indices and UV-VIS spectra (diode array detection). J Chromatogr 1987; 404:195–214.
42. Paterson RRM, Bridge PD. Biochemical Techniques for Filamentous Fungi. Wallingford: CAB International, 1994.
43. Frisvad JC, Filtenborg O, Thrane U. Analysis and screening for mycotoxins and other secondary metabolites in fungal cultures by thin-layer chromatography and high-performance liquid chromatography. Arch Environ Contam Toxicol 1989; 18: 331–335.
44. Onji Y, Aoki Y, Tani N. Analysis of *Fusarium* mycotoxins by gas chromatography/mass spectrometry without chemical derivatization. Mycotoxins 1994; 40:49–51.
45. Santesson J. Chemical studies on lichens 10. Mass spectrometry on lichens. Arkiv Chem 1969; 30:363–377.
46. Santesson J. Chemical studies on lichens 28. The pigments of some foliicolous lichens. Acta Chem Scand 1970; 24:371–373.
47. Van Eijk GW, Roeijmans HJ. Separation and identification of naturally occurring anthraquinones by capillary gas chromatography and gas chromatography–mass spectrometry. J Chromatogr 1984; 295:497–502.
48. Ichinoe M, Kurata H, Sugiura Y, Ueno Y. Chemotaxonomy of *Gibberella zeae* with special reference to production of trichothecenes and zearalenone. Appl Environ Microbiol 1983; 46:1364–1369.
49. Logrieco A, Bottalico A, Altomare C. Chemotaxonomic observations on zearalenone and trichothecene production by *Gibberella zeae* from cereals in southern Italy. Mycologia 1988; 80:892–895.
50. Mirocha CJ, Abbas HK, Windels CE, Xie W. Variation in deoxynivalenol, 15-acetyldeoxynivalenol, 3-acetyldeoxynivalenol, and zearalenone production by *Fusarium graminearum* isolates. Appl Environ Microbiol 1989; 55:1315–1316.
51. Altomare C, Bottalico A, Logrieco A. Alcuni aspetti chemiotassonomici del genere *Fusarium* in relazione alla produzione di micotossine. Mic Ital 1989; 3:21–24.
52. Sugiura Y, Watanabe Y, Tanaka T, Yamamoto S, Ueno Y. Occurrence of *Gibberella zeae* strains that produce both nivalenol and deoxynivalenol. Appl Environ Microbiol 1990; 56:3047–3051.

53. Miller JD, Greenhalgh R, Wang Y, Lu, M. Trichothecene chemotypes of three *Fusarium* species. Mycologia 1991; 83:121–130.
54. Vesonder RF, Golinski P, Plattner R, Zietkiewicz DL. Mycotoxin formation by different geographic isolates of *Fusarium crookwellense*. Mycopathologia 1991; 113:11–14.
55. Lauren DR, Sayer ST, di Menna ME. Trichothecene production by *Fusarium* isolates from grain and pasture throughout New Zealand. Mycopathologia 1992; 120: 167–176.
56. Szécsi A, Bartók T. Trichothecene chemotypes of *Fusarium graminearum* isolated from corn in Hungary. Mycotox Res 1995; 11:85–92.
57. Frisvad JC, Thrane U. Application of high performance liquid chromatography. In: Betina V, ed. Chromatography of Mycotoxins: Techniques and Applications. Amsterdam: Elsevier, 1993:253–372.
58. Cole RJ, Cox RH. Handbook of Toxic Fungal Metabolites. New York: Academic Press, 1981.
59. Smedsgaard J. Micro-scale extraction procedure for standardized screening of fungal metabolite production in cultures. J Chromatogr 1997; 760:264–270.
60. Nielsen MS, Nielsen PV, Frisvad JC. Micellar electrokinetic capillary chromatography of fungal metabolites. Resolution optimized by experimental design. J Chromatogr 1996; 721:337–344.
61. Taquler R, Smilde A, Kowalski B. Selectivity, local rank, three-way data analysis and ambiguity in multivariate curve resolution. J Chemometrics 1995; 9:31–58.
62. Bruins AP. Electrospray, technique and applications. J Chim Phys Phys-Chim Biol 1993; 90:1335–1344.
63. Smith RD, Loo JA, Edmonds CG, Barinaga CJ, Udseth HR. New developments in biological mass spectrometry: electrospray ionization. Anal Chem 1990; 62: 882–889.
64. Evershead EP, Robertson DHL, Beynon RJ, Green BN. Application of electrospray ionization mass spectrometry with maximum entropy analysis to allelic 'fingerprinting' of major urinary proteins. Rapid Commun Mass Spectrom 1993; 7:882–886.
65. Perkins JR, Smith B, Gallagher RT, et al. Application of electrospray mass spectrometry and matrix-assisted laser desorbtion ionization time-of-flight mass spectrometry for molecular weight assignment of peptides in complex mixtures. J Am Soc Mass Spectrom 1993; 4:670–684.
66. Poon GK, Bisset GMF, Mistry P. Electrospray ionization mass spectrometry for analysis of low-molecular-weight anticancer drugs and their analogues. J Am Soc Mass Spectrom 1993; 4:588–595.
67. Korfmacher WA, Bloom J, Churchwell MI, et al. Characterization of three rifamycins via electrospray mass spectrometry and HPLC-thermospray mass spectrometry. J Chromatogr Sci 1993; 31:498–501.
68. Smedsgaard J, Frisvad JC. Using direct electrospray mass spectrometry in taxonomy and secondary metabolite profiling of crude fungal extracts. J Microbiol Meth 1996; 25:5–17.
69. Smedsgaard J. Terverticillate penicillia studies by direct electrospray mass spectrometric profiling of crude extracts. II. Database and identification. Biochem Syst Ecol 1997; 25:65–71.

70. Filtenborg O, Frisvad JC, Thrane U, Lund F. Screening methods for secondary metabolites produced by fungi in pure cultures. In: Samson RA, Hoekstra ES, Frisvad JC, Filtenborg O, eds. Introduction fo Food-Borne Fungi. Baarn: Centraalbureau voor Schimmelcultures, 1995:270–274.
71. Frisvad JC, Filtenborg O. Secondary metabolites as consistent criteria in *Penicillium* taxonomy and a synoptic key to *Penicillium* subgenus *Penicillium*. In: Samson RA, Pitt JI, eds. Modern Concepts in *Penicillium* and *Aspergillus* Classification. New York: Plenum Press, 1990:373–384.
72. Andersen B. Consistent production of phenolic compounds by *Penicillium brevicompactum* for chemotaxonomic characterization. Antonie Leeuwenhoek 1991; 60: 115–123.
73. Delpassand ES, Charl MV, Stager CE, Morrisett JD, Ford JJ, Romazi M. Rapid identification of common human pathogens by high-resolution proton magnetic resonance spectroscopy. J Clin Microbiol 1994; 33:1258–1262.
74. Vogels JTWE, Tas AC, van den Berg F, van der Greef J. A new method for classification of wines based on photon and carbon-13 NMR spectroscopy in combination with pattern recognition techniques. Chemom Intell Lab Syst Lab Inf Manage 1993; 21:249–258.
75. Hollands MV, Bernreuther A, Reneiro F. The use of amino acids as a fingerprint for the monitoring of European wines. In: Belton PS, Delgadillo I, Gil AM, Webb GA, eds. Magnetic Resonance in Food Science. Cambridge: Royal Society of Chemistry, 1995:136–145.
76. Frisvad JC. Chemometrics and chemotaxonomy: a comparison of multivariate statistical methods for the evaluation of binary fungal secondary metabolite data. Chemom Intell Lab Syst 1992; 14:253–269.
77. Frisvad JC. Correspondence, principal coordinate, and redundancy analysis used on mixed chemotaxonomical qualitative and quantitative data. Chemom Intell Lab Syst 1994; 23:213–229.
78. Vogt NB. Soft modelling and chemosystematics. Chemom Intell Lab Syst 1987; 1: 213–231.
79. Magee JT. Analytical fingerprinting methods. In: Goodfellow M, O'Donnell AG, eds. Chemical Methods in Procaryotic Systematics. Chichester: John Wiley and Sons, 1994:523–553.
80. Sneath PHA, Sokal RR. Numerical Taxonomy. San Fransisco: Freeman, 1973.
81. Vining LC. Roles of secondary metabolites from microbes. In: Chadwick DJ, Whelan J, eds. Secondary Metabolites: Their Function and Evolution. Ciba Foundation Symposium 171. Chichester: John Wiley and Sons, 1992:184–198.
82. Engel G, von Milczewski KE, Prokopek D, Teuber M. Strain-specific synthesis of mycophenolic acid by *Penicillium roqueforti* in blue-veined cheese. Appl Environ Microbiol 1982; 43:1034–1040.
83. Bu'Lock JD. Mycotoxins as secondary metabolites. In: Steyn PS, ed. The Biosynthesis of Mycotoxins. A Study in Secondary Metabolism. New York: Academic Press, 1980:1–16.
84. Moss MO. The mycelial habit and secondary metabolite production. In: Jennings DH, Rayner ADM, eds. The Ecology and Physiology of the Fungal Mycelium. Cambridge: Cambridge University Press, 1984:127–142.

85. Brunner F, Petrini O. Taxonomy of some *Xylaria* species and xylariaceous endophytes by isozyme electrophoresis. Mycol Res 1992; 96:723–733.
86. Monaghan RL, Polishook JD, Pecore VJ, Bills GF, Nallin-Olmstead M, Streicher SL. Discovery of novel secondary metabolites from fungi—is it really a random walk through a random forest? Can J Bot 1995; 73:S925–S931.
87. Culberson WL, Culberson CF. Secondary metabolites as a tool in ascomycete systematics: lichenized fungi. In: Hawksworth DL, ed. Ascomycete Systematics: Problems and Perspectives in the Nineties. New York: Plenum Press, 1994:155–163.
88. Thrane U. Grouping *Fusarium* section *Discolor* isolates by statistical analysis of quantitative high performance liquid chromatographic data on secondary metabolite production. J Microbiol Meth 1990; 12:23–39.
89. Thrane U. *Fusarium* species and their specific profiles of secondary metabolites. In: Chelkowski J, ed. *Fusarium*: Mycotoxins, Taxonomy and Pathogenicity. Amsterdam: Elsevier, 1993:199–225.
90. Thrane U, Hansen U. Chemical and physiological characterization of taxa in the *Fusarium sambucinum* complex. Mycopathologia 1995; 129:183–190.
91. Lund F, Frisvad JC. Chemotaxonomy of *Penicillium aurantiogriseum* and related species. Mycol Res 1994; 98:481–492.
92. Larsen TO, Frisvad JC. Characterization of volatile metabolites from 47 *Penicillium* taxa. Mycol Res 1995; 99:1153–1166.
93. Appel O, Wollenweber HW. Grundlagen einer Monographie der gattung *Fusarium*. Arb Kais Biol Anst Land-Forstwirtsch 1910; 8:1–207.
94. Hawksworth DL, Kirk PM, Sutton BC, Pegler DN. Ainsworth & Bisby's Dictionary of the Fungi. 8th ed. Wallingford: CAB International, 1995.
95. Andary C, Cosson L, Haupert R. Role des polyols and des acides amines dans la différenciation des bolets. Crypt Mycol 1988; 9:277–288.
96. Laatsch H, Matthies L, Schwibbe M. Mustererkennung von Aminosäure-Profilen—Chemotaxonomische Klassifizierung von Pilzen. Z Mykologie 1993; 59:99–112.
97. Krzeczkowska I, Burzynski S, Czerniak Z. Investigation on the possibility of the determination of mushroom species on the basis of the composition of their amino acids. Ann Univ Mariae Curie-Sklodowska (Lublin, Poland) Sect D 1965; 20: 221–229.
98. Augustyn OPH, Kock JLF, Ferraira D. Differentiation between yeast species, and strains within a species, by cellular fatty acid analysis. 5. A feasible technique? Syst Appl Microbiol 1992; 15:105–115.
99. Wauthoz P, El Lioui M, Decallone J. Gas chromatographic analysis of cellular fatty acids in the identification of foodborne bacteria. J Food Prot 1995; 58:1234–1240.
100. Van der Westhuizen JPJ, Kock JLF, Botha L, Botes PJ. The distribution of the ω3- and ω6-series of cellular long-chain fatty acids in fungi. System Appl Microbiol 1994; 17:327–345.
101. Embley TM, Wait R. Structural lipids of Eubacteria. In: Goodfellow M, O'Donnell AG, eds. Chemical Methods in Procaryotic Systematics. Chichester: John Wiley and Sons, 1994:121–161.
102. Jones GJ, Nichols PD, Shaw PM. Analysis of microbial sterols and hopanoids. In: Goodfellow M, O'Donnell AG, eds. Chemical Methods in Procaryotic Systematics. Chichester: John Wiley and Sons, 1994:163–195.

103. De Rosa M, Gambacorta A. Archeal lipids. In: Goodfellow M, O'Donnell AG, eds. Chemical Methods in Procaryotic Systematics. Chichester: John Wiley and Sons, 1994:197–264.
104. Zuckerkandl E, Pauling L. Molecules as documents of evolutionary history. J Theoret Biol 1965; 8:357–366.
105. Frelin C, Vuilleumier F. Biochemical methods and reasoning in systematics. Z Zool Syst Evolut-Forsch 1979; 17:1–10.
106. Davis JI, Nixon KC. Populations, genetic variation, and the delimitation of phylogenetic species. Syst Biol 1992; 41:421–435.
107. Bills GF, Peláez F, Polishook JD, et al. Distribution of zaragozic acids (squalestatins) among filamentous ascomycetes. Myc Res 1994; 98:733–739.
108. Agosti G, Birkinshaw JH, Chaplan P. Studies in the biochemistry of microorganisms 112. Anthraquinone pigments of strains of *Cladosporium fulvum* Cooke. Biochem J 1962; 85:528–530.
109. Aldridge DC, Burrows BF, Turner WB. The structures of the fungal metabolites cytochalasin E and F. J Chem Soc Chem Commun 1972; 3:148–149.
110. Anke H, Kolthoum I, Zähner H, Laatsch H. Metabolic products of microorganisms. 185. The anthraquinones of the *Aspergillus glaucus* group. 1. Occurrence, isolation, identification and antimicrobial activity. Arch Microbiol 1980; 126:223–230.
111. Anzai K, Suzuki S. A new antibiotic bovicidin, identified as 3-nitropropionic acid. J Antibiotics 1960; 13:133–136.
112. Ayer WA, Pena-Rodriguez L, Vederas JC. Identification of sterigmatocystin as a metabolite of *Monocillium nordinii*. Can J Microbiol 1981; 27:846–847.
113. Blight MM, Grove JF. New metabolic products of *Fusarium culmorum*: toxic trichothec-9-en-8-ones and acetylquinazolon-4(3H)-one. J Chem Soc Perkin Trans 1974; 1974:1691–1693.
114. Brian PW, Curtis PJ, Hemming HG. A substance causing abnormal development of fungal hyphae produced by *Penicillium janczewskii* Zal. Trans Br Mycol Soc 1949; 32:30–33.
115. Broadbent B. Antibiotics produced by fungi. Bot Rev 1966; 32:219–242.
116. Broadbent D, Hemming HG, Lehan M. Production of griseofulvin by *Khuskia oryzae*. Trans Br Mycol Soc 1974; 62:625–626.
117. Bush MT, Touster O, Brockman JE. The production of 3-nitropropionic acid by a strain of *Aspergillus flavus*. J Biol Chem 1951; 188:685–694.
118. Canigueral S, Vila R, Vucario G, Tomas X, Adzet T. Chemometrics and essential oil analysis: chemical polymorphism in two *Thymus* species. Biochem Syst Ecol 1994; 22:307–315.
119. Carter CL, Karabin, the glucoside of *Corynecarpus laevigata* and hiptagenic acid. J Soc Ind Chem 1943; 62:238–240.
120. Chadwick DJ, Easton IW. 2-acetyl-4(3H)-quinazolone, $C_{10}H_8N_2O_2$. Acta Cryst 1983; C39:454–456.
121. Chien MM, Schiff PL, Slatkin DJ Jr, Knapp JE. Metabolites of aspergilli. III. The isolation of citrinin, dihydrocitrinone and sclerin from *Aspergillus carneus*. J Nat Prod Lloydia 1977; 40:301–302.
122. Cooke AR. The toxic constituents of *Indigophera endecaphylla*. Arch Biochem Biophys 1955; 55:114–120.

123. Davies JE, Kirkaldy D, Roberst JC. Studies in mycological chemistry. Part VII. Sterigmatocystin, a metabolite of *Aspergillus versicolor* (Vuillemin) Tiraboschi. J Chem Soc 1960; 1960:2169–2178.
124. Demain AL, Hunt NA, Malık V, et al. Improved procedure for production of cytochalasin E by *Aspergillus clavatus*. Appl Environ Microbiol 1976; 31:138–140.
125. Eckert JW, Ratnayake M. Role of volatile compounds from wounded oranges in induction of germination of *Penicillium digitatum* conidia. Phytopathology 1994; 84:746–750.
126. Endo A, Kurata M. Citrinin, an inhibitor of cholesterol synthesis. J Antibiot 1976; 29:841–843.
127. Ewert AJ. The presence of citrinin in *Crotolaria crispata*. Ann Bot 1933; 47: 913–915.
128. Furuya K, Emokita R, Shirasaka M. Studies on the antibiotics from fungi. II. A new griseofulvin producer *Nigrospora oryzae*. Ann Sakkyo Res Lab 1967; 19:91–95.
129. Groter K. Hiptagin, a new glucoside from *Hiptage madablota*. Gaertu Bull Jordin Bt Buitenzorg 1920; 2:187–202.
130. Hamilton JGC, Gough AJE, Staddon BW, Games DE. Multichemical defence of plant bug *Hotea gambiae* (Westwood) (Heteroptera: Scutelleridae): (E)-2-hexenol from abdominal gland in adults. J Chem Ecol 1985; 11:1399–1409.
131. Hetherington AC, Raistrick H. Studies in the biochemistry of microorganisms. Part XIV. On the production and chemical constitution of a new yellow colouring matter, citrinin, produced from glucose by *Penicillium citrinum* Thom. Phil Trans R Soc Ser B 1931; 220B:269–295.
132. Hikino H, Nabetani S, Takemoto T. Structure and biosynthesis of chrysogine, a metabolite of *Penicillium chrysogenum*. Yakugaku Zasshi 1973; 93:619–623.
133. Holzapfel CW, Purchase IHF, Steyn PS, Gows L. The toxicity and chemical assay of sterigmatocystin, a carcinogenic mycotoxin, and its isolation from two new fungal sources. S Afr Med J 1966; 40:1100–1101.
134. Jarvis BB, Zhou Y, Jiang J, et al. Toxigenic molds in water-damaged buildings: dechlorogriseofulvins from *Memnoniella echinata*. J Nat Prod 1996; 59:553–554.
135. Kingston D, Chen P, Vercellotti JR. Metabolites of *Aspergillus versicolor*: 6,8-di-O-methylnidurufin, griseofulvin, dechlorogriseofulvin, and 3,8-dihydroxy-6-methoxy-1-methylxanthone. Phytochemistry 1976; 15:1037–1039.
136. Kögl F, Potovsky JJ. Untersuchungen über Pilzfarbstoffe II. Über die farbstoffe des blutenrote Hautkorpfs (*Dermocybe sanguinea* Wulf.). Just Lieb Ann Chem 1925; 444:1–7.
137. Liu XJ, Luo XY, Hu WJ. *Arthrinium* spp. and the etiology of deteriorated sugercane poisoning. In: Natori S, Hashimoto K, Ueno Y, eds. Mycotoxins and Phycotoxins '88. Amsterdam: Elsevier, 1989:109–118.
138. Li H, Madden JL, Potts BM. Variation in volatile leaf oils of the Tasmanian *Eucalyptus* species. 1. Subgenus *Monocalyptus*. Biochem Syst Ecol 1995; 23:299–318.
139. Nakano H, Komiya T, Shibata S. Anthraquinones of the lichens *Xanthoria* and *Caloplaca* and their cultivated mycobionts. Phytochemistry 1972; 11:3505–3508.
140. Natori S, Sato F, Udagawa S. Anthraquinone metabolites of *Talaromyces avellaneus* (Thom et Turreson) C.R. Benjamin and *Preussia multispora* (Saito et Minoura) Cain. Chem Pharm Bull 1965; 13:385–389.

141. Niederer D, Tamm C, Zürcher W. Nitrogen containing metabolites of *Fusarium sambucinum*. Tetrahedron Lett 1992; 33:3997–4000.
142. Olagbemiro TO, Staddon BW. Isoprenoids from metathoracic scent gland of cotton seed bug. *Oxycarenus hyalinipennis* (Costa) (Heteroptera: Lygaeidae). J Chem Ecol 1983; 9:1397–1412.
143. Oxford AE, Raistrick H, Simonart P. Studies in the biochemistry of microorganisms 60. Griseofulvin $C_{17}H_{17}O_8Cl$ a metabolic product of *Penicillium griseofulvum* Dieckx. Biochem J 1939; 33:240–248.
144. Raistrick H, Smith G. Studies in the biochemistry of microorganisms. XLII. The metabolic product of *Aspergillus terreus* Thom. New metabolic products, terrein. Biochem J 1935; 29:606–611.
145. Raistrick H, Stössl A. Studies in the biochemistry of microorganisms 104. Metabolites of *Penicillium atrovenetum* G. Smith: 3-nitropropionic acid, a major metabolite. Biochem J 1958; 68:647–653.
146. Schroeder HW, Kelton WH. Production of sterigmatocystin by some species of the genus *Aspergillus* and its toxicity to chicken embryos. Appl Environ Microbiol 1975; 30:589–591.
147. Stierle A, Strobel G, Stierle D. Taxol and taxane production by *Taxomyces andreanea*, an endophyte fungus of Pacific Yew. Science 1993; 260:214–216.
148. Shibata S, Udagawa S. Metabolic products of fungi. XIX. Isolation or rugulosin from *Penicillium brunneum* Udagawa. Chem Pharm Bull 1963; 11:402–403.
149. Shibata S, Shoji J, Ohta A, Watanabe M. Metabolic products of fungi. XI. Some observation on the occurrence of skyrin and rugulosin in mold metabolites, with a reference to structural relationships between penicilliopsin and skyrin. Pharm Bull 1957; 5:380–383.
150. Tatsuno T, Kobayashi N, Okubo K, Tsunoda H. Recherches toxicologique sur les substances toxique de *Penicillium tardum* I. Isolement et identification des substances cytotoxique. Chem Pharm Bull 1975; 23:351–354.
151. Udagawa S, Muroi T, Kurata H, et al. The production of chaetoglobosins, sterigmatocystin, O-methylsterigmatocystin, and chaetocin by *Chaetomium* spp. and related fungi. Can J Microbiol 1979; 25:170–177.
152. Udagawa S, Muroi T, Kurata H, Sekita S, Yoshihira K, Natori S. *Chaetomium udagawae*, a new producer of sterigmatocystin. Trans Mycol Soc Jpn 1979; 20: 475–480.
153. Wells JM, Cole RJ, Kirksey J. Emodin, a toxic metabolite of *Aspergillus wentii* isolated from weevil-damaged chestnuts. Appl Microbiol 1975; 30:26–28.
154. Baxter RL, Hanley AB, Chan HW-S, et al. Fungal biosynthesis of 3-nitropropionic acid. J Chem Soc Perkin Trans 1992; I:2495–2502.
155. Tétényi P. Homology of biosynthetic routes: the base in chemotaxonomy. In: Bendz G, Santesson J, eds. Chemistry in Botanical Classification. New York: Academic Press, 1973:67–77.
156. Courtecuisse R. Mycologie traditionelle et nouvelles technologies: quelle taxonomie pour demain? Cryptogam Mycol 1995; 16:1–16.
157. Crawford DJ. Flavonoid chemistry and angiosperm evolution. Bot Rev 1978; 44: 431–456.
158. Rogers RW. Chemical variation and the species concept in lichenized ascomycetes. Bot J Linn Soc 1989; 101:229–239.

159. Gould BS, Raistrick H. Studies in the biochemistry of microorganisms. XL. The crystalline pigments of species in the *Aspergillus glaucus* series. Biochem J 1934; 28:1640–1656.
160. Dettner K. Chemosystematics and evolution of beetle chemical defences. Annu Rev Entomol 1987; 32:17–48.
161. Aldrich JR. Chemical ecology of the heteroptera. Annu Rev Entomol 1988; 33: 211–238.
162. Wicklow DT. Metabolites in the coevolution of fungal chemical defence systems. In: Pirozynski KA, Hawksworth DL, eds. Coevolution of Fungi with Plants and Animals New York: Academic Press 1988:173–201.
163. Williams DH, Stone MJ, Hauck PR, Rahman SK. Why are secondary metabolites (natural products) biosynthesized? J Nat Prod 1989; 52:1189–1208.
164. Lawrey JD. Lichen secondary compounds: evidence for a correspondence between antiherbivores and antimicrobial function. Bryologist 1989; 92:326–328.
165. Christophersen C. Evolution in molecular structure and adaptive variance in metabolism. Comp Biochem Physiol 1991; 98B:427–432.
166. Dowd PF. Insect interactions with mycotoxin-producing fungi and their hosts. In: Bhatnagar D, Lillehoj EB, Arora DK, eds. Handbook of Applied Mycology. Vol. 5. Mycotoxins in Ecological Systems. New York: Marcel Dekker, 1992:137–155.
167. Barnea A, Harborne JB, Pannell C. What parts of fleshy fruits contain secondary compounds toxic to birds and why? Biochem Syst Ecol 1993; 21:421–429.
168. Hawksworth DL. The recent evolution of lichenology: a science for our times. Crypt Bot 1994; 4:117–129.
169. Sterner O. Toxic terpenoids from higher fungi and their possible role in chemical defence systems. Crypt Mycol 1995; 16:47–57.
170. El-Refai A-M, Sallam NR, Naim N. The alkaloids of fungi. 1. The formation of ergoline alkaloids by representative mold fungi. Jpn J Microbiol 1970; 14:91–97.
171. Steyn PS, Tuinmann AA, Van Heerden FR, Van Rooyen PH, Wessels PL, Rabie CJ. The isolation, structure, and absolute configuration of the mycotoxin rhizonin, a novel cyclic heptapeptide containing N-methyl-3-(3-furyl)alanine, produced by *Rhizopus microsporus*. J Chem Soc Chem Commun 1983; 1983:47.
172. Rabie CJ, Lubben A, Schipper MAA, Van Heerden FR, Fincham JE. Toxigenicity of *Rhizopus* species. Int J Food Microbiol 1985; 1:263–270.
173. Powell MJ. Looking at mycology with a Janus face: a glimpse at Chytridiomycetes active in the environment. Mycologia 1993; 85:1–20.
174. Mantle PG. Secondary metabolites of some non-lichenized ascomycetes. In: Hawksworth DL, ed. Ascomycete Systematics: Problems and Perspectives in the Nineties. New York: Plenum Press 1994:145–153.
175. Sheard JW. The taxonomy of the *Ramalina siliquosa* species aggregate (lichenized Ascomycetes). Can J Bot 1977; 56:916–938.
176. Samson RA, Nielsen PV, Frisvad JC. The genus *Neosartorya*: differentiation by scanning electron microscopy and mycotoxin profiles. In: Samson RA, Pitt JI, eds. Modern Concepts in *Penicillium* and *Aspergillus* Classification. New York: Plenum Press, 1990:455–467.
177. Wang H, Gloer JB, Wicklow DT, Dowd PF. Aflavinines and other antiinsectan metabolites from the ascostromata of *Eupenicillium crustaceum* and related species. Appl Environ Microbiol 1995; 61:4429–4435.

178. Elix JA. Biochemistry and secondary metabolites. In: Nash TH III, ed. Lichen Biology. Cambridge: Cambridge University Press, 1996:154–180.
179. Andersen B. Characterization of cereal-borne *Alternaria* and *Stemphylium*. Ph.D. thesis. Department of Biotechnology, Lyngby, 1995.
180. Andersen B, Thrane U. Differentiation of *Alternaria infectoria* and *Alternaria alternata* based on morphology, metabolite profiles, and cultural characteristics. Can J Microbiol 1996; 42:685–689.
181. Andersen B, Solfrizzo M, Visconti A. Metabolite profiles of common *Stemphylium* species. Mycol Res 1995; 99:672–676.
182. Kachlicki P. Metabolites of Helminthosporia. In: Chelkowski J, ed. Helminthosporia Metabolites, Biology Plant Diseases. Bipolaris, Drechslera, Exserophilum. Poznan: Institute of Plant Genetics, 1995:1–26.
183. Sivanesan A. The taxonomy and biology of dematiaceous hyphomycetes and their mycotoxins. In: Fungi and Mycotoxins in Stored Products. ACIAR Proceedings 36. Canberra: Australian Centre for International Agricultural Research, 1991:47–64.
184. Montemurro N, Visconti A. *Alternaria* metabolites—chemical and biological data. In: Chelkowski J, Visconti A, eds. *Alternaria* Biology, Plant Diseases and Metabolites. Amsterdam: Elsevier, 1992:449–557.
185. Frisvad JC, Samson RA. Chemotaxonomy and morphology of *Aspergillus fumigatus* and related taxa. In: Samson RA, Pitt JI, eds. Modern Concepts in *Penicillium* and *Aspergillus* Classification. New York: Plenum Press, 1990:201–208.
186. Leslie JF, Plattner RD, Desjardins AE, Klittich JR. Fumonisin B1 production by strains from different mating populations of *Gibberella fujikuroi* (*Fusarium* section *Liseola*). Phytopathology 1992; 82:341–345.
187. Moretti A, Logrieco A, Bottalico A, Fogliano V, Randazzo G. Diversity in beauvericin and fusaproliferin production of *Gibberella fujikuroi* (*Fusarium* section *Liseola*). Sydowia 1996; 48:44–56.
188. Burgess LW, Summerell BA, Backhouse D, Benyon F, Levic J. Biodiversity and population studies in *Fusarium*. Sydowia 1996; 48:1–11.
189. Udagawa S. Taxonomy of mycotoxin producing *Chaetomium*. In: Kurata H, Veno Y, eds. Toxigenic Fungi—Their Toxins and Health Hazard. Developments in Food Science 7. Amsterdam: Elsevier, 1984:139–147.
190. Sekita S, Yoshihira K, Natori S, et al. Mycotoxin production by *Chaetomium* spp. and related fungi. Can J Microbiol 1981; 27:766–772.
191. Gill M, Steglich W. Pigment of fungi (Macromycetes). In: Herz W, Grisebach H, Kirby GW, Tamm C, eds. Progress in the Chemistry of Organic Natural Products 51 Wien: Springer Verlag, 1987:1–317.
192. Høiland K. The genus *Gymnopilus* in Norway. Mycotaxon 1990; 39:257–279.
193. Benedict RG, Brady LR. Taxonomic status of *Pholiota aurea*. Mycologia 1972; 64: 1167–1169.
194. Garz J. Occurrence of psilocybin, psilocin and baeocystin in *Gymnopilus purpuratus*. Persoonia 1984; 14:19–22.
195. Kreisel H, Lindequist U. *G. purpuratus*, ein psilocybinhaltiger Pilz adventiv in Bezirk Rostock. Z Mykol 1988; 54:73–76.
196. Stijve T, Kuyper TW. Absence of psilocybin in species of fungi previously reported to contain psilocybin and related tryptamine derivatives. Persoonia 1988; 13:463–465.

197. Döpp H, Musso H. Die konstitution des Muscaflavins aus *Amanita muscaria* und über betalaminsäure. Naturwiss 1973; 60:477.
198. Döpp H, Musso H. Eine chromatographische Analysenmethode für Betalainfarbstoffe in Pilzen und Höheren Pflanzen. Z Naturforsch 1974; 29C:640.
199. Von Ardenne R, Döpp H, Musso H, Steglich W. Über das Vorkommen von Muscaflavin bei Hygrocyben (Agaricales) und seine Dihydroazepin-Struktur. Z Naturforsch 1974; 29C:637.
200. Besl H, Bresinsky A, Kronawitter I. Notizen über Vorkommen und systematische Bewertung von Pigmenten in Höheren Pilzen (1). Z Pilzk 1975; 41:81–98.
201. Besl H, Bresinsky A. Notizen über Vorkommen und systematische Bewertung von Pigmenten in Höheren Pilzen (2). Z Pilzk 1977; 43:311–322.
202. Bresinsky A, Besl H. Notizen über Vorkommen und systematische Bewertung von Pigmenten in Höheren Pilzen (3). Untersuchungen an Boletales aus America. Z Mykol 1979; 45:247–264.
203. Keller G, Ammirati JF. Chemotaxonomic significance of anthraquinone derivatives in North American species of *Dermocybe*, section *Dermocybe*. Mycotaxon 1983; 18: 357–377.
204. Besl H, Bresinsky A, Kämmerer A. Chemosystematik der Coniophoraceae. Z Mykol 1986; 52:277–286.
205. Arnold N, Besl H, Bresinsky A, Kemmer H. Notizen xur Chemotaxonomie der gattung *Dermocybe* (Agaricales) und zu ihrem Vorkommen in Bayern. Z Mykol 1987; 53:187–194.
206. Andary C, Cosson L, Bourrier MJ, Wylde R, Heitz A. Chimiotaxonomie des bolets de la section *Luridi*. Crypt Mycol 1992; 13:103–114.
207. Beutler JA, der Marderosian AH. Chemical variation in *Amanita*. J Nat Prod 1981; 44:422–431.
208. Camazine S, Resch J, Eisner T, Mainwald J. Mushroom chemical defence: pungent sesquiterpenoid dialdehyde antifeedant to opossum. J Chem Ecol 1983; 9:1439–1442.

13

Special Metabolites in Relation to Conditions of Growth

J. M. Frank
University of Surrey, Guildford, Surrey, United Kingdom

I. INTRODUCTION

The morphological characters of classical taxonomy are phenotypic and influenced by the conditions of growth. This is true whether such characters are of obvious selective significance, as are reproductive features, or whether they lack any such dimension, as does colony texture. The same sensitivity to growth conditions can be seen in the expression of secondary metabolites, which have been defined as biochemical products serving no function to the producing organism. In either case the problem is reducible to one of understanding the factors that influence variation in expression and accounting for this variation. The central concept of "species" is, of course, that they have different properties. It is also clear that two isolates of one species are likely to show some differences and this fact is accounted for in the breadth of species' concepts. What is less clear is why the same strain grown under "identical" conditions should vary, a property associated with secondary metabolites in particular.

The fungal cell produces few biosynthetic building blocks, compounds such as acetyl CoA, malonyl CoA, pyruvate, and α-ketoglutarate. The cofactors, principally ATP/ADP, NADH/NAD, and NADPH/NADP, are also few. Collectively these intermediary compounds are the substrate for the vast diversity of secondary metabolites as well as for the energy-deriving and replicatory activities of cell growth. It must be recognized that their principal *quantitative* significance to the fungus lies in their role as intermediaries of "primary" metabolism and as such will always be present in a functioning mycelium. The implication of this simple fact is that the nonoccurrence of a feature or metabolite in a strain that is capable of its production is due to regulatory function, not the unavailability of constituents.

The substrate for the synthesis of secondary metabolites, although always present, is not uniformly distributed. This can be seen as a property of a differentiated hyphal structure (1). There is no doubt that the idiosyncratic appearance of secondary products can be understood in some cases to result from the process of differentiation. Over time the mycelium displays functional specialization, spatial separation of activities, and integration of these activities in producing a life cycle that is clear and unmistakable if more subtle than that which occurs in species capable of producing muscle and nerve, liver and lung.

A priori, then, one is led to expect that any environmental (extrinsic) factor capable of affecting either growth or development could also affect the expression of "secondary" chemical characters. Where the genetic or regulatory (intrinsic) properties of different species or isolates are not equivalent we would not expect to observe the same response from different fungi exposed to a given environmental constraint. In other words, there is no single set of growth conditions best suited for the expression of all fungal metabolites and the expression of these features is a dual function of intrinsic and extrinsic parameters. This is of course exactly the situation for classical taxonomic features.

After a brief review of some theoretical considerations associated with the concept of secondary metabolism and the development of interpretive models, this chapter will selectively review reports relating environmental parameters to the production of secondary metabolites.

II. SECONDARY OR SPECIAL METABOLISM

Bennett and Bentley (2) have suggested that the term "special metabolism" or "specific metabolism" should be preferred to "secondary metabolism." This, it is argued, would avoid the implication that secondary metabolites lack importance by being referred to as secondary. There may well be some truth to this observation although secondary products are, as a matter of fact, quantitatively and in terms of energy balance less important than primary metabolites. The author considers that two other points relating to the term "secondary metabolism" carry more serious implications and require clarification. Bennett and Bentley's analysis appears to overlook the point that the use of either of the terms "secondary metabolism" or "special metabolism" strongly implies the existence of a discrete *process* and one that is exclusive of and distinct from primary (sometimes called "general") metabolism. Primary and secondary metabolites both require the same limited number of substrates for their elaboration, which suggests that this distinction is untenable. No set of conditions emerges as requisite for the production of these metabolites generally, and in programs screening fungal isolates for novel antibiotic metabolites, doubling the number of media has been reported to be as effective as doubling the number of isolates (3).

The assumption of the existence of a process can affect the interpretation of data. For example, in a study of pigment production by species of *Monascus* it is stated (4): "It has already been shown by Bu'Lock (1961), that balanced but limited growth, as well as nitrogen exhaustion is required for the induction of secondary metabolites." The data published clearly show that the key factor is not nitrogen concentration at all but whether nitrogen is supplied as amino N, ammonium N, or nitrate N. The data are marshaled in an attempt to describe the "switch" that the term "secondary metabolism" implies instead of leading to legitimate doubt regarding the existence of such a switch. It has been pointed out that the trophophase/idiophase model is a doubtful validity applied to filamentous organisms and that several regulatory schemes can be identified including those where production occurs during maximum growth (5).

A second semantic problem with the term "secondary metabolism" arises due to the adoption of what is essentially a botanical term for the fungi. In the original botanical sense it relates to a process of differentiation (lignification) that is unified, tightly regulated, and essential to the plant's survival. Latterly it has also been used to represent metabolites induced by the attack of plant pathogens (6). If the analogy with plants were rigorous, accommodating the facts that in both plants and fungi there are different tissues with different metabolic activity, that nutritional factors move freely throughout such organisms, and that the diversion of carbon from vegetative growth to other fates is part of a developmental program, the concept would have traveled more accurately to the fungi. We understand there to be a simpler form of integration, organization, and development in fungi with vegetative hyphae, aerial hyphae, perithecia, sclerotia, conidiophores, phiallides, and so forth.

The model for secondary metabolite production in fungi seems to have arisen from single-celled organisms that might well differentiate exclusively as a population of cells, not as a multicellular organism, and the two cases must necessarily be different. It may be that production of secondary metabolites cannot take place at the same site as primary metabolism due to functional specialization of tissue but since fungi are differentiated organisms the two processes can take place at the same time. A tradition of confusing culture development, viz. "trophophase" to "idiophase," with morphological development has led, in the author's view, to an erroneous model of the relationship between environmental factors and the production of secondary metabolites. Correlation between some measurable time-dependent changes in batch culture and some biochemical event or events is routinely give causal status; i.e., the appearance of a product is said to be caused by the disappearance of a nutrient. That the appearance of the product is due to the normal developmental cycle is often at least as good an explanation. It is clear that secondary metabolism in the fungal kingdom has not been shown to be a process, per se, much less a single process. Indeed, it would be astonishing if the expression of secondary metabolites as disparate as cyclic peptides and isoprenoids, for example, proved to be subject to a single or even to similar control systems.

Bennett and Bentley (2) have suggested that the universally recognized term "secondary metabolism" be replaced by the term "special metabolism" or "specific metabolism" but have offered no significant redefinition. In the author's view, the proposed change could only be justified if a refined definition for the new term, aimed at rectifying the problems discussed above, could be offered. As we have argued, secondary metabolism corresponds to a concept and not to a mechanism. This simple point may explain why a satisfactory diagnostic definition has yet to be developed and may be unattainable: the concept, though useful, does not correspond to an actual physical entity.

The following descriptive definition is offered. *Special metabolism* is a concept applied to filamentous fungi and actinomycetes representing the various schemes whereby special metabolites are produced. *Special metabolites* are compounds of terpenoid, polyketide oligopeptide, or mixed biosynthetic origin, of limited molecular weight (generally less than about 1000), which occur irregularly. Typically, such a compound will be known from some families and genera only. Many are peculiar to a limited number of species within a genus and within those species may only be produced by certain strains. Likewise, the narrowness of their occurrence with respect to timing during the growth cycle, the environmental requirements for their expression, and sometimes their association with differentiated cells distinguish these compounds from those small intermediates of primary (or general metabolism) that are universal, or nearly so, in their occurrence.

Many attempts at developing a single model and an adequate definition for special metabolism are available in the literature (5,7–16). A brief summary of some notable theories is offered here.

A. Internal Necessity

An often-suggested view is derived from the older idea of shunt metabolism (17) that "secondary metabolism ... serves to maintain mechanisms essential to cell multiplication in operative order when that cell multiplication is no longer possible" (7). The products are inconsequential but their synthesis is essential. This proposal readily appeals to a logic that suggests that if there is no synthesis there must be deterioration. However, there is no experimental evidence that the operation of the biosynthetic pathways of special metabolism would "maintain mechanisms essential to cell multiplication." Neither have the biosynthetic intermediates been shown to be generally toxic or deleterious, thus necessitating disposal, as the original concept of shunt metabolism requires. The fact that both hyphae and perennating structures maintain the potential for growth for periods up to years without producing special metabolites in doing so tends to refute any link between special metabolism and the maintenance of cellular growth potential.

There is a parallel with the operation of photo respiration in C-3 plants (those using the ribulose diphosphate carboxylase system for fixing carbon) but it is superficial and in no way analogous to the situation in molds or other heterotrophic organisms. In photo respiration O_2 is fixed in place of CO_2 when CO_2 availability becomes limiting. This allows the light-capturing apparatus to effectively discharge and along with cyclic photophosphorylation helps minimize destruction by photo-oxidation under conditions where light energy is more available than is the sink for this energy, the reduction of CO_2. There is no indication of a similar threat to cellular integrity under conditions of excess carbohydrate in proportion to nitrogen, for example.

B. Response to Stress

Perhaps partly inspired by the phytoalexins of plants that are induced by pathogen attack or mechanical injury (6), the suggestion has been made that stress in some form is required to initiate or augment the production of special metabolites. The occasional correlation of these products with the so-called idiophase (nutrient stress) is taken to be a manifestation of this requirement.

Although a useful concept, even a basic definition of stress is difficult to construct since adaptation is a fundamental property of organisms. So high salt levels are not stressful to halophilic organisms, high temperatures do not constitute a stress in thermophilic species, and an antibiotic does not stress an organism that carries resistance to it.

Stress induced by the application of biocides does not, as a rule, enhance special metabolite production (18). In this light, normal laboratory practice, relying as it does on rich media and axenic culture, may even exert stress on fungi, particularly soil fungi, which have evolved in diverse communities adapted to conditions that are oligotrophic to dystrophic.

If the effect of stress can result in a breakdown in intracellular compartmentation then perhaps Zamir's "derailment metabolites" (19) are examples of stress-induced products formed when enzymes of low specificity act on substrates they would not encounter under normal conditions but for which they have some affinity. Frequently fungi readily incorporate exogenously supplied intermediates or analogs into special metabolites, which implies a low enzyme specificity and a low degree of complexing in these biosynthetic pathways. But as a general explanation of special metabolic activity this scenario is far too narrow.

C. Ecological Explanation

As Campbell has emphasized (16), special metabolites, in some cases at least, have adaptive functions. The antibiotic activity of some metabolites has long been cited in arguing that competitive exclusion is the primary function that these substances provide (20,21) but this is clearly too narrow an interpretation. There are many mechanisms of natural selection aside from competition. In particular, group selection, "soft" or density-dependent selection, coevolution, and the adaptive advantage of genetic diversity (genetic load; see Ref. 22) have largely been neglected in evaluating potential for selective advantage derived from special metabolites. The apparent long-term maintenance in the gene pool of the ability to produce these compounds could simply reflect an ability for adaptation, and it is this, a potential, that is selected for. In an environment that is variable and offers a great variety of stable niches, variability is adaptive. This represents one form of genetic load (23) and can be understood as the cost of adaptability, which is actively beneficial only under certain circumstances.

Group selection is selection that takes place between consortia of organisms leading to the maintenance of traits favorable to a consortium but not necessarily to an individual species outside of that context (24). A characteristic of phytoplankton, for example, is the large proportion of photosynthate and nitrogen that they lose to the water column (25–27). This tendency enhances nutrient recycling (C and N) but does not help the organism that has to maintain the apparatus for fixing carbon (and in some species nitrogen). In this example, leakiness of the plasmalemma benefits the group in niche creation, which encourages diversity, thereby stability, but also maximizes energy flow in the ecosystem. Thus it is not imperative that every characteristic be of direct advantage to a species.

Mutualistic associations such as that between *Lolium perenne* and *Acremonium* (28,29) or *Baccharis* and *Myrothecium* (30) are more obvious cases where fungal special metabolites have no value in culture but confer an unmistakable adaptive advantage in situ—in these cases, a defense against herbivory. Mycotoxins associated with perennating structures, such as the ergot alkaloids in sclerotia of *Claviceps* (31), the sporidesmins in the spores of *Pithomyces* (32), or ochratoxin A in the sclerotia of *Aspergillus carbonarius* (33), may also serve a protective function evident only when the natural context is considered.

D. Manifestation of Differentiation

Some metabolites are produced in specific differentiated structures as noted above or as mycophenolic acid is produced in conidiophores of *Penicillium brevicompactum* (34,35). Other metabolites may play a role in controlling differentiation, such as trisporic acid and carotenoid biosynthesis in the sexual cycle of some mucoraceous fungi (36). Such integration into developmental programs also serves to emphasize the adaptive value of some special metabolites.

Although the origin of this idea probably lies with such facts of natural history, it undoubtedly developed as a theory of special metabolism through analysis of the dynamics of antibiotic production in relation to pelleting and branching frequency in submerged culture (8). In such experimental designs it is particularly difficult to identify causal relationships between the large number of time-dependent parameters that can be measured such as nutrient depletion, product formation, pH change, and gas exchange. Correlations that may arise can be merely coincident with observed morphological changes such as pellet size progression or sporulation, which will tend to proceed with the growth of the organism as the "normal" developmental program.

The interpretation of fermenter studies is further hampered by the fact that such conditions are a long way from any that a fungal regulatory system could have evolved to cope with. Campbell et al. (37) have forcefully argued that the developmental organization and control of metabolite production patterns that they have documented in solid culture cannot be expressed in liquid culture. It is a clear principle that any regulatory system can be expected to regulate effectively only within a limited domain of conditions to which it could conceivably be exposed. Outside of this domain control is ceded to other phenomena. In general terms, one is entitled to predict that this domain should reflect the conditions under which the control system was developed or, in the case of biological systems, under which it evolved. It remains a moot point as to whether any laboratory system, not just the extreme case of submerged liquid culture, relates to normal conditions well enough to allow extrapolation to the field.

One can imagine that even within one set of stable conditions, such as the forest floor of a hardwood climax sere, there is scope for many adaptive strategies that can be exploited to yield survival success. This could affect the relationship between morphogenesis and other "programmed" events; thus some fungi relying on rapid dispersal will produce conidia quickly under almost any conditions whereas others that rely on slow growth and thorough exploitation of local resources may be expected to form extracellular products very quickly but conidia perhaps only tardily.

III. METABOLIC INTEGRATION

A thorough discussion of metabolic control mechanisms is beyond the scope of this review because the extent of recent progress in the genetics and enzymology of special metabolism is such that the subject requires a review on its own (see, for example, Refs. 38–40). A brief discussion of some selected aspects of the organization and regulation of metabolic activity in filamentous fungi is offered to develop the context and provide a basis for rationalizing how and why environmental conditions invariably affect internal dynamics.

It has been recognized for some decades that biological control mechanisms differ fundamentally from electromechanical systems although there is a strong superficial similarity that invites analogy (41). A principal difference lies in the ability of biological control mechanisms to adapt to ambient mean signal levels so that the response to a stimulus from one ambient level elicits a different response than the same stimulus applied after the system has adapted to a different ambient level. This is an element of the fundamental adaptive property of biological systems, which also includes adaptation mechanisms such as enzyme induction and possibly selection of nuclei in a population of genotypically different nuclei in heterokaryotic organisms.

Figure 1 recapitulates some selected features of fungal metabolic activity. These pathways are often represented, as here, as an interconnected single entity. It is crucial to understand, however, that these elements are separated in space, within and between cells, and sometimes in time, associated with events of differentiation. Contemporaneous activities are connected by cyclosis and mass flow through incomplete septae and membrane transport systems (42), an organization representing an important layer of structure and regulation.

Two observations that bear on mass balance should be emphasized. The carbon for lipid special metabolites is drawn from the acetyl-CoA pool and could be said to be in competition with the Krebs cycle and with animo acid production for substrate. The availability of carbon for the acetyl-CoA pool could be said to be in competition with the other CO_2-generating process, the pentose phosphate pathway. The source of NADPH, the principal cofactor for biosynthesis, is produced by the pentose phosphate pathway; thus more reducing power for biosynthesis requires that there is less carbon available for substrate. This suggests a separation of processes so that there is more than one pool of acetyl CoA. The mobilization of storage lipid could also balance the conflicting demands implied by this relationship. The mannitol cycle (not shown), where in effect NADH reduces NADP (43,44), could also contribute to balancing cellular demand for biosynthetic reducing power against structural carbon in fungi where it is found. It is clear that such an interrelated structure with compartmentalization superimposed provides a potent regulatory framework that would be sensitive to factors affecting carbon or energy availability.

Other aspects that Figure 1 develops are these. Two enzymes are named of which malonyl CoA reductase is the rate-limiting step in the polyketide/fatty acid branch and HMG (3-hydroxy-3-methyl-glutaryl) CoA reductase is the first committed step to isoprenoid biosynthesis. The cyclases involved in the folding step of isoprenoid products may be relatively nonspecific since geranyl-geranyl pyrophosphate provided to yeast squalene synthetase (the steroid cyclizing enzyme) instead of farnesyl-farnesyl pyrophosphate was able to produce lycopene, the initial carotenoid precursor, which demonstrates a lack of chain length specificity (47).

In Figure 2 a model is proposed that relates growth and differentiation to

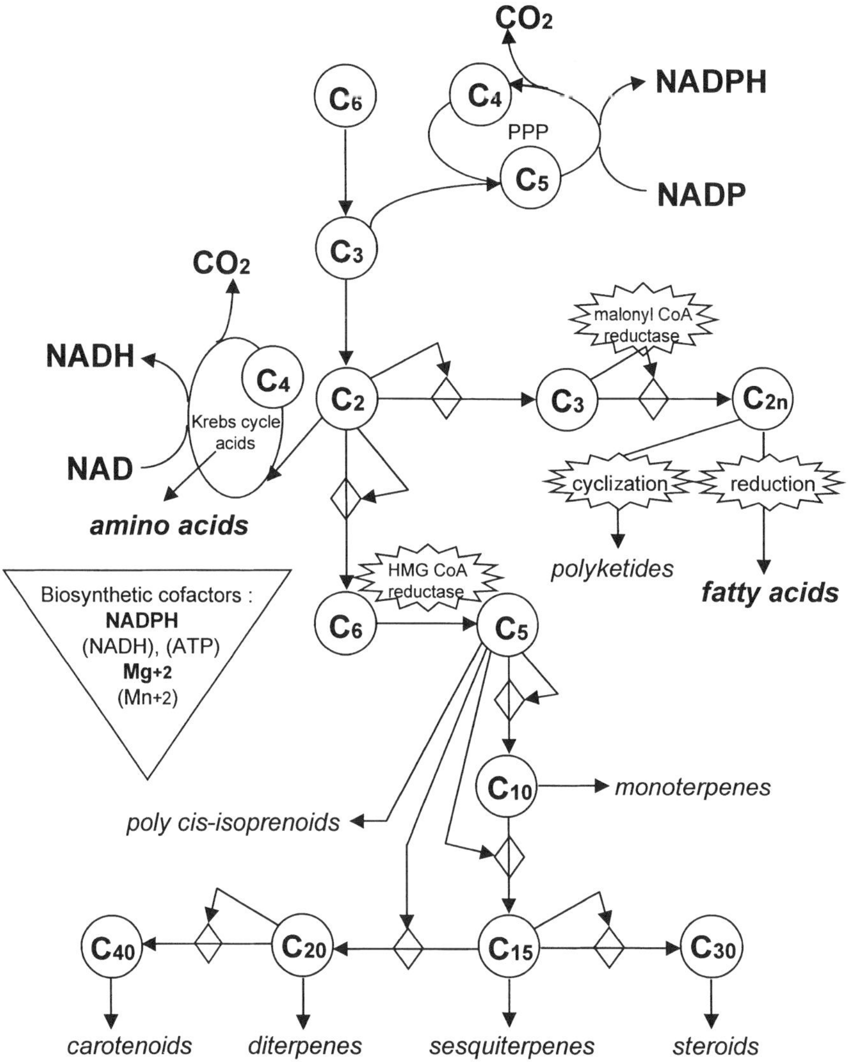

Figure 1 Flow diagram of selected aspects of metabolic relations relevant to the production and regulation of special metabolites. (Developed from Refs. 45 and 46.)

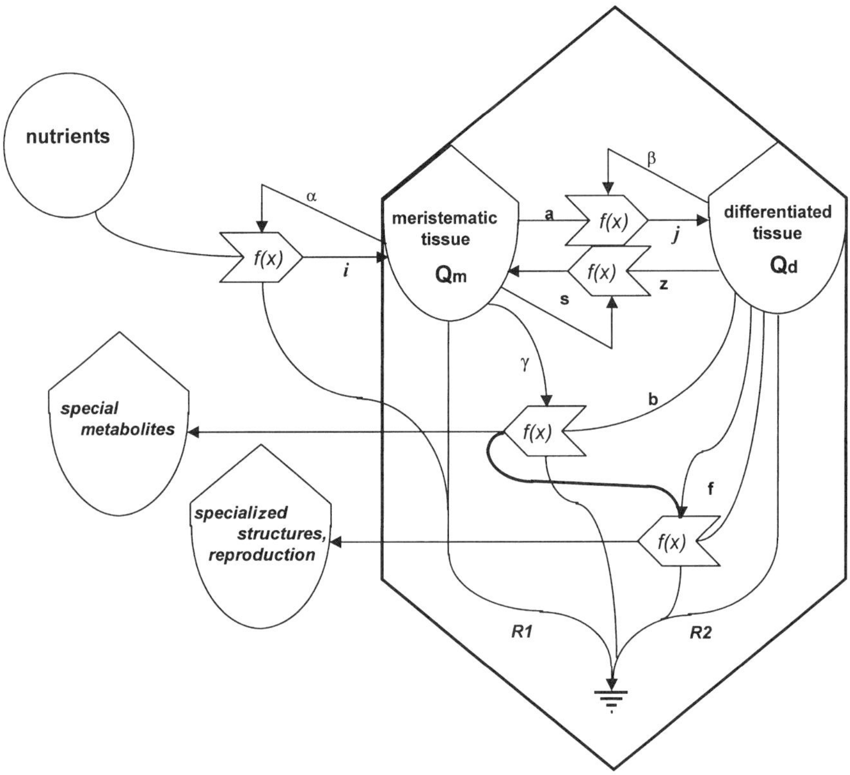

Figure 2 A two component model for special metabolism where meristematic biomass (Qm) is functionally distinct from differentiated biomass (Qd) and the interaction of the two is a key regulatory feature. Symbols are defined as follows (from a modeling language called "energese") with their electronic equivalence, which could be applied in an analog computer model: ⬯ self-replicating heterotroph—amplifier and capacitor; ⪢ work gate relates two inputs to one output—transistor; ○ forcing function, "infinite" source—potentiometer; ⬯ temporary storage—capacitor; ⏚ loss to the system—common ground. (After Ref. 48.)

the production and regulation of special metabolites. It is intended to provide an interpretive framework based on the assumptions that special metabolism is associated with differentiated tissue and that the organism is compartmentalized but functionally integrated. Regulation results from source/sink or push/pull balance but a more sophisticated version incorporating other forms of regulation could be developed. Such an interaction is suggested by the line connecting special metabolism with sporulation and could represent the role of trisporic acid in some mucoraceous fungi (36,49) for example. To simplify the diagram, not all of the losses to the system (*R1* and *R2*) are shown.

During a brief initial period all of the biomass is actively growing ("meristematic" biomass = *Qm*) and there is little or no special metabolite production (*b* and γ are small) nor respiratory demands from differentiated tissue (*a* and β are small). As the mycelium develops, tissue that had been meristematic differentiates (*Qd* increases) with the passage of time according to an evolved program of differentiation. The precise dynamics of this program is constrained by the prevailing environmental conditions, including nutrient availability, water availability, and temperature, and, in vivo, features relating to the activity of other strains, mating types, and species. This is one means by which environmental conditions affect special metabolism. Because differentiation is a continuous process, only the young hyphae of the margin of the hyphal body are undifferentiated and this tissue exploits nutrients and supports differentiated biomass (*Qd*) over the now exhausted substrate. The differentiated tissue, although presumably less metabolically active, will begin to predominate. As the mycelium grows out and as the flows $R1 + a + \gamma$ tend toward equality with *i*, the model predicts that the mycelium will stall as is observed in field infections (21) or produce a cyclical stalling pattern as is often observed on agar in the laboratory with the formation of concentric rings. The flow α is readily understood as the metabolic apparatus relating to uptake and utilization of nutrients while the loss from the work gate (part of *R1*) includes protein turnover and the normal thermodynamic losses associated with metabolism.

The feedback amplifier cycle *s* is constrained by both large *Qm* and small *Qd*. When *i* becomes small, however (nutrient exhaustion), lipid storage in differentiated tissue is mobilized. This activity (*z*) would necessarily compete with sporulation (amplifier *f*) and special metabolism (*b*). We can identify an adaptive choice in this thermodynamic imperative. Different species could be variously adapted to strategies of meristematic development or of spore production to reach unexploited resources, and arguably special metabolites could have a regulatory role, potentially even an interspecific interactive role, to play here.

β controls the redistribution of assimilate from meristematic regions to differentiated regions and will be larger as *Qd* is larger. Likewise, *b* will tend to increase with *Qd* and increase γ by demand "pulling." This is the second aspect of the model where growth conditions exert influence on special metabolism since availability of γ has a clear function in regulating the special metabolite product. The controlling flow γ can be understood as reducing power while *b* is carbon skeleton and the enzymes of special metabolic biosynthesis exerting pull demand on γ.

The principal role of models is to simplify complex systems. Superficially they invariably appear to constitute a process but here special metabolism is represented as a part of interdependent cellular activity, not as separate or mutually exclusive of primary metabolism and not as a discrete process. The model emphasizes how different adaptive strategies can produce different regulatory outcomes within one general framework. The models presented above show one way in which the separation of function, differentiation, and competition for common substrate can be resolved in keeping with our general understanding of the organization and development of fungi. It seems likely that another tier of regulation operating at the level of specific biochemical steps acts to refine the general tendencies that these speculative models outline and predict. Biochemical inhibitors with relatively well-defined modes of action may help to elucidate this class of regulatory activity, as indeed the ergosterol biosynthesis inhibitors have with the sterol pathways (50,51).

IV. ENVIRONMENTAL FACTORS AFFECTING SPECIAL METABOLITE PRODUCTION

Useful generalizations concerning how special metabolite production is likely to be affected by growth conditions are difficult to develop—a logical extension of special metabolism not representing a process, per se. Different strains of a fungal species can show different responses to the same variable (52); the production of different metabolites from a single strain is not necessarily influenced in unison by a variable (53). Initial conditions such as inoculation density and the resident variability of single-spore lines exert an influence apparently in addition to the conditions prevailing during the period of product formation itself (52,54). One general feature that can be identified is that the conditions supporting the production of a particular special metabolite (or sporulation or sexual reproduction) are a subset of those required to support growth.

Any environmental feature that is capable of affecting assimilation or development could also affect special metabolite production. On the other hand, special metabolite production may be radically affected while assimilation is affected little, if at all (55–57). This is because a unit of biomass also has important developmental and compositional dimensions, as emphasized in the models described above. Environmental factors work in concert and interact with a self-regulating, adaptive entity, the mycelial fungus. A change in one pervasive environmental feature can change the response of the fungus to another; thus, in principle, a change in incubation temperature could elicit a different response at one water activity than at another. This self-adjusting property of homeostatic systems tends to obscure regulatory function: if an environmental variable is capable of altering both morphological and biochemical expression, are these changes coincident or causal?

Almost since the discovery of penicillin much of the interest in the physiology of special metabolite production was motivated by industrial production of exploitable compounds (58,59). Since the recognition that the cause of turkey-X disease in the early 1960s (60) was a fungal metabolite, additional impetus has been provided by concern about the impact of mycotoxins on human and animal health. For these reasons, much of the work cited below addresses certain specific high-profile compounds often in highly artificial production systems. It is fair to say that a systematic and comprehensive study of the spectrum of special products of any fungus has yet to be reported.

A. Biotic Interactions

This subject area is somewhat off the point of the taxonomic theme of this book but a very brief treatment is offered for the interested reader (see also Ref. 61). Fungi of agricultural significance interact with a responsive host (not a substrate) sensitive to fungal activity (6) as well as with a community of co-occurring microorganisms. It has been forcefully argued (62) that the rationale of special metabolites cannot be understood outside of this context and without taking account of specific, often mutualistic interactions between organisms that have coevolved to exploit specific niches. Even strain differences in the host will produce markedly different yields of mycotoxins (53,63). The mechanisms responsible are not known but there are plant products such as phytic acid (64) and caffeine (65) that affect the production of special metabolites. There is no doubt that in axenic culture, temperature and water activity are important factors influencing special metabolite production. They have even greater significance when two or more organisms are in association (66–68). The possibility of using microbial interactions to improve industrial production of microbial products has not gone unnoticed (69).

B. Physicochemical Factors

Unlike nutritional factors, which often produce narrow and idiosyncratic effects, temperature, pH, ambient light, water activity, and atmospheric composition produce more general responses. Studies usually focus on one or a few special products and many have been conducted on autoclaved natural substrates due to the interest directed toward how physical factors influence mycotoxin accumulation in commodities. The substrate exerts a controlling influence on the outcome of such experiments. The water content optimum for zearalenone production by *Fusarium roseum* is reported to be 45% in corn and 60% in rice at the temperature optimum of 15°C. There was no production in soya beans and peas and little in oats and barley (70). The difference between rice and corn is also seen in deoxynivalenol (DON) production by strains of *Fusarium graminearum* but here, although corn is always better than rice for zearalenone production in the strains tested, some of them produce more DON on rice (71). Sometimes a consistent response is reported. All of fourteen toxigenic strains from species of the Sporotrichiella section (*Fusarium*) showed maximum toxicity grown at 8°C with an initial pH of 5.6 in seven different liquid media, some of them defined (72). Ground nut fodder when inoculated with *Aspergillus flavus* shows both the greatest weight loss and aflatoxin accumulation at an equilibrium relative humidity above 90% and temperatures between 25 and 30°C (73), which represent typical values for the *A. flavus*/aflatoxin production system in other substrates. In laboratory media water activity can be controlled by the use of solute concentration. The interpretation of such results is difficult when sugars are used as controlling solutes. Using this approach it appears that the optimal water activity for the formation of fumitremorgins A and C on sugar-supplemented Czapek Yeast agar is 0.98 when glucose or fructose is used as the controlling solute but 0.99 when sucrose is employed (74). Verruculogen is optimally produced at a water activity of 0.97 on glucose, 0.98 on fructose, and 0.99 on sucrose.

Citreoviridin was reported to be optimally produced by *Penicillium citreoviride* in a defined liquid medium at 20–24°C and the relative metabolite production levels shadowed assimilation over the range of temperatures tested (75). This optimum is in good agreement with that determined in rice and luteoskyrin production by *Penicillium islandicum* and fusarenon-X production by *Fusarium nivale* (FN2-B = *Fusarium sporotrichioides*) show optima at 24 and 27°C, respectively (75).

Usually the effect of pH is assessed by setting the initial pH of the medium, a value that changes as growth occurs. The gradient plate method is a perhaps underused application of this approach but has been used to show that the optimal initial pH for aflatoxin production by *Aspergillus parasiticus* is below 3.5 (76).

The widespread storage of fruit in modified atmospheres has generated some studies on the effect of a reduction of O_2 and elevation of CO_2 on the accumulation of some mycotoxins. Patulin seems to be produced, although in somewhat reduced amounts, even in extreme atmospheres such as 7.5% CO_2 and 2% O_2 (77,78). There is a theoretical requirement for oxygen gas in the biosynthesis of oxygenated trichothecenes (79) and presumably any metabolite the synthesis of which requires a P450 mono-oxygenase. That the concentration of this requirement can be low is suggested by the often reported production of trichothecenes in oxygen-limited fermentation conditions. *Trichoderma viride*, however, requires abundant oxygen for high yields of the oxygenated metabolites gliotoxin and viridiol (80), which may indicate oxygen regulates as a nutrient rather than as a biosynthetic substrate in this case. The availability of oxygen influences the proportion of carbon assimilated via the pentose phosphate pathway (PPP) as opposed to glycolysis and the Krebs cycle. Using C-1-radiolabeled glucose, a smaller proportion of label was incorporated and less aflatoxin produced in aerated conditions (81). The interpretation of this result, that increased PPP activity leads to rapid loss of label and thus lower incorporation and that therefore a higher PPP activity leads to reduced toxin formation, was tested by blocking terminal electron transfer in the formation of CO_2 with NaN_3. This produced higher toxin and lipid production, a result interpreted as showing that loss of the ability to oxidize NADH leads to an accumulation of acetyl CoA and a stimulation of polyketide and lipid biosynthesis (81).

Light is well recognized to sometimes affect sporulation, production of sexual stages, sclerotia, and pigment formation among other features. All of these may, as discussed above, bear a relationship to special products insofar as special metabolites may be associated with differentiated tissues. Generally, mycotoxins seem to be less influenced by light than these developmental events. There was little effect of a light treatment on aflatoxin and anthraquinone accumulation in Adye and Mateles medium by several strains of *A. flavus* although sclerotia formation was markedly affected (82). Likewise, little effect of light was reported on trichothecene production by *F. sporotrichioides* in shaken liquid culture (54). Carotenoid production, on the other hand, is influenced by light in the mucorales (49) in conjunction with the sexual cycle.

In comparison of results from solid and liquid culture, the differences in the form that vegetative growth takes is generally neglected. Incubation of *Fusarium tricinctum* on vermiculite impregnated with defined liquid media gives a higher yield of trichothecenes than liquid medium alone (83), suggesting either that attached growth performs differently or that the microclimate of such conditions is superior for the production of this special metabolite.

C. Micronutrients and Phosphate

This subject has been reviewed (84) and only a few comments are offered here. Magnesium and manganese are the most common cofactors for the enzymes of general metabolism. They are rarely limiting under natural conditions and are uncommonly associated with regulatory activity in special metabolism outside of what might be attributable to their effect on growth and development. Specific regulatory impact of manganese is reported in patulin biosynthesis however (85). Of copper, cadmium, calcium, iron, molybdenum, zinc, cobalt, magnesium, and manganese, only the lack of manganese leads to the accumulation of the patulin precursor 6-methylsalicylic acid and the prevention of patulin accumulation. The effect was attributed to a requirement at the transcription stage of one or more of the late patulin biosynthetic enzymes (86). By implication this suggests that RNA polymerases can have specific requirements for inorganic cofactors.

Zinc has been shown to inhibit glucose-6-phosphate dehydrogenase (87) and therefore the pentose phosphate pathway, which may explain why it appears to be more involved than most micronutrients in the regulation of special metabolism. It is a requirement for aflatoxin and versicolorin accumulation and can, when present at concentrations superoptimal for growth, stimulate higher yields of these products (88).

The regulatory impact of calcium is rarely evaluated but an interesting case has been reported where the presence of calcium at a rate of 2% or more produced a hundredfold increase in verruculogen yield and three orders of magnitude increase in sporulation (89). Obviously this effect is demonstrated at a concentration far from micronutrient thresholds.

Phosphate limitation is often implicated in the regulation of special metabolite production by actinomycetes in submerged culture but in the author's experience is of little importance in special metabolite regulation in fungi. *Claviceps*, in producing ergot alkaloids in submerged culture, however, is reported to be generally inhibited by phosphate in the medium (90) but the better-producing strains are ones that have overcome this inhibition to some extent (91). Likewise, initial concentrations of phosphate greater than 0.5 mM inhibited abscisic acid accumulation by *Cercospora rosicola*, a concentration below which growth is inhibited (92).

D. Carbon and Nitrogen Sources

The case-by-case nature of responses to environmental features has been a consistent theme of this discussion and the responses of fungi to carbon and nitrogen nutrition are no exception. One might expect, and the models set out above predict, that regulation of products of mixed lipid–amino acid biosynthetic origin or of oligopeptides is likely to be separate and distinct from that of pure acetyl/malonyl CoA–derived products. In addition to this fundamental difference, the "best" nitrogen source with a given sugar may be a poor substrate when used with a different sugar. Different species and strains perform differently and often the best substrate for one metabolite is poor for the expression of some other products from the same fungus.

There are undoubtedly many reasons for this close relationship between the two principal nutrients. Amino acids do not appear to be generally utilized by a reversal of biosynthetic pathways and their uptake in some cases is regulated by specific uptake mechanisms. They can be deaminated and oxidized to NO_3, an ability that can be correlated to special metabolite production (93), as well as simply deaminated. In products containing amino acid moieties the distinction between nitrogen nutrition and precursor feeding becomes ambiguous (94,95). Oxidized nitrogen sources can have a direct effect on cellular energetics in at least two guises. There is a substantial implicit requirement in the reduction of NO_3 to NH_4, which is mediated by the NADPH-requiring nitrogenase enzyme complex; nitrate is reported to have a direct influence on the activities of several of the enzymes of the pentose phosphate pathway (96). The constituent amino acids of complex mixtures are not utilized in tandom. *Fusarium roseum* utilized most amino acids of a peptone–yeast extract to the extent of 60–70% but aspartate, proline, and glycine were used to about half of this extent (97). The mixture was poor for trichothecene production.

Some sugars under certain regimes of nitrogen nutrition show catabolite repression, but not under others (98), which may be one mechanism through which production kinetics are distorted such that production of a special metabolite begins only after growth ceases whereas production accompanies growth when a simple source is supplied (99). The "optimal" carbon source can be strain and nitrogen-source dependent (99,100). The effect of a mixture of amino acids can be dependent on other conditions as in the instance where cassein supports high aflatoxin production in stirred culture but very little in stationary culture (101). In this case *A. flavus* and *A. parasiticus* behaved similarly and in stationary culture proline supported the best production while aspartate was markedly inferior to asparagine.

Supplied with aspartate, glutamate, or ammonium nitrate alone and in combinations, *Claviceps purpurea* produces ergot alkaloids in an additive fashion during the growth phase. Asparagine alone caused this fungus to continue to produce these products into stationary phase producing the best yield (102). The relationship between initial nitrate concentration and enniatin production by *Fusarium oxysporum* shows a maximum at 25 mM with a superoptimal reduction at higher concentrations. Feeding with branched-chain amino acids further enhances both the spectrum of products and their amounts. Ammonium alone supports little production and is inhibitory when added in concert with other nitrogen sources (99).

The overall picture with carbon nutrition is similar to that of nitrogen. Table 1 shows that sucrose produces a different performance than the mean of glucose and fructose and that the performance on glucose is distinct from that on maltose starch or cellobiose. The two fungi, although producing the same metabolites, show different properties in their response to the different substrates.

Ueno et al. found the same kind of pattern in their survey of 13 isolates of toxigenic fusaria (100) in that the response to different carbon substrates was not

Table 1 Performance of *Fusarium roseum*/*Fusarium graminearum* on Different Carbon Sources with Respect to the Production of Three Special Metabolites

Carbon source	Deoxynivalenol	3- or 15-acetyl deoxynivalenol	Zearalenone
Sucrose	460/n.d.	21250/5350	n.d./n.d.
Fructose	tr/tr	3900/1370	523/n.d.
Glucose	500/410	21800/1480	52/24
Maltose	310/n.d.	927/4150	n.d./2800
Starch	124/280	n.d./tr	n.d./150
Cellobiose	387/140	116/tr	130/150

n.d. = not detected; tr = trace.
Source: Ref. 97.

uniform across the taxa tested. Nitrate stimulated the greatest yields of extracellular lipid and although these fractions contained less trichothecenes, toxicity in bioassay was not necessarily less. This stimulation of lipid biosynthesis and inhibition of trichothecene production was also observed in *F. sporotrichioides* (54). Sucrose is a good substrate for these products in *Fusarium* but for enniatin production lactose and glycerol are reported to be the best substrates for *Fusarium sambucinum* although there was little product from *F. oxysporum* using those carbon sources (99). One way to avoid this idiosyncratic response pattern in trying to establish special metabolite profiles is to combine extracts from several different media for analysis (103).

V. CONCLUSION

Whether they are referred to as "special," "specific," or "secondary" metabolites, an effective understanding of the metabolism that the terms refer to has eluded research efforts stretching back some four or five decades. It is an empirical fact that the relationship between growth conditions and special metabolites is a case-by-case matter, which is not also to say that the relationship is random or without pattern. It is likely that models can be developed to guide inquiries and aid the interpretation of these patterns, a simple one of which is proposed above.

Control of the production of special metabolites is the experimental dimension of the rationale of their production. Perhaps understanding this rationale could in itself allow the extrapolation of known results to predict the behavior of unstudied systems. More likely, interest in the question reflects a natural desire to *unify*: a logical extension of the controlling influence that selection and evolution is seen to exert on biological systems. It seems likely that there is a unitary basis at some very fundamental level, but the subtleties of the relationship of fitness and selection to variability, variation, and adaptation obscure it. Additionally, there has been a tendency to confuse intellectual constructs that aid thought and communication with biological processes.

The idea that there is a single set of conditions for demonstrating the complete special metabolite profile of an isolate is almost certainly untenable. If the application is taxonomic, the complete profile is not a necessity, only the diagnostic elements need to be seen. The absence of a metabolite, however, will always beg the question of whether growth conditions were correct for its expression—this is a conundrum. The development of a model capable of producing testable hypotheses and the application of defined media with systematic alterations of constituents to inform the model focused on carbon and nitrogen sources is the way forward here. This approach may even provide other information of taxonomic value since in an evolutionary context the condition under which a property is expressed is a feature of equal significance to the ability to express the feature. The production of secondary metabolites is a dynamic process in every sense, and once we develop a methodology capable of coping with many interrelated layers or dimensions, progress will no doubt be rapid.

REFERENCES

1. Moss MO. The mycelial habit and secondary metabolite production. In: Jennings DH, Rayner ADM, eds. The Ecology and Physiology of the Fungal Mycelium. Cambridge, UK: Cambridge University Press, 1984.
2. Bennett JW Bentley R. What's in a name?—Microbial secondary metabolism. Adv Appl Microbiol 1989; 34:1–28.
3. Wildman H. Use of a resampling procedure to analyse the contributions of fungal diversity and growth medium diversity to an industrial screening campaign for secondary metabolites. Univ of Liverpool, UK: British Mycological Society Tropical Mycology Symposium, 1992.
4. Shepherd, D. The relationship between pigment production and sporulation in *Monascus*. In: Meyrath J, Bu'lock JD, eds. Biotechnology and Fungal Differentiation, FEMS Symposium No. 4. London: Academic Press 1977:103–118.
5. Demain AL. Biology of antibiotic formation. Drugs Pharm Sci 1984; 22:33–42.
6. Haard NF. Stress metabolites. In: Lieberman M, ed. Post-Harvest Physiology and Crop Preservation. London: Plenum Press, 1981:299–314.
7. Bu'Lock JD. Intermediary metabolism and antibiotic biosynthesis. Adv Appl Microbiol 1961; 3:293–342.
8. Smith JE, Berry DR. Differentiation, secondary metabolism and industrial mycology. In: Smith JE, Berry DR, eds. An Introduction to Biochemistry of Fungal Development. London: Academic Press 1974:282–308.
9. Bull AT, Trinci AJP. The physiology and metabolic control of fungal growth. Adv Microbiol Physiol 1977; 15:1–84.
10. Martin JF, Demain AL. Fungal development and metabolite formation. In: Smith JE, Berry DR, eds. The Filamentous Fungi, Vol. III. London: Edward Arnold, 1978.
11. Bu'Lock JD. Chapter 1. Mycotoxins as secondary metabolites. In: Steyn PS ed. The Biosynthesis of Mycotoxins, a Study in Secondary Metabolism. London: Academic Press, 1980.
12. Bennett JW. Secondary metabolism as differentiation. J Food Safety 1983; 5:1–11.
13. Luckner M, Nover L and Böhm H. Secondary Metabolism and Cell Differentiation. Berlin: Springer-Verlag 1977.
14. Malik VS. Microbial secondary metabolism. Trends Biochem Sci 1980; 5:68–72.
15. Woodruff HB. Natural products form microorganisms. Science 1980; 208:1225–1229.
16. Campbell IM. Secondary metabolism and microbial physiology. Adv Microbiol Physiol 1984; 25:1–60.
17. Foster JW. Chemical Activities of Fungi. New York: Academic Press, 1949.
18. Moss MO, Frank JM. Prevention: effects of biocides and other agents on mycotoxin production. In: Watson DH, ed. Natural Toxicants in Food. Chichester, UK: Ellis Horwood Ltd, 1987:231–251.
19. Zamir LO. Chapter 7. The biosynthesis of patulin and penicillic acid. In: Steyn PS

ed. The Biosynthesis of Mycotoxins, a Study in Secondary Metabolism. London: Academic Press, 1980:223–268.

20. Rayner ADM, Webber JF. Inter-specific mycelial interactions—an overview. In: Jennings DH, Rayner ADM, eds. The Ecology and Physiology of the Fungal Mycelium. Cambridge, UK: Cambridge University Press, 1984:383–417.
21. Wicklow DT, Hesseltine CW, Shotwell OL, Adams GL. Interference, competition and aflatoxin levels in corn. Phytopathology 1980; 70:761–764.
22. Wallace B. Genetic Load. Its Biological and Conceptual Aspects. Englewood Cliffs, NJ: Prentice-Hall, 1970.
23. Bryant EH. Habitat selection in a variable environment. J Theor Biol 1973; 41: 421–429.
24. Odum EP. Fundamentals of Ecology. London: WB Saunders, 1971:274–275.
25. Fogg GE, Westlake DF. The importance of extracellular products of algae in freshwater. Int. Vereingung fur Theoretische und Angewandte Limnologie Verhandlungen 1955; 12:219–232.
26. Saunders GW. Carbon flow in the aquatic system. In: Cairns J, ed. The Structure and Function of Fresh Water Microbial Communities. Blacksburg, VA: Virginia Polytechnic Institute Press, 1970.
27. Hellebust JA. Excretion of some organic compounds by marine phytoplankton. Limnol Oceanogr 1965; 10:192–206.
28. Culvenor CCL, et al. Isolation of toxic metabolites of *Phomopsis leptostromiformis* responsible for lupinosis. Aust J Biol Sci 1977:30:269–277.
29. Latch GCM. Endophytes and ryegrass staggers. In: Lacy J, ed. Trichothecenes and Other Mycotoxins. Chichester, UK: Wiley, 1985:135–140.
30. Jarvis BB, Midiwo JO, Tuthill D, Bean GA. Interaction between the antibiotic trichothecenes and the higher plant *Baccharis megopotamica*. Science 1981; 214: 460–461.
31. van Rensburg SJ, Altenkirk B. *Claviceps purpurea*—ergotism. In: Purchase FH, ed. Mycotoxins. Amsterdam: Elsevier, 1974:69–96.
32. Dodd DC. Facial eczema in ruminants. pp.. In: Wogan GN, ed. Mycotoxins in Foodstuffs. Cambridge, MA: MIT Press, 1965:105–110.
33. Wicklow DT, Dowd PF, Alfatafta AA, Gloer JB. Ochratoxin A: an antiinsectan metabolite from the sclerotia of *Aspergillus carbonarius* NRRL 369. Can J Microbiol 1996; 42:1100–1103.
34. Bartman CD, Doerfler DL, Bird BA, Remaley AT, Peace NJ, Campbell IM. Mycophenolic acid production by *Penicillium brevicompactum* on solid medium. Appl Environ Microbiol 1981; 41:729–736.
35. Bird BA, Campbell IM. Disposition of mycophenolic acid, brevianimide A, asperphenamate and ergosterol in solid cultures of *Penicillium brevicompactum*. Appl Environ Microbiol 1982; 43:345–348.
36. Rao S, Modi VV. Carotenogenesis: possible mechanism of action of trisporic acid in *Blakeslea trispora*. Experientia 1977; 33:31–33.
37. Campbell IM, Doerfler DL, Bird BA, Remaley AT, Rosato LM, Davis BN. 11. Secondary metabolism and colony development in solid cultures of *Penicillium brevicompactum* and *P. patulum*. In: Krumphanzl V, Sikyta B, Vanek Z, eds. Overproduction of Microbial Products. London: Academic Press, 1982.

38. Beck J, Ripka S, Siegner A, Schiltz E, Schweizer E. The multifunctional 6-methylsalisylic acid synthetase gene of *Penicillium patulum*. Eur J Biochem 1990; 192: 487–498.
39. Timberlake WE. Molecular genetics of *Aspergillus* development. Annu Rev Genet 1990; 24:5–36.
40. Mayorga ME, Timberlake WE. The developmentally regulated *Aspergillus nidulans* wA gene encodes a polypeptide homologous to polyketide and fatty acid synthetases. Mol Gen Genet 1992; 235:205–212.
41. Stark L. Defining biological feedback control systems. Ann NY Acad Sci 1963; 117: 426–442.
42. Jennings DH. Water flow through mycelia. In: Jennings DH, Rayner ADM, eds. The Ecology and Physiology of the Fungal Mycelium. Cambridge, UK: Cambridge University Press, 1984.
43. Hult K, Gatenbeck S. Production of NADPH in the mannitol cycle and its relation to polyketide formation in *Alternaria alternata*. Eur J Biochem 1978; 88:607–612.
44. Niehaus WG, Jiang W. Nitrate induces enzymes of the mannitol cycle and suppresses versicolorin synthesis in *Aspergillus parasiticus*. Mycopathology 1989; 107:131–137.
45. White A, Handler P, Smith EL. Principles of Biochemistry. New York: McGraw-Hill, 1973.
46. Weete JD. Lipid Biochemistry of Fungi and Other Organisms. London: Plenum Press, 1980.
47. Quershi AA, Barnes FJ, Semmler J, Porter JW. Biosynthesis of prelycopene pyrophosphate and lycopersene by squalene synthetase. J Biol Chem 1973; 248:2755–2767.
48. Odum HT. Systems Ecology: An Introduction. Chichester, UK: Wiley, 1983.
49. Dandekar S, Modi, VV, Jani UK. Chemical regulators of carotenogenesis by *Blakeslea trispora*. Phytochemistry 1980; 19:795–798.
50. Benveniste P. Sterol biosynthesis. Annu Rev Plant Physiol 1986; 37:275–308.
51. Desjardins AE, Plattner RD, Beremand MN. Ancymidol blocks trichothecene biosynthesis and leads to accumulation of trichodiene in *Fusarium sporotrichioides* and *Gibberella pulicaris*. Appl Environ Microbiol 1987; 53:1860–1865.
52. Moss MO, Frank JM. Variability in the production of trichothecenes by *Fusarium sporotrichiodes*. Int Biodet 1988; 24:445–453.
53. Hart LP, Braselton WE, Stebbins TC. Production of zearalenone and deoxynivalenol in commercial sweet corn varieties. Plant Dis 1982; 66:1133–1135.
54. Frank JM, Moss MO. Behaviour of a toxigenic *Fusarium sporotrichioides* in liquid culture. In: Moss MO, Frank JM, eds. 5th Meeting on Mycotoxins in Animal and Human Health. Guildford, UK: University of Surrey Press, 1984:155–168.
55. Moss MO, Badii F. Increased production of aflatoxins by *Aspergillus parasisticus* Speare in the presence of rubratoxin B. Appl Environ Microbiol 1982; 43:895–898.
56. Moss MO, Frank JM. Influence of the fungicide tridemorph on T-2 toxin production by *Fusarium sporotrichioides*. Trans Br Mycol Soc 1985; 84:585–590.
57. Bahtnagar D, McCormick SP. The inhibitory effect of neem (*Azadirachta indica*) leaf extracts on aflatoxin synthesis in *Aspergillus parasiticus*. JAOCS 1988; 65: 1166–1168.

58. Borrow A, Jefferies EG, Lessell RHJ, LLoyd EC. Metabolism of *Gibberella fujikouroi* in stirred culture. Can J Microbiol 1961; 7:227–276.
59. van Suijdam JC, Metz B. Influence of engineering variables upon the morphology of filamentous moulds. *Biotechnol Bioeng* 1981; 23:111–148.
60. Sergeant K, O'Kelly J, Carnaghan RBA, Allcroft R. The assay of the toxic principle in certain groundnut meal. Vet Rec 1961; 73:1219—1222.
61. Moss MO, Frank JM. The influence on mycotoxin production of interactions between fungi and their environment. In: Lacy J, ed. Trichothecenes and Other Mycotoxins. Chichester, UK: Wiley, 1985:257–268.
62. Wicklow DT. Ecological approaches to the study of mycotoxigenic fungi. In: Kurata H, Ueno Y, eds. Toxigenic Fungi—Their Toxins and Health Hazard. Amsterdam: Elsevier Press, 1984:33–43.
63. Shannon GM, Shotwell O, Lyons AJL, White DG, et al. Laboratory screening for zearalenone formation in corn hybrids and inbreds. J Assoc Off Anal Chem 1980; 63:1275–1277.
64. Ehrich K, Ciegler. Effect of phytate on aflatoxin formation by *Aspergillus parasiticus* grown on different grains. Mycopathologia 1985; 92:3–6.
65. Buchanan RL, Harry MA, Gealt MA. Caffeine inhibition of sterigmatocystin, citrinin and patulin production. J Food Res 1983; 48:1226–1228.
66. Cuero RG, Smith JE, Lacey J. Interaction of water activity, temperature and substrate on mycotoxin production by *Aspergillus flavus*, *Penicillium viridicatum* and *Fusarium graminearum* in irradiated grains. Trans Br Mycol Soc 1987; 89: 221–226.
67. Boller RA, Schroeder HW. Influence of *Aspergillus chevalieri* on production of aflatoxin in rice by *Aspergillus parasiticus*. Phytopathology 1973; 63:1507–1510.
68. Boller RA, Schroeder HW. Influence of *Aspergillus candidus* on production of aflatoxin in rice by *Aspergillus parasiticus*. Phytopathology 1974; 64:121–123.
69. Harrison DEF. Mixed cultures in industrial fermentation processes. Adv Appl Microbiol 1978; 24:129–164.
70. Eugenio CP, Christensen CM, Mirocha CJ. Factors affecting production of the mycotoxin F-2 by *Fusarium roseum*. Phytopathology 1970; 60:1055–1058.
71. Greenhalgh R, Neish GA, Miller JD. Deoxynivalenol, acetyl deoxynivalenol and zearalenone formation by Canadian isolates of *Fusarium graminearum* on solid substrates. Appl Environ Microbiol 1983; 46:625–629.
72. Joffe AZ. Growth and toxigenicity of fusaria of the sporotrichiella section as related to environmental factors and culture substrates. Mycopathology 1974; 54:35–46.
73. Surekha M, Reddy M. Influence of temperature and humidity on biodeterioration and aflatoxin production in groundnut fodder by *Aspergillus flavus*. J Toxicol Toxin Rev 1989; 8:291–297.
74. Nielsen PV, Beuchat LR, Frisvad JC. Growth of and fumitremorgin production by *Neosartorya fischeri* as affected by temperature, light and water activity. Appl Environ Microbiol 1988; 54:1504–1510.
75. Ueno Y. Temperature-dependent production of citreoviridin, a neurotoxin of *Penicillium citreo-viride* Biourge. Jpn J Exp Med 1972; 42:107–114.
76. Sacks LE, King AD, Schade JE. A note on pH gradient plates for fungal growth studies. J Appl Bacterial 1986; 61:235–238.

77. Sommer NF, Buchanan JR, Fortlage RJ. Production of patulin by *Penicillium expansum*. Appl Microbiol 1974; 28:589–593.
78. Paster N, Huppert D, Barkai-Golan R. Production of patulin by different strains of *Penicillium expansum* in pear and apple cultivars stored at different temperatures and modified atmospheres. Food Addit Contam 1995; 12:51–58.
79. Desjardins AE, Plattner RD, Vanmiddlesworth F. Trichothecene biosynthesis in *Fusarium sporotrichiodes*: origin of the oxygen atoms of T-2 toxin. Appl Environ Microbiol 1986; 51:493–497.
80. Bu'lock JD, Yuen TH. Oxygen requirements for secondary metabolism in *Trichoderme viride* and the effect of barbiturate. Phytochemistry 1971; 10:1835–1836.
81. Shih C-N, Marth EH. Aflatoxin formation, lipid synthesis and glucose metabolism by *Aspergillus parasiticus* during incubation with and without agitation. Biochim Biophys Acta 1974; 338:286–296.
82. Bennett JW, Fernholz FA, Lee LS. Effect of light on aflatoxins, anthraquinones and sclerotia in *Aspergillus flavus* and *A. parasiticus*. Mycologia 1978; 70:104–116.
83. Cullen D, Smalley EB, Caldwell RW. New process for T-2 toxin production. Appl Environ Microbiol 1982; 44:371–375.
84. Weinberg ED. Biosynthesis of microbial metabolites—regulation by mineral elements and temperature. In: Krumphanzl V, Sikyta B, Vanek Z, eds. Overproduction of Microbial Products. New York: Academic Press, 1982:181–194.
85. Scott RE, Jones A, Lam KS, Gaucher GM. Manganese and antibiotic biosynthesis. I. A specific manganese requirement for patulin production in *Penicillium urticae*. Can J Microbiol 1986; 32:259–267.
86. Scott RE, Jones A, Gaucher GM. Manganese and antibiotic biosynthesis. III. The site of manganese control of patulin production in *Penicillium urticae*. Can J Microbiol 1986; 32:273–279.
87. Niehaus WG, Dilts RP. Purification and characterization of glucose-6-phosphate dehydrogenase from *Aspergillus parasiticus*. Arch Biochem Biophys 1984; 228: 113–119.
88. Niehaus WG, Failla LJ. Effect of zinc on versicolorin production by a mutant strain of *Aspergillus parasiticus*. Exp Mycol 1984; 8:80–84.
89. Day JB, Mantle PG, Shaw BI. Production of verruculogen by *Penicillium estinogenum* in stirred fermenters. J Gen Microbiol 1980; 117:405–410.
90. Robbers JE, Robertson LW, Hornemann KM, Jindra A, Floss HG. Physiological studies on ergot: further studies on the induction of alkaloid synthesis by tryptophan and its inhibition by phosphate. J Bacteriol 1972; 112:791–796.
91. Arcamone F, Casinelli G, Ferni G, Penco S, Penella P, Pol C. Ergotamine production and metabolism of *Claviceps purpurea* strain 275 FI in stirred fermenters. Can J Microbiol 1970; 16:923–931.
92. Griffin DH, Walton DC. Regulation of abscisic acid formation in *Mycosphaerella* (*Cercospora*) *rosicola* by phosphate. Mycologia 1982; 74:614–618.
93. White JP, Johnson GT. Aflatoxin production correlated with nitrification in *Aspergillus flavus* group species. Mycologia 1982; 74:718–723.
94. Gräfe U (1982). Relevance of microbial nitrogen metabolism to production of secondary metabolites. In: Krumphanzl V, Sikyta B, Vanek Z, eds. Overproduction of Microbial Products. New York: Academic Press, 1982:63–75.

95. Mendelowitz S, Aharonowitz Y. Regulation of cephamycin C synthesis, aspartokinase, dihyrodipicolinic acid synthetase and homoserine dehydrogenase by aspartic acid family amino acids in *Streptomyces clavuligerus*. Antimicrob Agents Chemother 1982; 21:74–84.
96. Hankinson O, Cove DJ. Regulation of the pentose phosphate pathway in the fungus *Aspergillus nidulans*—the effect of growth with nitrate. J Biol Chem 1974; 249: 2344–2353.
97. Miller JD, Greenhalgh. Nutrient effects on the biosynthesis of trichothecenes and other metabolites by *Fusarium graminearum*. Mycologia 1985; 77:130–136.
98. Vinning LC, Chatterjee S. Catabolite repression and the control of secondary metabolism. In: Krumphanzl V, Sikyta B, Vanek Z, eds. Overproduction of Microbial Products. New York: Academic Press, 1982:35–46.
99. Madry N, Zocher R, Kleinkauf H. Enniatin production by *Fusarium oxysporum* in chemically defined media. Eur J Microbiol Biotechnol 1983; 17:75–79.
100. Ueno Y, Sawano M, Ishii K. Production of trichothecene mycotoxins by *Fusarium* species in shake culture. Appl Microbiol 1975; 30:4–9.
101. Payne GA, Hagler WM. Effect of specific aminoacids on growth and aflatoxin production by *Aspergillus parasiticus* and *A. flavus* in defined media. Appl Environ Microbiol 1983; 46:805–812.
102. Rehacek Z, Desai JD, Sajdl P, Pazoutova S. The cellular role of nitrogen in the biosynthesis of alkaloids by submerged culture of *Claviceps purpurea*. Can J Microbiol 1977; 23:586–600.
103. Miller JD, Greenhalgh R, Wang YZ, Lu M. Trichothecene chemotypes of three *Fusarium* species. Mycologia 1991; 83:121–130.

14

Taxonomic Use of Metabolic Data in Lichen-Forming Fungi

Helge Thorsten Lumbsch
Botanical Institute, University of Essen, Essen, Germany

I. INTRODUCTION

Chemistry has had an important role in the taxonomy of lichenized ascomycetes since the early years of lichen systematics, when lichens were distinguished by obvious differences in color which are caused by different secondary metabolites (e.g., the yellowish-orange genus *Xanthoria* and the gray *Physcia*), or characterized by their bitter taste due to the presence of some secondary metabolites (e.g., *Pertusaria amara* with picrolichenic acid). Nowadays, as Hawksworth (1) points out, any taxonomic revision in lichenology which does not consider chemical data is likely to be regarded as incomplete. In lichen chemotaxonomy, secondary metabolites have mainly been used, while polyols, free fatty acids, and other primary products have played only a marginal role.

Lichens are symbiotic organisms typically comprised of a fungus and a photosynthetic partner, which may be a cyanobacterium or green alga or both. It is now generally accepted, and has been recently supported by molecular data (2), that lichens are only a nutritional group of fungi associated with photosynthetic organisms and not a monophyletic group. The usual situation of the lichen symbiosis can become far more complex with the involvement of additional partners, and it is therefore more helpful to interpret the lichen thallus as an ecosystem rather than an organism. Fungi or other lichens can be found associated with lichens and at least some of these associations seem to be mutualistic rather than parasitic (3,4), thus extending the symbioses to include more than two or three partners. The name given to a lichen refers to its mycobiont, and approximately 15,000 to 20,000 species are known. Most of the fungal partners belong to the Ascomycetes; only few are Basidiomycetes. Since lichenized Basidiomycetes are not known to contain lichen substances and secondary metabolites have not been employed in taxonomy of these organisms, they are not treated here.

Morphologically three main growth forms of lichens can be distinguished: crustose, foliose, and fruticose. While the thallus is closely attached to its substratum in crustose lichens, foliose lichens resemble the shape of leaves in higher plants and the thallus is attached to its substratum by rhizinae, an umbilicus, or other means. Fruticose lichens have a shrubby thallus. Several intermediate and special types exist, as well as combinations of different growth types within a single thallus (e.g., dimorphic thallus in *Cladonia*). For further details, the reader is referred to reviews on the morphology of lichens (5,6). Anatomically different types can be distinguished, but the thallus is generally stratified in most foliose and fruticose lichens, and different layers can be distinguished. The photosynthetic partner is concentrated in an algal layer situated between the upper cortex and the medulla. The lower part of the thallus may or may not be covered by a lower cortex. In most crustose lichens the algal layer is covered by an ahyphal polysaccharide layer, known as the epinecral layer. The fruiting bodies of the lichens are generally similar to those of nonlichenized Ascomycetes; i.e., they may be apothecia, perithecia, hysterothecia, or pseudothecia. In the apothecial margin of some apothecia, algal cells are incorporated; this type is usually called lecanorine. Further details of the anatomy of lichenized ascomycetes can be found in textbooks (7–9).

The secondary metabolites usually employed in lichen chemotaxonomy are exclusively produced by the fungal partner. Several "lichen products" are produced in pure cultures of mycobionts (10–13), but differences occur as well. A discussion of these phenomena is found in Ahmadjian's textbook (13). A doubtful record of the presence of lichen substances in algal cells has been published (14); we were unable to reproduce those results in our laboratory.

More than 650 secondary products have been found in lichenized ascomycetes, and further substances will almost certainly be found, as exemplified by the structural elucidation of simonyellin. Simonyellin was found to be a naphthopyran, a class not previously known in lichens (15). Secondary metabolites may belong to different classes (16–19) and are produced via the acetate-polymalonate, mevalonic acid, and shikimic acid pathways (20,21), as shown in Fig. 1. These metabolites accumulate on the outside walls of the mycobiont hyphae. The vast majority of secondary metabolites in lichens are synthethized by the acetate-polymalonate pathway and belong to the depsides and depsidones. Substances unique to lichens usually belong to these groups, while substances in other groups—or closely related substances—might occur in other groups of plants or fungi as well. A list of the classes of secondary metabolites with reported

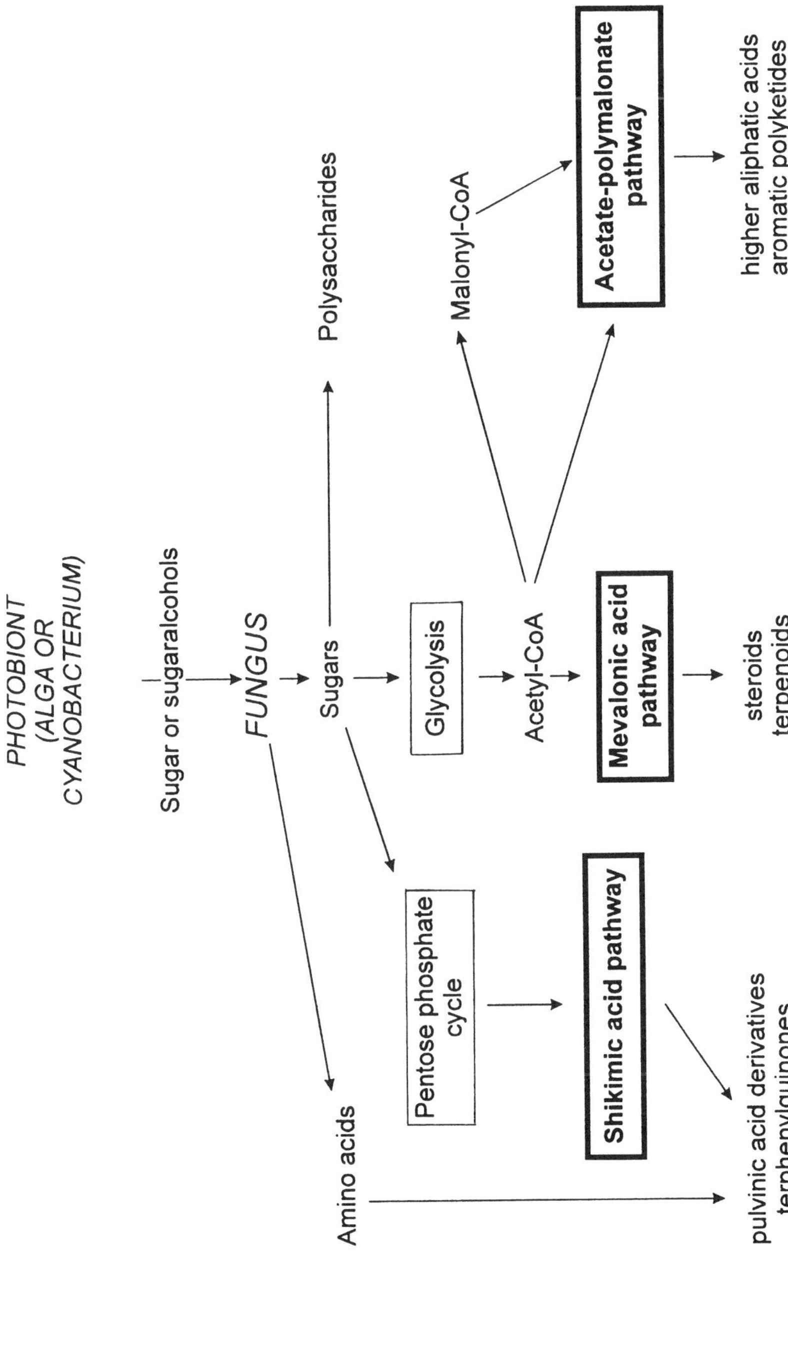

Figure 1 Simplified scheme of probable pathways leading to the major groups of lichen substances.

Table 1 Major Classes of Secondary Metabolites in Lichenized Ascomycotina

A. Acetate-polymalonate pathway
   I. Higher aliphatic acids and related compounds (Fig. 2)
   II. Phenolic compounds
      1. Monocyclic compounds (Fig. 3)
      2. Depsides (Fig. 4)
      3. Tridepsides (Fig. 5)
      4. Depsidones (Fig. 6)
      5. Depsones (Fig. 7)
      6. Benzyl esters (Fig. 8)
      7. Diphenyl ethers (Fig. 9)
      8. Dibenzofuranes, usnic acids and related compounds (Figs. 10, 11)
      9. Chromones (Fig. 12)
      10. Xanthones (Fig. 13)
      11. Naphthoquinones (Fig. 14)
      12. Anthraquinones (Fig. 15)
      13. Napthopyran (Fig. 16)
B. Mevalonic acid pathway
   I. Terpenoids (Figs. 17, 18)
   II. Steroids (Fig. 19)
C. Shikimic acid pathway
   I. Terphenylquinones (Fig. 20)
   II. Pulvinic acid derivatives (Fig. 21)

chemotaxonomic value in lichenized Ascomycetes is given in Table 1 and examples of structural formulas are presented in Figures 2 through 21.

Chemotaxonomy in lichens has focused on those products that are crystallized in situ. Our knowledge of the chemistry of amorphous pigments, which occur widely in lichens, is very limited (22). Although they are poorly known, these substances may have some taxonomic significance, as shown by observations in "brown Parmeliae" or *Micarea* species (23–25).

Cell-wall polysaccharides were recently employed as a further character in

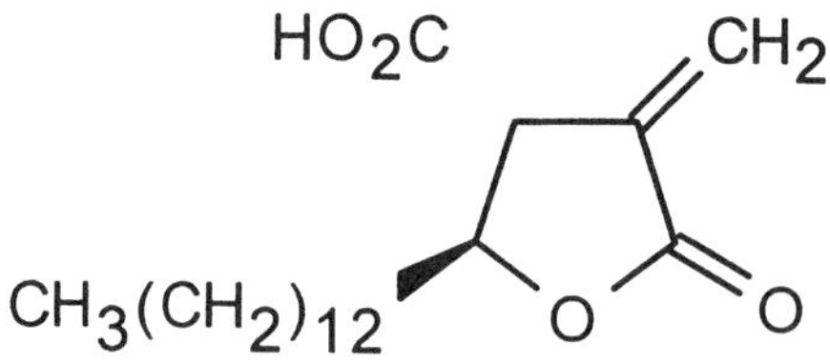

Figure 2 Structural formula of (+)-protolichesterinic acid.

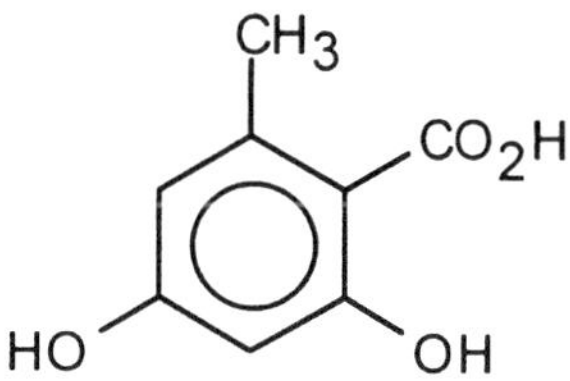

Figure 3 Structural formula of orsellinic acid.

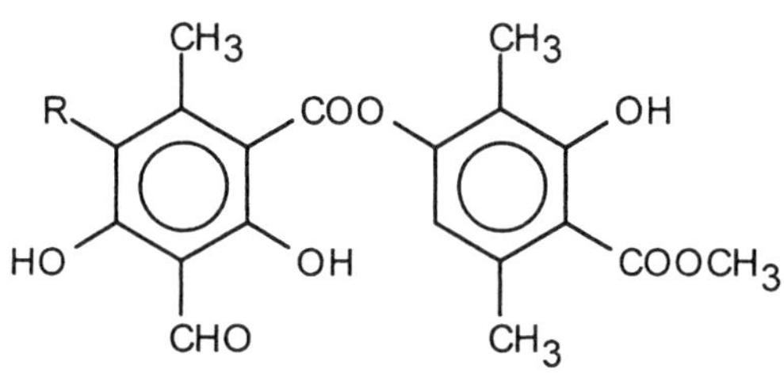

Substance	*R*
Atranorin	*H*
Chloroatranorin	*Cl*

Figure 4 Structural formulas of atranorin and chloroatranorin.

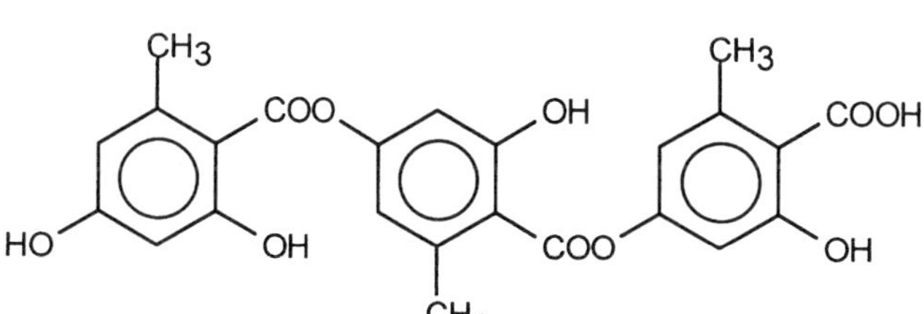

Figure 5 Structural formula of gyrophoric acid.

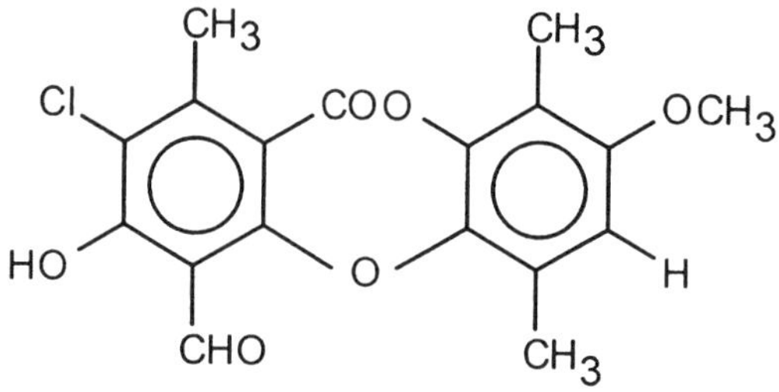

Figure 6 Structural formula of pannarin.

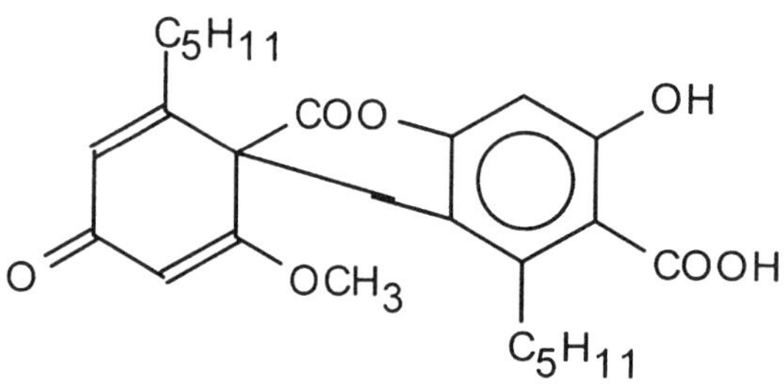

Figure 7 Structural formula of picrolichenic acid.

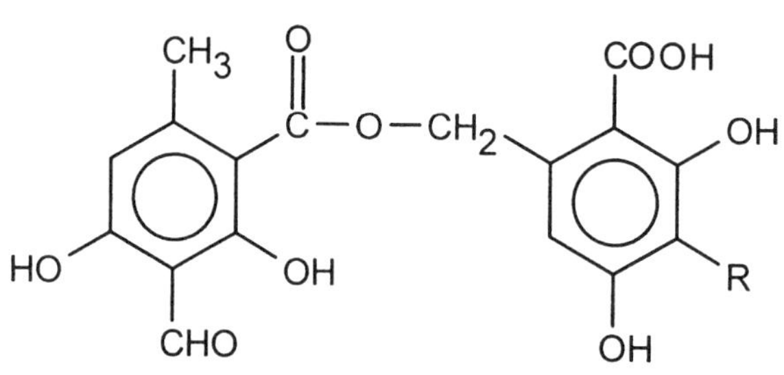

Substance	*R*
Alectorialic acid	CH_3
Barbatolic acid	*CHO*

Figure 8 Structural formulas of alectorialic and barbatolic acids.

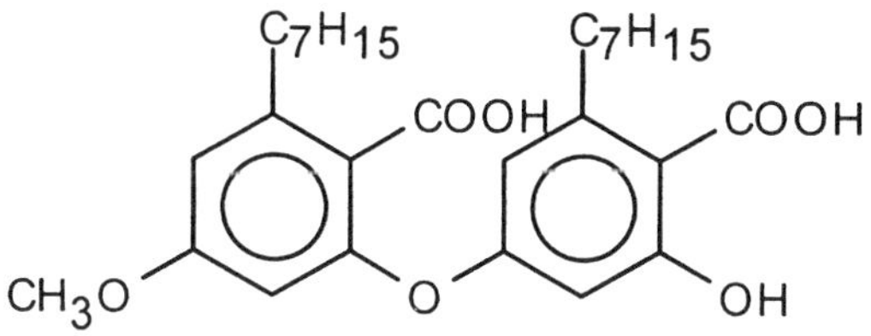

Figure 9 Structural formula of micareic acid.

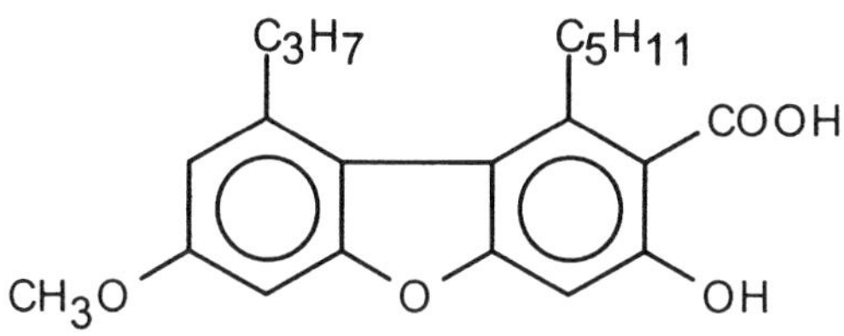

Figure 10 Structural formula of didymic acid.

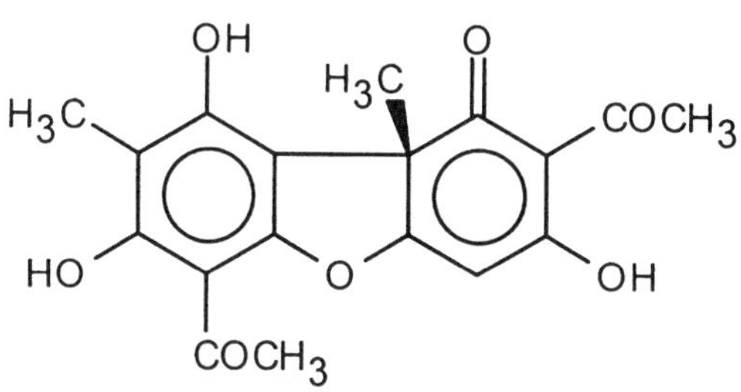

Figure 11 Structural formula of usnic acid.

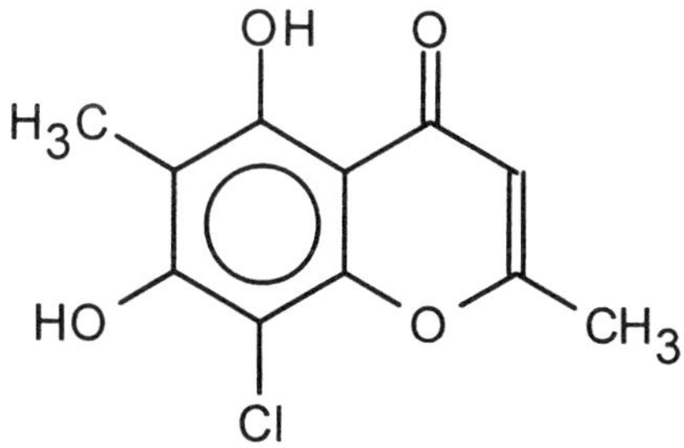

Figure 12 Structural formula of sordidone.

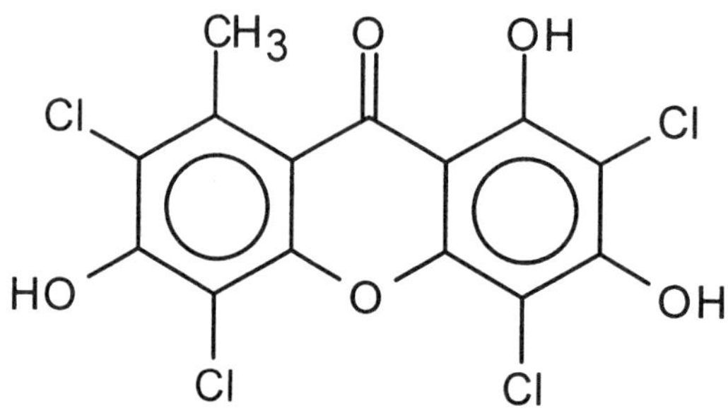

Figure 13 Structural formula of thiophanic acid.

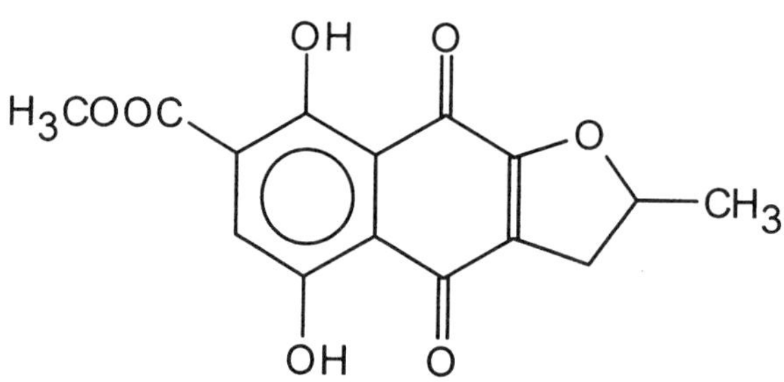

Figure 14 Structural formula of haemaventosin.

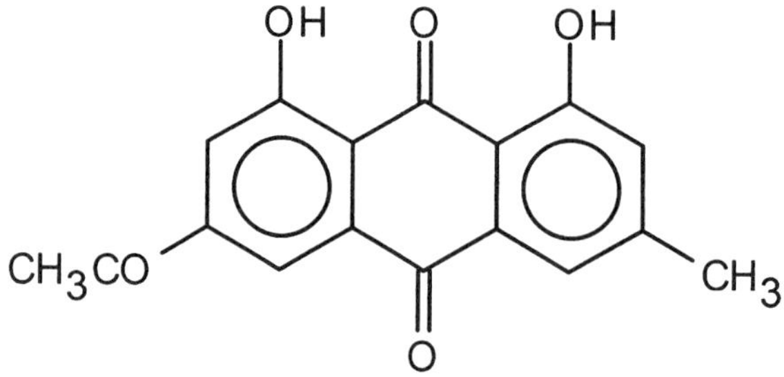

Figure 15 Structural formula of physcion (= parietin).

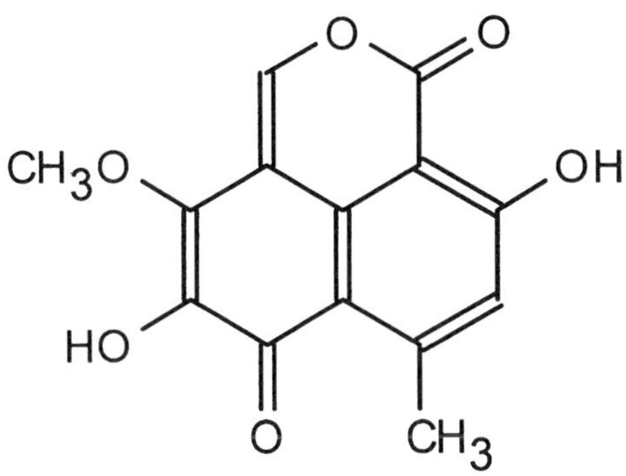

Figure 16 Structural formula of simonyellin (naphtopyran)

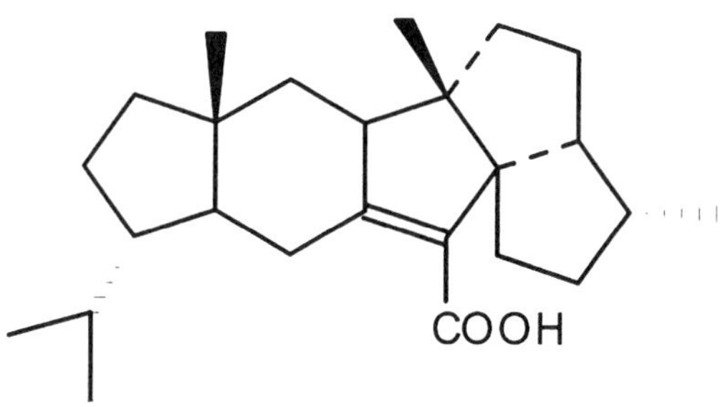

Figure 17 Structural formula of retigeranic acid.

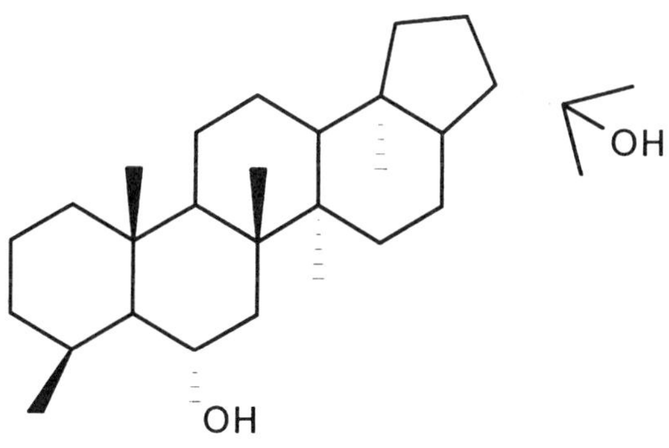

Figure 18 Structural formula of zeorin.

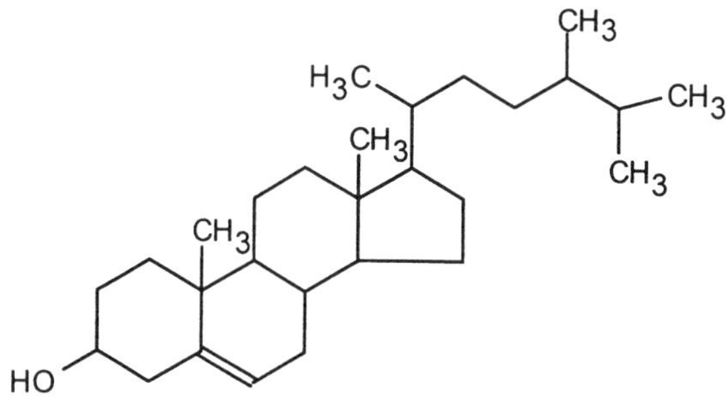

Figure 19 Structural formula of β-sitosterol.

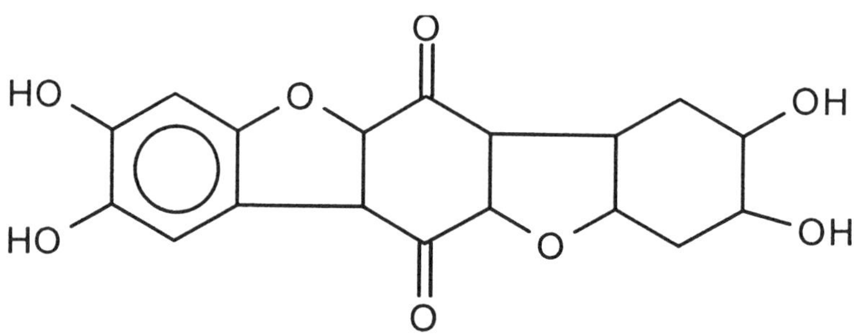

Figure 20 Structural formula of thelephoric acid.

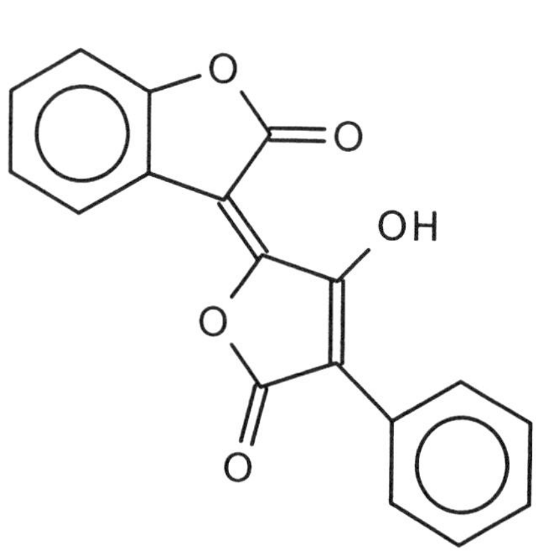

Figure 21 Structural formula of calycin.

distinguishing species groups in Parmeliaceae sensu lato (26). Four main groups of polysaccharides were distinguished based on the stereochemistry of the glycosidic bonds, and this character set correlates quite well with other characters in that group (27,28).

In this review we can only discuss some examples of chemical variation and their significance to taxonomy in lichenized fungi. Exhaustive treatments of the chemistry of lichen substances have been published since the early compilation by Zopf (29,16–18,30–38). Lichen chemotaxonomy has been treated in detail by numerous authors (30–48).

II. IDENTIFICATION OF LICHEN SUBSTANCES

Here we can only briefly refer to the methods used in the identification of lichen substances. The methods are discussed in more detail by other authors (37,49–52).

A. Microchemical Tests

Secondary metabolites are produced in large amounts by lichens, which has made it possible to detect them easily by spot tests with KOH and bleaching powder (C), which were first employed in the 1860s by Nylander (53,54). With these reagents, the thallus and apothecia may give characteristic yellow, red, or green color reactions. Later, Asahina (55) added a further spot test with p-phenylendiamine which gives yellow to red reactions with secondary metabolites containing a free aldehyde group. These methods were supplemented by the introduction of microcrystal tests (56).

B. Thin-Layer Chromatography

The introduction of chromatographic methods has made chemical investigations an integral part of any modern taxonomic study in lichenology. In the 1950s, paper chromatography was used for the identification of lichen substances (57), but this method was rapidly superseded by thin-layer chromatography (TLC).

A standardized method to identify lichen products by TLC was developed by Chicita Culberson and co-workers with subsequent additions and modifications (58–60). In this method, three main solvent systems are used (Table 2) and two

Table 2 Solvent Systems in the Standardized TLC System by Culberson (58–60)

Solvent A:	Toluene (180 ml)	Dioxane (45 ml)	Acetic acid (5 ml)
Solvent B:	Hexane (140 ml)	*tert.*-butyl ether (72 ml)	Formic acid (18 ml)
Solvent C:	Toluene (170 ml)	Acetic acid (30 ml)	

internal standards (atranorin and norstictic acid) to which the actual R_f data are compared. Besides the R_f values, the colors of the spots in daylight and UV are used before and after spraying with sulfuric acid. Alternative reagents to sulfuric acid have been used for the development of the TLC plates in special cases: a stabilized PD reagent (61); anisaldehyde with sulphuric acid (62); and 3-methyl-2-benzothiazolone hydrazone hydrochloride (MBTH) (63).

Additional solvent systems were applied to some groups of lichen substances which were difficult to separate under standardized conditions, such as some β-orcinol depsidones, chlorinated xanthones, or depsides with long side chains in *Pertusaria* (64–66). For complex mixtures a two-dimensional TLC system has been developed (67). When depsides with long side chains are examined, as perlatolic acid or sphaerophorin, the chromatography of hydrolysis products can prove helpful (59,68). Arup et al. (70) provide a list of R_f values differing when HPTLC plates are used in a developing chamber instead of the originally proposed larger TLC plates in the standardized system of Culberson (69).

C. High-Performance Liquid Chromatography

Because most lichen substances are very thermolabile or nonvolatile, gas chromatography has never been widely used in lichen chemistry. Instead, high-performance liquid chromatography (HPLC) has become an ideal tool for the detection of lichen substances, especially when only trace amounts are involved. Early attempts were made with normal-phase silica columns (71,72), but better results were obtained using reversed-phase columns. Both isocratic elution (73–76) and gradient elution have been used.

Since lichen extracts often contain compounds with different hydrophobicities, gradient elution is ideal for the HPLC analyses of lichens. Strack and coworkers (77) separated 13 selected lichen products with a 70-min linear gradient. More recently, standardized systems for the identification of lichen compounds using gradient elution HPLC were proposed by Huovinen et al. (78) for the genera *Cladina* and *Cladonia*, and by Feige et al. (79) for over 300 aromatic lichen substances present in different groups of lichenized Ascomycotina, both using internal standards.

When transformed into their strongly UV absorbing phenacyl esters, aliphatic acids can be examined by HPLC as shown by Huneck and colleagues (80). An HPLC method using a photodiode detection array has been proposed by Yoshimura et al. (81). HPLC is now more regularly used in taxonomic analysis, which is especially important in the analysis of chemosyndromes as discussed below.

D. Mass Spectrometry

Usually, relatively few secondary metabolites are present in lichenized ascomycetes. These substances may sublime if the lichen is heated at very low pressure in mass spectrometry. Lichen mass spectrometry (LMS) was developed by Santesson (82), where small lichen samples are introduced into a mass spectrometer by a direct inlet system. The sample is heated, and many lichen compounds sublime readily at low pressure. Mass spectra of the subliming substance may then be recorded and are usually similar to the spectra of the corresponding pure compounds.

E. Computer-Assisted Identifications

The chromatographic and mass spectrometry data are now available for both Apple and Dos operating systems (83–85). The computer programs use a database which includes R_f values from six standard solvent systems, the HPLC RI value in the standard system of Feige et al. (79), visibility and color of TLC spots before and after sulfuric acid and MBTH, and major peaks in the mass spectrum as well as other data.

III. TAXONOMIC USE OF METABOLIC DATA

A. Metabolic Data at Species and Subspecies Levels

Most discussions of the use of chemistry in lichen taxonomy have centered on its application at species and subspecific level. When introducing the color tests, Nylander (53) used them to distinguish species based on different color reactions of the thallus. While some contemporaries, such as Leighton (86), were quick in adapting the new method, others, such as Lindsay (87), were skeptical or critical. There is still disagreement on the rank that should be used to distinguish chemically different populations. Only a brief overview on the history of lichen chemotaxonomy can be given here; the reader is referred to Hawksworth (41) for a more in-depth historical account. The main discussion has centered on the problem of how to rank those populations where no morphological characters can be correlated with the chemical differences. While some workers uncritically described new taxa solely on the basis of different color reactions (88,89), others preferred to treat chemical races at variety or forma level (90), or proposed that chemical characters should not be used at any taxonomic rank and only chemical strains should be accepted (91).

Chemotaxonomy was given a much sounder basis by the extensive studies of Asahina and co-workers in the 1930s to 1960s, but the main disagreement on the ranking of chemical variants remained. With the introduction of new methods (microcrystallization), it was possible to determine the lichen substances in herbarium specimens more accurately than previously. In his early papers, Asahina (92) regarded chemical differences as sufficient to distinguish species, although in later works he himself more often used subspecific rank to classify chemically different populations (93). The formal attempt to solve the ranking problem by introducing the term "chemovar." for chemical races, originally proposed for vascular plants (94), was only adopted by a few lichenologists (95). The implicit idea that chemical characters somehow differ from other characters and therefore deserve special terms must be rejected.

Subsequently, chemical races have been accepted at various ranks by different authors. Some authors prefer to include these chemotypes within one single, morphologically defined species, even when differences in ecology or geography can be shown (44); others prefer a distinction at species (96,97) or subspecies level (41,98–100). The discussion on this subject seems to be endless and cannot be reiterated here. An excellent review of that discussion was compiled by Hawksworth (41).

It is difficult to understand how the taxonomic value of chemical variation can be established a priori. The significance of morphological and anatomical characters can vary, and some characters which are useful to separate larger groups (e.g., multispored asci in *Acarospora*) may be of little value in other groups (e.g., eight- and multispored asci in *Candelariella faginea* [101]). A character in itself—e.g., number of spores per ascus—has no taxonomic value a priori, but may have importance when correlated with other independent characters. This, however, can only be evaluated a posteriori, at the end of an examination. Chemical characters do not fundamentally differ from morphological and anatomical characters, and it is therefore not surprising to find that chemical characters may be of importance at different ranks in different groups.

This position includes the situation where some types of observed chemical variation do not deserve any taxonomic recognition. One example of this type of variation is where it may be caused by a simple mutation. Dickhäuser et al. (102) found chemical features to be important characters for distinguishing taxa in the *Lecanora subcarnea* group, an assemblage of crustose lichens with heavily pruinose apothecial disks which occur on siliceous rocks in all continents, except Antarctica. Most species are characterized by chemical differences in addition to morphological ones (Table 3). However, in one species, *L. subcarnea*, some specimens occur with virensic acid instead of protocetraric acid. This chemical race was collected in Austria, France, Germany, Sweden, Switzerland, and Turkey, and was always rare among the collections for these countries, while the protocetraric acid race is widely distributed throughout Europe and the Mediterranean. Virensic acid is closely related to protocetraric acid (Fig. 22), differing only in the absence of a hydroxy-group in the 9 position of the B-ring, and the virensic acid chemotype occurs sporadically within the distribution area of the protocetraric acid chemotype. This indicates that the virensic acid chemotype is a simple mutant of *L. subcarnea*, which lacks the ability to oxidize the methyl-group at position 9, and that the chemotype is polyphyletic and does not merit taxonomic recognition.

Similar cases are known from *Candelariella* species. In this crustose genus the thallus is yellow due to the presence of pulvinic acid derivatives. However,

Table 3 Chemical Patterns of Taxa in the *Lecanora subcarnea* Group[a]

Taxon	1	2	3	4	5	6
L. farinacea	+		±		±	
L. ochroidea	+		+			
L. rhodi	+	+	+			
L. sanctae-helenae	+		+	+	+	
L. subcarnea chemical race I	+				+	
L. subcarnea chemical race II	+					+
L. subcarnea var. *soralifera*	+				+	

[a]Only major substances mentioned: 1 = atranorin (Fig. 4); 2 = 2′-*O*-methylperlatolic acid (Fig. 24); 3 = norstictic acid (Fig. 23); 4 = placodiolic acid (Fig. 25); 5 = protocetraric acid (Fig. 22); 6 = virensic acid (Fig. 22).

citrine-yellow thalli occur sporadically in populations of three species, which result from the suppression of calycin (Fig. 21) production (103). In this case these mutants were treated at forma level.

A careful examination of two chemotypes in the genus *Arctoparmelia* has shown that the races cannot be separated as different species (104), as previously suggested (105). The chemotypes differ in the presence or absence of usnic acid (Fig. 11), the usnic acid-deficient strain occurs in Fennoscandia and Canada. Both chemotypes grow side by side in Canada and show no detectable differences in morphology or epiflora of lichenicolous fungi. The Fennoscandian and Canadian populations of the pigment-deficient strain, however, contain different combina-

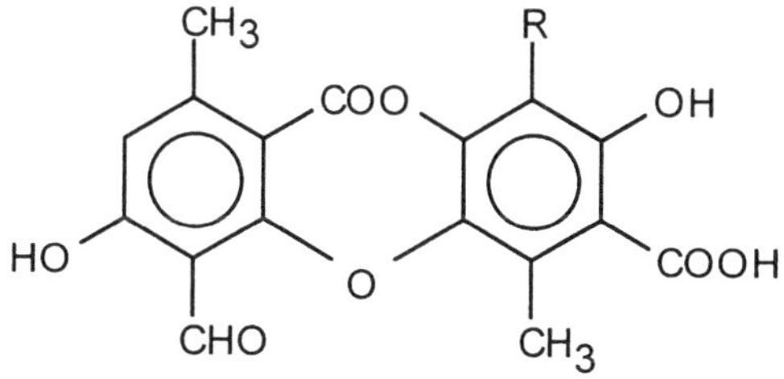

Protocetraric acid: $R = CH_2OH$

Virensic acid: $R = CH_3$

Figure 22 Structural formulas of protocetraric and virensic acids.

tions of fatty acids. The pigment-deficient strains probably originated separately on the two continents, rather than from a common ancestor.

Follmann and Schulz (106) found remarkable changes in the secondary products of lichen thalli due to environmental stress or infection by lichen parasites. This included alterations in the amounts of specific compounds, displacements from major to minor products, and vice versa. The authors pointed out that great care should be taken when describing new taxa, especially from small samples, since such alterations and deviations in the secondary compound patterns are considered to be not as scarce as previously assumed. A change in secondary metabolite pattern was also found by Feige and co-workers (107) in *Roccella* species parasitized by *Lecanographa grumulosa* (as *Lecanactis grumulosa*). However, it is not clear whether the change in secondary metabolites in this case is due to a reaction of the host, or whether lichenicolous fungi or lichens produce or use lichen substances. In this connection it is of interest to note that in five species of lichenicolous fungi 13 metabolites have been detected (108), including two also known from lichenized Ascomycotina, viz. gyrophoric acid (Fig. 5) in two species and pannarin (Fig. 6) in *Milospium graphideorum*.

In most cases metabolic data correlate well with other independent characters and can help to sort out confusing assemblages of difficult species complexes. A good example is those populations that were filed under the collective name *Lecanora campestris*, a saxicolous species. A combination of anatomical, morphological, and chemical characters identifies five taxa in this group in North America (43). The same was true for a revision of the Australasian taxa. In this case, six species could be distinguished, but interestingly, *L. campestris* s.str. does not occur in that region; it is now known to be restricted to Europe and Western North America (109). In the morphologically difficult and anatomically variable *Lecanora dispersa* group, Poelt and Leuckert (110) found the presence of chlorinated xanthones to be useful in the distinction of taxa.

A study on the sociology and ecology of two morphologically similar species of *Cladonia* (*C. rei* and *C. subulata*) revealed two chemically defined taxa. The two differ in their ecology and sociology; although morphology was not reliable in this case, different morphological tendencies were observed (111).

Many studies have concentrated on phenolic substances and their significance in the distinction of species. However, in *Lecanora* (112), triterpenoids have also been used successfully as species level. In *Pyxine*, different patterns of triterpenoids on a TLC plate were used as taxonomic characters ("fingerprints") (113).

The problem of the taxonomic estimation of chemical variation is complicated by the occurrence of fertile and sterile populations of taxa which are otherwise identical morphologically, the so-called species pairs (114–118). The taxonomomic status of these species pairs has also been disputed. Some authors prefer to treat them as species (118), others as forms (115) or at different taxonomic levels (116). An example in the foliose genus *Parmotrema*, where a parallel evolution of secondary metabolites and development of sterile taxa derived from fertile ancestors occurs, was discussed by the Culbersons (119,120). The asexual *Parmotrema* ("*Parmelia*") *hypotropa* has two chemical races that differ in their distribution and appear to have been derived by morphological parallelism from chemically identical races of the closely related *Parmotrema perforata*, which is sexual. It was proposed that each chemical type be accepted at species level for each morphotype, resulting in four species in that group (119).

1. Distribution of Secondary Metabolites in a Lichen

The distribution of secondary metabolites in the thallus and ascomata of lichens may also provide taxonomically important information. The presence of substances in particular areas are characteristic features of some species, such as norstictic acid (Fig. 23) which is restricted to the apothecial discs in *Letharia columbiana* (121), or pulvinic acid derivatives which are found in parts of the medulla surrounding pseudocyphellae in *Pseudocyphellaria* species (122). The different localization of chromones and xanthones was used as a taxonomic character in the distinction of taxa in the *Lecanora rupicola* group (99), as summarized in Table 4. Further examples of the localized occurrence of lichen substances are mentioned by Brodo (42).

In lichens, the substances which accumulate in the cortex and the medulla are usually different. Cortical compounds have an ecological importance and must therefore also have an evolutionary significance. It is generally agreed that

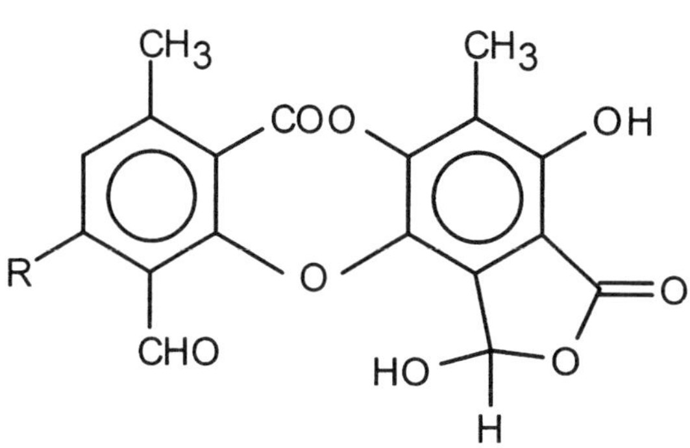

Substance	*R*
Norstictic acid	*OH*
Stictic acid	OCH_3

Figure 23 Structural formulas of norstictic and stictic acids.

Table 4 Localization of Some Phenolic Compounds in Selected Taxa of the *Lecanora rupicola* Group[a]

	Sordidone (Fig. 12)				Thiophanic acid (Fig. 13)			
Taxon	Disk	Apoth. margin	Thallus cortex	Thallus medulla	Disk	Apoth. margin	Thallus cortex	Thallus medulla
L. bicincta var. *bicincta*	+	?	–	–	–	–/?	–/+	–/+
L. bicincta var. *sorediata*	+	?	–	–	–	–	–	–
L. rupicola ssp. *rupicola*	+	?	–	–	–	–	–	–
L. rupicola ssp. *subplanata*	+	?	–	–	–	?	?	+
L. rupicola ssp. *sulphurata*	+	?	–	–	–	+	+	+
L. rupicola ssp. *arctoa*	+	+	+	?	–	–	–	–
L. lojkeana	+	+	+	?	–	–	–	–
L. swartzii ssp. *swartzii*	+	+	+	?	–	–	–	–
L. swartzii ssp. *nylanderi*	+	?	–	–	–	–	–	–

Source: Ref. 90.
[a]Atranorin present in the apothecial margin and thallus of all taxa.

some cortical substances are important at higher taxonomic rank, either to circumscribe genera (e.g., *Letharia* with vulpinic acid [Fig. 26]) or species groups (e.g., *Lecanora polytropa* group with usnic acid [Fig. 11]). However, in some cases the presence of a cortical substance varies at species level, as in *Parmeliopsis ambigua* (yellowish with usnic acid) and *P. hyperopta* (gray, lacking usnic acid, but containing atranorin [Fig. 4]). This case and the general significance of cortical metabolites are discussed at length in a stimulating review by Rikkinen (123).

Variation in medullary constituents is mostly used as a discriminator at species level. However, certain compounds may also be characteristic for species groups. For example, orcinol derivatives are common in *Cetrelia*, while the closely related genus *Platismatia* contains aliphatic acids or β-orcinol derivatives (124).

2. Simple Chemical Variation

In classical chemical variation, chemotypes show simple replacements of one or a few substances. The substances are present solitarily, being the only representa-

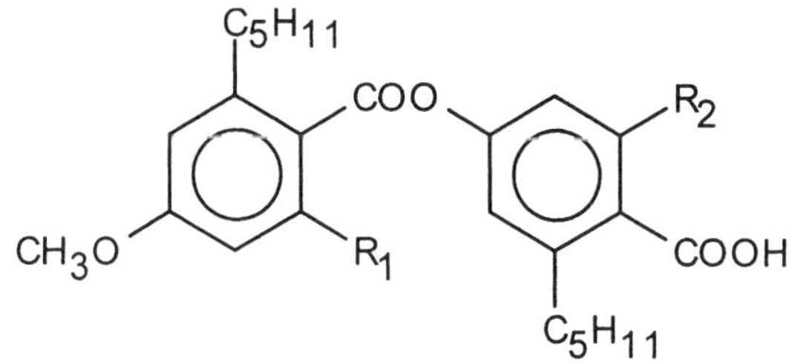

Substance	*R_1*	*R_2*
2-O-methylperlatolic acid	*OCH_3*	*OH*
2'-O-methylperlatolic acid	*OH*	*OCH_3*
perlatolic acid	*OH*	*OH*

Figure 24 Structural formulas of perlatolic acid and related substances.

tive of a type of compound. They can co-occur with unrelated substances in different combinations; e.g. the triterpenoid zeorin (Fig. 18) occurs with the β-orcinol depside atranorin in *Lecanora melacarpella*, and with lichexanthone (Fig. 27) in *Lecidella stigmatea*. In these cases each substance can be used as a taxonomic character.

A well-known example of replacement compounds in morphologically

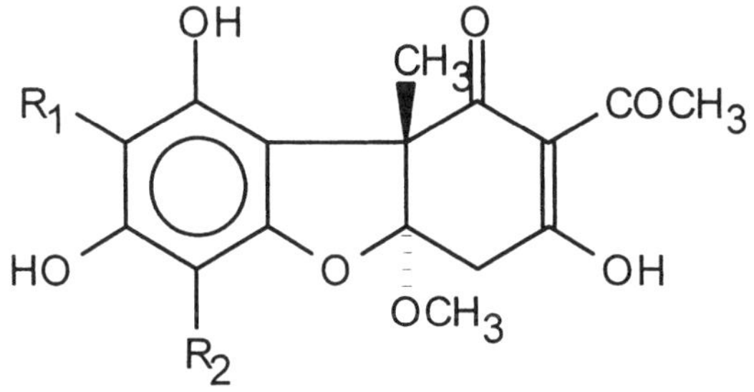

Substance	*R_1*	*R_2*
Placodiolic acid	*$COCH_3$*	*CH_3*
Pseudoplacodiolic acid	*CH_3*	*$COCH_3$*

Figure 25 Structural formulas of placodiolic and pseudoplacodioloic acids.

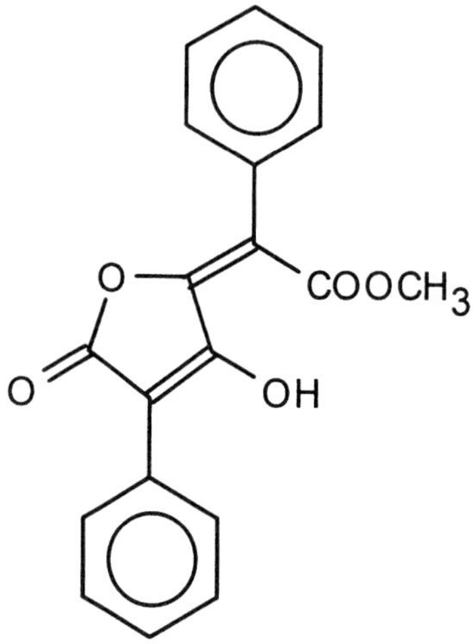

Figure 26 Structural formula of vulpinic acid.

identical populations of lichens, is in *Pseudevernia furfuracea*, which belongs in the Parmeliaceae (125–129). In this morphotype three chemical races occur. A lecanoric acid (Fig. 28) strain occurs in North America, while in Europe a chemical strain with olivetoric acid (Fig. 29) and a second one with physodic acid (Fig. 30) occur. The geographical distribution of the European chemical races differs significantly, although the two races occur together throughout most parts of Europe. Hale (126) distinguished the chemical races at species level, while others preferred to regard them as varieties (41). Biogenetically, olivetoric and physodic acids are closely related, and a small percentage of populations contain both substances (128,129). Therefore, it is nowadays generally accepted that the two European strains represent a single species which shows some genetic variation, while the allopatrically distributed North American strain is regarded as a

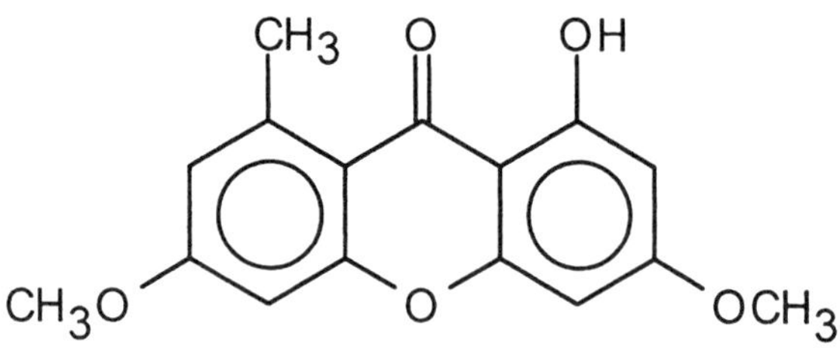

Figure 27 Structural formula of lichexanthone.

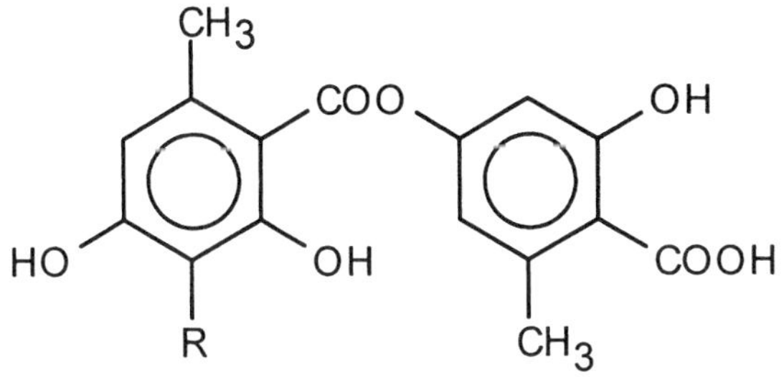

Substance	*R*
Diploschistesic acid	*OH*
Lecanoric acid	*H*

Figure 28 Structural formulas of diploschistesic and lecanoric acids.

species of its own, *P. consocians*. Although the chemistry is not important at any taxonomic rank in the case of the European strains of *Pseudevernia furfuracea*, the secondary metabolites provide us with useful information regarding development and distribution of the population of this species.

Sometimes, differences in ecology or distribution are regarded as important indicators which support the taxonomic recognition of chemical strains. This is generally acceptable, as in the case of *P. consocians* discussed above. However, it should not be forgotten that different mutations and strains can have different histories. Therefore, they must differ in ecology or distribution when they are monophyletic. Only when a chemical strain is an offspring of the common chemodeme, as in *Arctoparmelia centrifuga* (see above), the distribution of the two chemical strains does not differ fundamentally. It has been further suggested

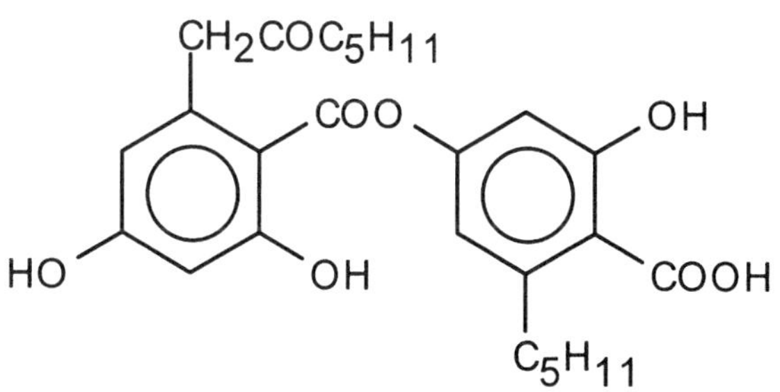

Figure 29 Structural formula of olivetoric acid.

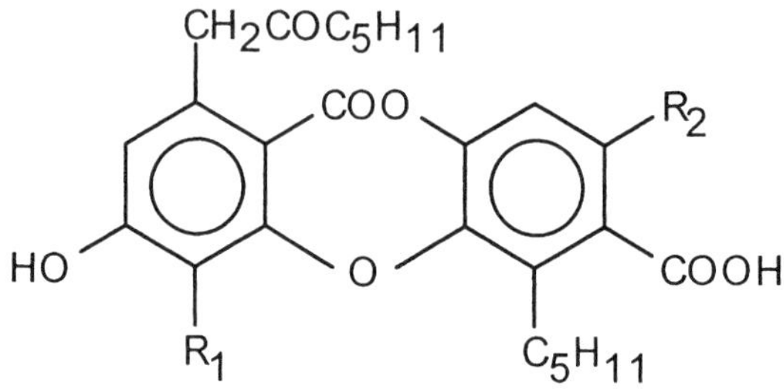

Substance	R_1	R_2
3-Hydroxyphysodic acid	*OH*	*OH*
2'-O-Methylphysodic acid	*H*	OCH_3
Physodic acid	*H*	*OH*

Figure 30 Structural formulas of physodic acid and related substances.

that the best evidence that chemical variation is under genetic control rather than being environmentally determined is the sympatric occurrence of chemical races, which maintain their integrity (130–132). Whether genetic control implies taxonomic recognition should also be considered.

3. *Chemosyndromic Variation*

A special kind of chemical variation is chemosyndromic variation. A chemosyndrome is a set of biogenetically closely related substances. It was first described in *Cetrelia* (133), and subsequently found in further Parmeliaceae, as the *Parmelia pulla* group (now regarded as *Neofuscelia*) and *Xanthoparmelia* species (74,134). Since then it has been found to occur in numerous groups of lichens (109,135,136).

In chemosyndromic variation, each taxon contains a characteristic set of products that appears to show progressive chemical change. An example of chemosyndromic variation in a group of species of *Relicina*, a foliose genus belonging to the Parmeliaceae is given in Table 5. In chemosyndromic variation the major constituents of each taxon are accompanied by biogenetically related minor constituents. The compounds that are the major substances in some species may be minor compounds in others. Thus, not only is the presence of lichen compounds important, but also the relative amounts of the substances and their relation to each other.

In a study on some *Hypogymnia* species from Turkey, Zeybek et al. (137) showed that the chemistry varied between the species in this group of foliose lichens, but the major and minor compounds were constant within every species

Table 5 Chemosyndromic Variation in Some *Relicina* Species

Species	Echinocarpic acid	Conechinocarpic acid	Hirtifructic acid	Gyrophoric acid (Fig. 5)	Fatty acids
R. samoensis	Major	Minor	—	—	—
R. terricrocodila	Major	Minor	Trace	—	—
R. fijiensis	—	—	Major	—	—
R. niuginiensis	—	—	Major	Trace	Minor
R. relicinula	—	—	—	—	Major

Source: Adapted from Ref. 147.

(Table 6). In this genus the species are well defined morphologically and their taxonomic status is not disputed. It is therefore interesting to see the constancy of the relative amount of secondary metabolites in this group as a model which indicates that the quantity of lichen compounds, as well as the quality, can supply taxonomic information.

In a character analysis of the secondary products of the crustose Porpidiaceae, Gowan (136) sorted 18 compounds into eight series, where they occurred in particular, relative concentrations. Each set was interpreted as a taxonomic character state. Each character set may represent a single recurring chemical pathway, and similarities in the pathways offer a basis for proposing evolutionary homology. Convergences, proposed on the basis of dissimilarity, were corroborated by co-occurrence. Hence the presence of chemosyndromes could not only be used as a character to define taxa, but also to infer phylogenetic hypotheses in a certain group of lichens.

4. *Patterns of Occurrence of Secondary Metabolites*

It is evident that for some reasons, chemical races without morphological or anatomical correlations are usually restricted to some species groups and do not occur randomly among lichenized Ascomycetes. In the genus *Cladonia*, morphology generally correlates quite well with chemical characters, but not in the *C. chlorophaea* group, which is notoriously variable chemically. In this group the 14 known chemotypes are either interpreted as sibling species (40,138) or as chemical races classified in two subspecies (139). New data have been added in this species complex recently (140–143), and gene flow was demonstrated by the analysis of secondary products in the progeny of individuals from natural populations of mixed chemotypes (140). While all chemotypes were formerly interpreted as sibling species by the Culbersons (40,138), they found both the *grayi* and *merochlorophaea* chemotypes within a single interbreeding population that was reproductively isolated from the *Cryptochlorophaea* chemotype. In a different

Table 6 Occurrence of Some Lichen Substances Within Selected *Hypogymnia* Species from Turkey[a]

Species	Atranorin (Fig. 4)	Chloro-atranorin (Fig. 4)	Physodalic acid (Fig. 31)	Proto-cetraric acid (Fig. 22)	Alectoronic acid (Fig. 32)	3-Hydroxy-physodic acid (Fig. 30)	2′-*O*-Methyl-physodic acid (Fig. 30)	Physodic acid (Fig. 30)	Vittatolic acid (Fig. 33)
H. bitteri	M	m	—	—	—	M	—	M	—
H. farinacea	m	M	—	—	—	M	m	M	—
H. laminisorediata	m	M	—	—	m	M	m	M	—
H. physodes	M	m	M	m	m	M	m	M	—
H. tubulosa	M	m	—	—	m	M	m	M	—
H. vittata	M	m	—	—	—	M	—	M	M

[a]According to Ref. 137.

[b]M = major substance; m = minor substance or trace; — = not detectable.

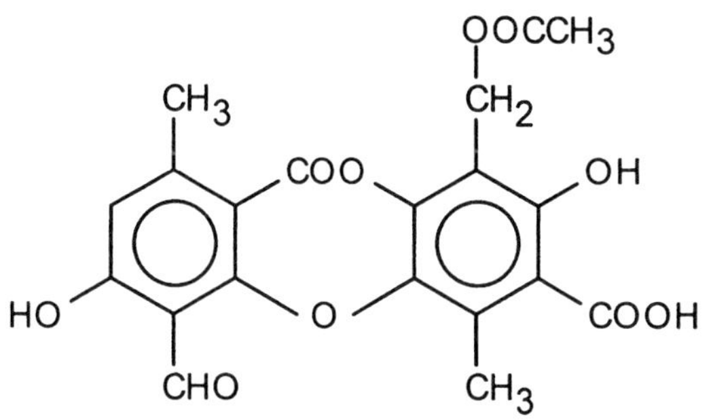

Figure 31 Structural formula of physodalic acid.

population, the *Cryptochlorophaea* chemotype hybridized with the local endemic *perlomera* chemotype.

Ribosomal DNA variation in a population of this group showed that three of the chemotypes had multiple rDNA restriction-fragment patterns, and two patterns were shared among chemotypes. The large number of restriction-fragment patterns suggested polymorphism within some interbreeding groups, and the sharing of patterns among chemotypes suggested that some chemotypes may be polymorphisms of a single species (143).

A much discussed case of chemical variation in a morphologically uniform species is the *Ramalina siliquosa* group, which includes chemical races occurring on coastal rocks, mainly in Western Europe. Where the races are sympatric, they populate different habitats (130,132). Recently, this case was reexamined (144) in order to study gene flow and reproductive isolation in the complex through an analysis of the progeny of maternals from nature. The chemotypes were shown to be reproductively isolated, the progeny tending to be chemically identical to their respective maternal individuals. However, one pair of chemotypes (stictic acid [Fig. 23] and acid-deficient strain) showed polymorphism.

In the Australasian taxa of the *Lecanora subfusca* group, chemical differences are generally associated with morphological, anatomical, and geographical

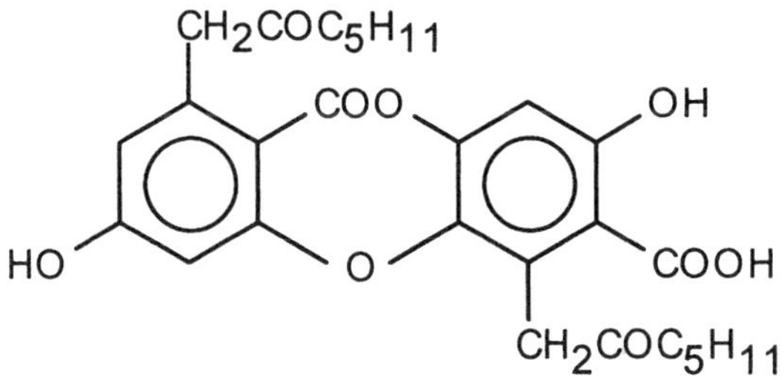

Figure 32 Structural formula of alectoronic acid.

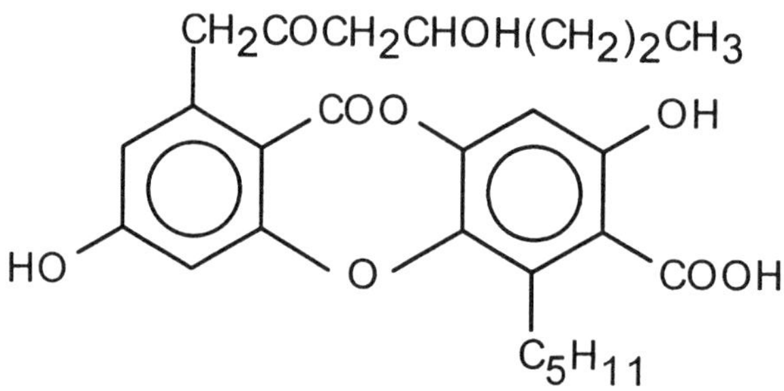

Figure 33 Structural formula of vittatolic acid.

differences, and chemistry was of considerable assistance in sorting the different groups (109). All but two morphological taxa have a constant chemistry where the presence of chemosyndromes was invariable. In the other two species, a high chemical diversity was found. When only the presence of chemosyndromes, not the substances themselves, was considered, two, resp. three chemical races were found, which were accepted as subspecies. The restricted occurrence of chemical races in lichens suggests that polyploidy or related processes might be involved.

In his studies of the genus *Pertusaria*, Archer (145) found that a species in this genus may contain (as major compounds) a combination of, or a selection from, three different classes of compounds—viz., a xanthone (A), an orcinol depside or depsone (B), and a β-orcinol depsidone or depside (C). This leads to seven theoretical possibilities of substance combination: ABC, AB, AC, BC, A, B, and C. Examples of each are given in Table 7. In addition, some taxa may lack lichen substances or contain aliphatic compounds.

Lecanora vacillans is a saxicolous lichen which grows on overhung rocks in southern Sweden. Morphologically it is reminiscent of two groups in *Lecanora*

Table 7 *Pertusaria* Taxa Selected to Show Possible Chemical Combinations

Combination	*Taxon*	Chemistry
ABC	*P. subventosa*	Lichexanthone (Fig. 27), picrolichenic (Fig. 7), and thamnolic (Fig. 34) acids
AB	*P. commutans*	Lichexanthone and lecanoric (Fig. 28) acid
AC	*P. thiospoda*	Thiophaninic (Fig. 35) and stictic (Fig. 23) acids
BC	*P. hartmannii*	Perlatolic (Fig. 24) and norstictic (Fig. 23) acids
A	*P. rigida*	4,5-dichlorolichexanthone (Fig. 35)
B	*P. mattogrossensis*	2-*O*-methylperlatolic (Fig. 24) acid
C	*P. novaezelandiae*	Hypothamnolic (Fig. 34) acid

Source: Adapted from Ref. 145.

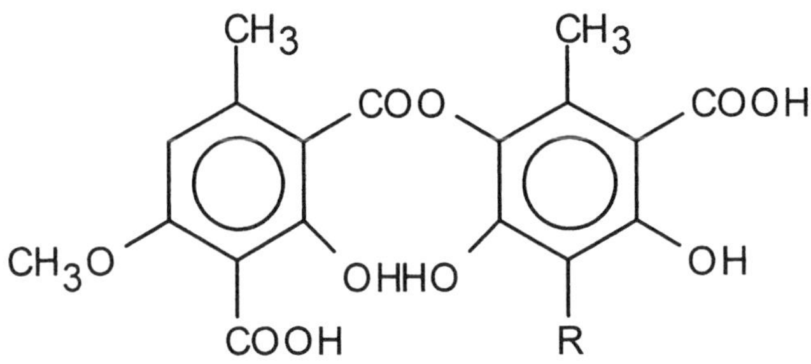

Substance	*R*
Hypothamnolic acid	CH_3
Thamnolic acid	*CHO*

Figure 34 Structural formulas of hypothamnnolic and thamnolic acids.

(*L. subcarnea* and *L. subfusca* groups), while the anatomy fits the *L. subfusca* group, and the chemistry the *L. subcarnea* group. It might be a rather basal offshot of the *L. subcarnea* phylogeny and, interestingly, contains nearly all substances present in the whole *L. subcarnea* group, thus having a more sophisticated chemistry than any other member of this group. These observations indicate that a pool of related substances can occur in species groups and some genetic factors may determine whether all or some products are accumulated.

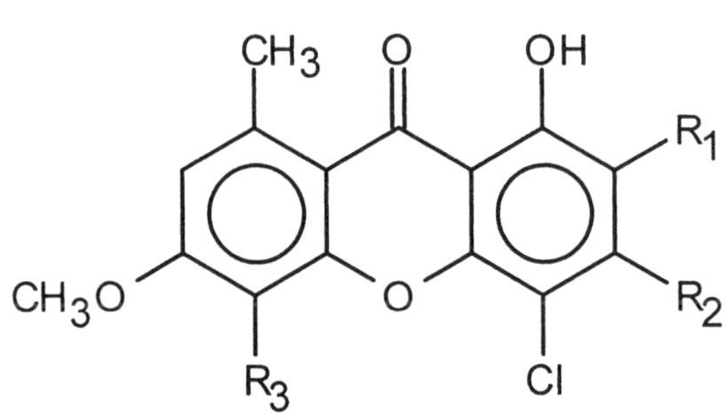

Substance	R_1	R_2	R_3
4,5-Dichlorolichexanthone	*H*	OCH_3	*Cl*
Thiophaninic acid	*Cl*	*OH*	*H*

Figure 35 Structural formulas of 4,5-dichlorolichexanthone and thiophaninic acid.

B. Metabolic Data at Genus and Subgenus Levels

Metabolic data have been widely used as characters at the generic and subgeneric levels. In the Physciaceae, Moberg (146) excluded the genus *Phaeophyscia* from *Physcia* mainly on the basis of the absence of atranorin (Fig. 4) and the presence of different conidia. Metabolic data have often used as important characters to circumscribe genera, particularly in the Parmeliaceae. Several genera were derived from the collective genus *Parmelia*, and chemistry has played an important role in the definition of new genera. A summary of this is given by Elix (28). In addition to secondary metabolites, the structure of the upper cortex, and type of cell wall polysaccharides, spore and conidial characters were also of special importance. Although these new genera are recognized by many lichenologists, they have not been universally accepted (139,147–149). Authors agree that most of the segregates are natural groupings, but the main disagreement in this debate concerns the ranking of these groups. An example of three genera distinguished by a correlation of morphological, geographical, and chemical characters is given in Table 8.

Chemical characters correlate to some extent with the new generic concept in cetrarioid lichens recently proposed by Kärnefelt, Thell, and co-workers (150–154), which is mainly based on ascomatal characters, such as ascus type, ascospores, and conidia. An overview of some genera in this group and their chemistry can be found in Table 9.

Triterpenoids can be important indicators for the detection of evolutionary relationships in the Peltigerineae, a suborder of Lecanorales with five families

Table 8 Comparison of Some Characters of *Parmelia* s.str., *Flavopunctelia*, and *Punctelia*

Characters	*Parmelia* s.str.	*Flavopunctelia*	*Punctelia*
Cortical chemistry	Atranorin	Usnic acid	Atranorin
Medullary chemistry	β-orcinol depsides	Orcinol depside	Orcinol depsides, fatty acids
Cell wall polysaccharides	Lichenin absent	Lichenin present	Lichenin absent
Main distribution	Arctic to temperate	Temperate-subtropical	Temperate-subtropical
Conidia	Bifusiform	Bifusiform	Filiform or unciform
Pseudocyphellae	Linear	Suborbicular	Suborbicular
Rhizines	± squarrose	Simple	Simple
Greatest diversity	East Asia and Australasia	America, Africa	America, Africa

Source: Adapted from Refs. 184–186.

Table 9 Secondary Metabolites Present in Some Genera of Cetrarioid Lichens

Genus	Fatty acids	Orcinol depsides and depsidones	β-Orcinol depsides and depsidones	Usnic acid	Quinones	Pulvinic acid derivatives
Allocetraria	+	−	−	−	+	−
Arctocetraria	+	−	−	−	−	−
Cetraria	+	−	+	−	−	−
Cetrariella	−	+	−	−	−	−
Cetrariopsis	+	+	−	+	−	−
Esslingeriana	+	−	+	−	+	−
Flavocetraria	+	−	−	+	+	−
Masonhalea	−	+	−	−	−	−
Nephromopsis	+	+	+	+	+	−
Tuckermannopsis	+	+	−	+	−	−
Vulpicida	−	−	−	+	−	+

Source: Adapted from Ref. 154.

(46,155). In the genus *Pseudocyphellaria*, numerous groups of triterpenoids occur. Hopan-22-ol derivatives, derived by direct H^+-initiated cyclization of squalene, occur only in white medulla species, whereas C_3-oxygenated ferenen, sticate, or lupeol derivatives, derived from H^+-initiated cyclization of 2,3-epoxysqualene, occur in species with a yellow medulla, due to the presence of pulvinic acid derivatives (155).

Examples where genera are defined by their chemistry are less obvious among crustose lichen families, but with the current trend of defining smaller natural units in crustose groups, some genera are also chemically well defined, such as *Miriquidica* with predominantly miriquidic acid (Fig. 36) (156) or *Rhizoplaca* with placodiolic or pseudoplacodiolic acids (Fig. 25) (157,158). At the subgeneric level, Archer (145) successfully used metabolic data in combination with morphology and spore characters to propose a new subdivision of *Pertusaria* (Table 10).

The chemistry of the crustose genus *Diploschistes* is rather uniform and few secondary metabolites have been found (159). However, among these is diploschistesic acid (Fig. 28), which has not been found in any other lichen genus. Two or three groups can be distinguished in the genus according to the opening of the ascomata. The presence of diploschistesic acid in both species with perithecioid and apothecioid ascomata supports the homogenity of the genus, despite the considerable morphological variability among the species.

The examples discussed above show that secondary metabolites in lichens

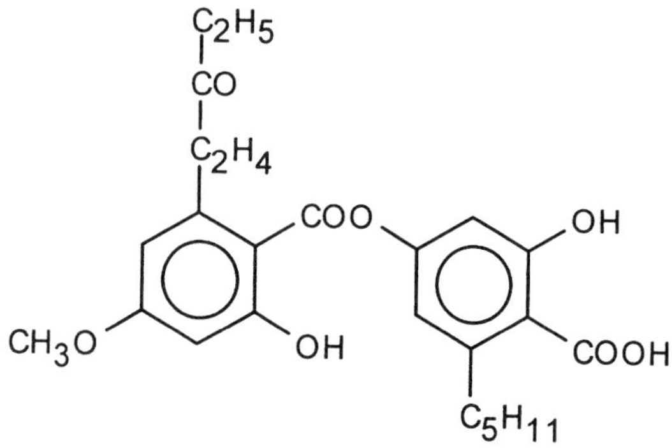

Figure 36 Structural formula of miriquidic acid.

Table 10 Main Characters to Distinguish Subgenera and Sections in *Pertusaria*

Subgenus:	*Pertusaria*	*Pionospora*	*Monomurata*	
Section:			*Monomurata*	*Digitatae*
Lichexanthone	+	–	+	–
Chlorinated xanthones	+	+	–	–
β-Orcinol depsidones	+	+	+	+
β-Orcinol depsides	–	–	+	+
Depsones	–	–	+	–
Orcinol-depsides with short side chains	–	+	+	–
Orcinol-depsides with long side chains	+	–	–	–
Apothecia	Verruciform	Disciform or pseudodisciform	Disciform	Disciform
Spore wall	Double, ± rough	Single or double, smooth	Single, smooth	Single, smooth
Spores per ascus	1–8	1–8	1–8	1–2

Source: Adapted from Ref. 145.

can be of immense importance in the definition of supraspecific categories and have helped to solve taxonomic problems. As pointed out by Brodo (43), "Chemistry sometimes acts as a red flag warning you 'take a closer look!' " (p. 136). An example he used was "*Ochrolechia geminipara*" which contains alectorialic and barbatolic acids (Fig. 8), substances otherwise unknown in *Ochrolechia*. This taxon was found to differ in anatomy, and belongs in the genus *Pertusaria* (160). Numerous additional examples could be added. However, in some cases secondary metabolites may lead to incorrect interpretations.

The genus *Lecanora* has many diverse species and clearly does not represent a natural group. Several entities exist within the genus, but the circumscription of monophyletic groups remains difficult. It is tempting to employ chemistry in this area as described earlier. Since the core group of *Lecanora* contains atranorin, the presence of atranorin (Fig. 4) and usnic acid (Fig. 11) usually exclude each other. It was suggested that the usnic acid-containing species formed a different group, for which the name *Straminella* was available (161). However, recent investigations showed that there appears to be no clear distinction among species in *Lecanora* containing atranorin and those containing usnic acid. Brodo and Elix (162) first reported a species containing both usnic acid and atranorin and having the anatomy of *Lecanora* s.str.; subsequently, further species containing usnic acid were found to be members of *Lecanora* s.str. (163,164), resulting in 18 taxa being placed in this group. This does not mean that all taxa in *Lecanora* with usnic acid will remain in the genus, but it is clear that there is not a simple dichotomy.

Another example where metabolic data did not help to resolve taxonomic problems is in the distinction of the genera *Ochrolechia* and *Pertusaria*. The two genera differ in the amyloidity of the hymenium, ascus type, and spore wall thickness. However, a number of taxa are known only from sterile material. Gyrophoric acid and related substances occur in numerous species of *Ochrolechia*, while this compound is rather rare in *Pertusaria*. Therefore, chemistry was thought to be helpful in sorting the sterile taxa. Almborn (165) and Hanko (166) proposed that most members containing gyrophoric acid (Fig. 5) and described as *Pertusaria* actually belonged to *Ochrolechia*, and that both genera can thus be distinguished by their chemistry as well. However, in a reexamination of this group, Schmitz et al. (167) found some fertile taxa containing gyrophoric acid which undoubtely belong in *Pertusaria*. Therefore, sterile taxa cannot be classified in either genus with certainty.

C. Metabolic Data at Suprageneric Level

While chemistry has been mainly applied at generic and species levels, some chemical data are also helpful in systematics at higher ranks. An overview of chemical data for phylogeny of lichenized Ascomycetes was given by the Culbersons (168). Since then, however, lichen systematics has seen a remarkable change of concepts in the circumscription of orders and families, especially since the observations of variation of ascus types in the Lecanoraceae and Lecideaceae sensu lato by Hafellner (169), and more metabolic data are available.

1. *Metabolic Data at Order Level*

While a great diversity of secondary metabolites can be found in some groups of lichens, such as the Lecanorales, others rarely produce any secondary metabolites. An overview of some major Ascomycetes orders with lichenized members and the metabolites found so far is given in Table 11. The occurrence of secondary metabolites shows clear patterns, and while orders such as the Arthoniales, Lecanorales, or Pertusariales contain a number of substances of various classes, others, such as the Gyalectales or Verrucariales, lack any compounds (cyanogenous substances were recently discovered in some genera of Verrucariales, but the structure of these compounds is not yet known [Huneck, in litt.]). The Lecanorales are by far the largest group of lichenized ascomycetes and include numerous families. On the other hand, the Pertusariales are a rather small order, containing only two families and fewer than 10 genera. Most of the chemical diversity seen in this group occurs in the single genus *Pertusaria*.

The distinction of the orders Gyalectales and Ostropales, which show striking convergences in their ascoma ontogeny (7,170), is supported by metabolic data. While the former order do not contain any metabolites, numerous substances have been detected in members of the latter. However, it should be noted that numerous genera with lichenized species in the Ostropales also lack metabolites.

The occurrence of particular chemical classes among the orders of lichenized Ascomycotina may also differ. While depsones have been found only in the Pertusariales, and diphenyl ether only in Lecanorales, depsides or quinones are present in numerous groups. In some cases it is helpful to distinguish groups within chemical classes—for example, xanthones are widely distributed among lichens. However, if one distinguishes between nonchlorinated and chlorinated xanthones, chlorinated xanthones are mostly found in the Lecanorales and Pertusariales, and seldom found in the Arthoniales and Caliciales.

2. *Metabolic Data at Family Level*

More information can be obtained from metabolic data for taxonomic decisions at the family level. When examining the systematic position of the Schaereriaceae, Lunke et al. (171) used, in addition to anatomical and ontogenetic characters, the presence of gyrophoric acid (Fig. 5) and related substances as a character to support the close relationship of this family and the Agyriaceae.

The Haematommaceae and Ophioparmaceae, two families in the Lecanorales whose members were formerly regarded as belonging to one genus, *Haematomma*, are now regarded as belonging to different families, mainly on the basis of their different ascus types (169,172). Both groups have similarly red-pigmented apothecial disks. The structure of the quinones present in both families, however,

Table 11 Overview of Occurrence of Secondary Metabolites in Orders of Ascomycotina with Lichenized Members

Order	Aliphatic acids	Depsides	Depsidones	Depsones	Benzyl ester	Diphenyl ether	Naphthopyran	Dibenzofuranes and usnic acids	Chromones	Xanthones	Quinones	Mevalonic acid pathway	Shikimic acid pathway
Arthoniales	+	+	+	–	–	–	+	+	+	+	+	+	–
Caliciales	–	+	+	–	–	–	–	+	–	+	+	+	+
Dothideales	–	–	–	–	–	–	–	–	–	–	–	–	–
Gyalectales	–	–	–	–	–	–	–	–	–	–	–	–	–
Lecanorales	+	+	+	–	+	+	–	+	+	+	+	+	+
Leotiales	–	+	+	–	–	–	–	–	–	–	–	–	–
Ostropales	–	+	+	–	–	–	–	–	–	+	+	+	–
Patellariales	–	+	+	–	–	–	–	–	–	–	–	–	+
Pertusariales	+	+	+	+	–	–	–	–	–	+	–	–	–
Pyrenulales	–	–	–	–	–	–	–	–	–	+	+	–	–
Trichotheliales	–	–	–	–	–	–	–	–	–	–	+	–	–
Verrucariales	–	–	–	–	–	–	–	–	–	–	–	–	–

Source: Systematic arrangement follows Ott and Lumbsch, in press.

was shown to be different (173,174), supporting the distinction of the two groups.

It is generally believed that most groups of foliose and fruticose lichens have derived from crustose ancestors, although evolution in the opposite direction might have occurred in some cases. The Ramalinaceae are a group of fruticose lichens where no crustose-related group was known. In this connection the endemic Australian genus *Ramalinora* is of special interest. This monotypic crustose genus was found on soil in tropical Queensland and was originally placed in *Lecanora*. It was, however, atypical for this genus due to the presence of metadepsides, which are common in the Ramalinaceae and Cladoniaceae. The ascus type, ascoma development, and chemistry suggested a close relationship to *Ramalina* and related genera, and it was therefore tentatively placed in the Ramalinaceae, as the first crustose genus in that family.

Lichens colonizing living leaves are rather small and were generally believed to lack secondary metabolites, mainly due to the insufficient sensitivity of the analytical methods applied to the low concentrations of metabolites in relatively small amounts of material. The application of HPLC surmounted these difficulties, and numerous foliicolous lichens were shown to contain metabolites (176–178). The genera *Badimia* and *Fellhanera* both contain numerous foliicolous species and have been placed in either the same or different families (179,180). A detailed anatomical and chemical reexamination showed that the two genera were closely related and that the chemistry correlated well with the anatomical data (177).

On the other hand, some families were mainly defined by their chemistry, and their circumscription remains uncertain. A group of genera with a usually yellowish-green thallus (due to the presence of pulvinic acid derivatives) was removed from the Lecanoraceae and placed in a separate family Candelariaceae (181). Although this family seems to be an natural group with development from crustose to foliose and fruticose genera (182), the group might not be so distant from the Lecanoraceae, since pulvinic acid derivatives have also been found in the *Lecanora subfusca* group, which represents the core group of *Lecanora* (183).

CONCLUSIONS

Chemotaxonomy has a long history in lichen taxonomy and has been widely used for the distinction of taxa, mainly at species and generic levels, although metabolic data can also be useful at higher ranks. However, no general statement on the taxonomic significance of metabolic data can be given; rather, each case should be evaluated individually. Some types of chemical variation are of secondary importance and do not deserve any taxonomic recognition, while others have been very helpful in taxonomic analysis. The recognition of chemosyndromic variation stimulated further investigation of chemical variation of lichenized Ascomycetes and might also be useful in other groups of fungi.

The introduction of molecular methods may also stimulate the examination of biosynthesis of lichen substances and the genetic background of chemical variation. It is likely that molecular methods and numerical and cladistic analyses, which have seldom been used in lichen chemotaxonomy until recently, will be applied more frequently in the future.

ACKNOWLEDGMENTS

I am grateful to Dr. William Sanders (Berkeley) for constructive criticism and linguistic correction of the text. Professor Benno Feige and Mr. Roland Guderley (both of Essen) are thanked for helpful comments on the manuscript.

REFERENCES

1. Hawksworth DL. Some advances in the study of lichens since the time of E.M. Holmes. Bot J Linn Soc (Lond) 1973; 67:3–31.
2. Gargas A, DePriest PT, Grube M, Tehler A. Multiple origins of lichen symbioses in fungi suggested by SSU rDNA phylogeny. Science 1995; 268:1492–1495.
3. Hawksworth DL. The variety of fungal-algal symbioses, their evolutionary significance, and the nature of lichens. Bot J Linn Soc 1988; 96:3–20.
4. Hawksworth DL. Secondary fungi in lichen symbioses: parasites, saprophytes and parasymbionts. J Hattori Bot Lab 1982; 52:357–366.
5. Jahns, HM. Anatomy, morphology and development. In: Ahmadjian V, Hale ME, eds. The Lichens. New York; Academic Press 1973:3–58.
6. Büdel B, Scheidegger C. Thallus morphology and anatomy. In: Nash TH, ed. Lichen Biology. Cambridge: Cambridge University Press, 1996:37–64.
7. Henssen A, Jahns A. Lichenes. Stuttgart: Georg Thieme Verlag, 1974.
8. Poelt J. Systematic evaluation of morphological characters. In: Ahmadjian V, Hale ME, eds. The Lichens. New York: Academic Press, 1973:91–115.
9. Nash TH. Lichen Biology. Cambridge: Cambridge University Press, 1996.
10. Hamada N, Miyagawa H. Secondary metabolites from isolated lichen mycobionts cultured under different osmotic conditions. Lichenologist 1995; 27:201–205.
11. Culberson CF, Armaleo D. Induction of a complete secondary-product pathway in a cultured lichen fungus. Exp Mycol 1992; 16:52–63.
12. Culberson CF, Culberson WL, Johnson A. Characteristic lichen products in cultures of chemotypes of the *Ramalina siliquosa* complex. Mycologia 1992; 84:705–714.
13. Ahmadjian V. The Lichen Symbiosis. New York: John Wiley & Sons, 1993.
14. Avalos A, Vicente C. The occurrence of lichen phenolics in the photobiont cells of *Evernia prunastri*. Plant Cell Reports 1987; 6:74–76.
15. Elix JA, Feige GB, Lumbsch HT, Mies B, Wardlaw JH, Willis AC. The structure determination of simonyellin—a new lichen naphthopyran. Aust J Chem 1995; 48: 2035–2039.
16. Culberson C. Chemical and Botanical Guide to Lichen Products. Chapel Hill: University of North Carolina Press, 1969.
17. Culberson C. Supplement to Chemical and Botanical Guide to Lichen Products. Bryologist 1970; 73:177–377.

18. Culberson CF, Culberson WL, Johnson A. Second Supplement to Chemical and Botanical Guide to Lichen Products. St. Louis: American Bryological and Lichenological Society, Missouri Botanical Garden, 1977.
19. Elix JA. Biochemistry and secondary metabolites. In: Nash TH, ed. Lichen Biology. Cambridge: Cambridge University Press, 1996:154–180.
20. Mosbach K. On the biosynthesis of lichen substances. Part 2. The pulvinic acid derivative vulpinic acid. Biochem Biophys Res Commun 1964; 17:363–367.
21. Mosbach K. Zur Biosynthese von Flechtenstoffen, Produkte einer symbiotischen Lebensgemeinschaft. Angew Chem 1969; 81:233–244.
22. Poelt, J Leuckert C. Substitution and supplementary addition of secondary products in the evolution of lichenized Ascomycotina. Bibl Lichenol 1993; 53:201–215.
23. Esslinger TL. A new status for the brown Parmeliae. Mycotaxon 1978; 7:45–54.
24. Coppins BJ. A taxonomic study of the lichen genus *Micarea* in Europe. Bull Br Mus (Nat Hist) Bot Ser 1983; 22:17–214.
25. Bachmann E. Über nichtkristallisierte Flechtenfarbstoffe. Ein Beitrag zur Chemie und Anatomie der Flechten. Jb Wiss Bot 1890; 21:1–61.
26. Common RS. The distribution and taxonomic significance of lichenan and isolichenan in the Parmeliaceae (lichenized Ascomycotina), as determined by iodine reactions. I. Introduction and methods. II. The genus *Alectoria* and associated taxa. Mycotaxon 1991; 41:67–112.
27. Elix JA, Nash TH III. A synopsis of the lichen genus *Psiloparmelia* (Ascomycotina, Parmeliaceae). Bryologist 1992; 95:377–391.
28. Elix JA. Progress in the generic delimitation of *Parmelia* sensu lato lichens (Ascomycotina: Parmeliaceae) and a synoptic key to the Parmeliaceae. Bryologist 1993; 96:359–383.
29. Zopf W. Die Flechtenstoffe in chemischer, botanischer, pharmokologischer und technischer Beziehung. Jena: Gustav Fischer Verlag, 1907.
30. Asahina Y, Shibata S. Chemistry of Lichen Substances. Tokyo: Japanese Society for the Promotion of Science in Tokyo, 1954.
31. Shibata S. Lichen substances. In: Modern Methods of Plant Analysis. Vol. VI. Berlin: Springer-Verlag, 1963:155–193.
32. Huneck S. Lichen substances. In: Reinhold L, Liwschitz Y, eds. Progress in Phytochemistry. Vol. 1. London: Interscience, 1968:223–346.
33. Huneck S. Chemie und Biosynthese der Flechtenstoffe. Fortschr Chem Organ Naturstoffe 1971; 29:209–306.
34. Huneck S. Nature of lichen substances. In: Ahmadjian V, Hale ME, eds. The Lichens. New York: Academic Press, 1973:495–522.
35. Huneck S. Fortschritte der Chemie von Flechtenstoffen. Beih Nova Hedwigia 1984; 79:793–838.
36. Huneck S. New results in the chemistry of lichens. Symbiosis 1991; 11:225–248.
37. Santesson J. Identification and isolation of lichen substances. In: Ahmadjian V, Hale ME, eds. The Lichens. New York: Academic Press, 1973:633–652.
38. Elix JA, Whitton AA, Sargent MV. Recent progress in the chemistry of lichen substances. Progr Chem Org Nat Prod 1984; 45:103–234.
39. Culberson W. The use of chemistry in the systematics of the lichens. Taxon 1969; 18:152–166.

40. Culberson WL. Chemistry and sibling speciation in the lichen-forming fungi: ecological and biological considerations. Bryologist 1986; 89:123–131.
41. Hawksworth DL. Lichen chemotaxonomy. In: Brown, DH, Hawksworth DL, Bailey RH, eds. Lichenology: Progress and Problems. London: Academic Press, 1976: 139–184.
42. Brodo IM. Changing concepts regarding chemical diversity in lichens. Lichenologist 1978; 10:1–11.
43. Brodo IM. Interpreting chemical variation in lichens for systematic purposes. Bryologist 1986; 89:132–138.
44. Rogers RW. Chemical variation and the species concept in lichenized ascomycetes. Bot J Linn Soc 1989; 101:229–239.
45. Elix JA. History of lichen chemistry and development of analytical techniques. Fl Aust 1992; 54:23–29.
46. Galloway DJ. Chemical evolution in the order Peltigerales: triterpenoids. Symbiosis 1991; 11:327–344.
47. Leuckert C. Probleme der Flechten-Chemotaxonomie—Stoffkombinationen und ihre taxonomische Wertung. Ber Dtsch Bot Ges 1985; 98:401–408.
48. Culberson W. Chemosystematics and ecology of lichen-forming fungi. Annu Rev Ecol Syst 1970; 1:153–170.
49. Culberson CF, Elix JA. Lichen substances. In: Harborne JB, ed. Methods in Plant Biochemistry. Vol. 1. Plant Phenolics. London: Academic Press, 1989:509–535.
50. Leuckert C. Die Identifizierung von Flechtenstoffen im Rahmen chemotaxonomischer Routineanalysen. Beih Nova Hedwigia 1984; 79:839–869.
51. White FJ, James PW. A new guide to microchemical techniques for the identification of lichen substances. Br Lich Soc Bull 1985; 57:1–41.
52. Huneck S, Yoshimura I. Identification of Lichen Substances. Berlin: Springer Verlag, 1996.
53. Nylander W. Circa novum in studio lichenum criterium chemicum. Flora 1866; 49: 198–201.
54. Nylander W. Quaedam addenda ad nova criteria chemica in studio lichenum. Flora 1866; 49:233–234.
55. Asahina Y. Über die Reaktion vom Flechten-Thallus. Acta Phytochim 1934; 8:47–64.
56. Asahina Y. Mikrochemischer Nachweis der Flechtenstoffe (I). J Jpn Bot 1936; 12: 516–525.
57. Wachtmeister C. Identification of lichen acids by paper chromatography. Bot Not 1956; 109:313–324.
58. Culberson CF, Kristinsson H-D. A standardized method for the identification of lichen products. J Chromatogr 1970; 46:85–93.
59. Culberson CF. Improved conditions and new data for the identification of lichen products by a standardized thin-layer chromatographic method. J Chromatogr 1972; 72:113–125.
60. Culberson CF, Johnson A. Substitution of methyl *tert.*-butyl ether for diethyl ether in standardized thin-layer chromatographic method for lichen products. J. Chromatogr 1982; 238:438–487.
61. Steiner M. Ein stabiles Diaminreagens für lichenologische Zwecke. Ber Dtsch Bot Gesell 1955; 63:35–40.

62. Leuckert C, Dolling K, Wolters W. Chemische Flechtenanalysen. Herzogia 1979; 5: 181–185.
63. Archer AW. 3-Methyl-2-benzothiazolone hydrazone hydrochloride as a spray reagent for phenolic lichen compounds. J Chromatogr 1978; 152:290–292.
64. Culberson CF, Culberson WL, Johnson A. A standardized TLC analysis of β-orcinol depsidones. Bryologist 1981; 84:16–29.
65. Leuckert C, Knoph J-G. European taxa of saxicolous *Lecidella* containing chloroxanthones: identification of patterns using thin layer chromatography. Lichenologist 1992; 24:383–397.
66. Archer AW. Identification of orcinol para-depsides in the lichen genus *Pertusaria* by thin layer chromatography. Mycotaxon 1993; 46:1–4.
67. Culberson CF, Johnson A. A standardized two-dimensional thin-layer chromatographic method for lichen products. J Chromatogr 1976; 128:253–259.
68. Esslinger TL. On the chemistry and distribution of *Cladonia petrophila*. Mycotaxon 1994; 51:101–105.
69. Culberson CF. Conditions for the use of Merck silica gel 60 F-254 plates in the standardized thin-layer chromatographic technique for lichen products. J Chromatogr 1974; 97:107–108.
70. Arup U, Ekman S, Lindblom L, Mattsson J-E. High performance thin layer chromatography (HPTLC), an improved technique for screening lichen substances. Lichenologist 1993; 25:61–71.
71. Culberson C. High-speed liquid chromatography of lichen extracts. Bryologist 1972; 75:54–62.
72. Nourish R, Oliver RWA. Chemotaxonomic studies on the *Cladonia chlorophaea-pyxidata* complex and some allied species in Britain. In: Brown DH, Hawksworth DL, Bailey RH, eds. Lichenology: Progress and Problems. London: Academic Press, 1976:185–214.
73. Culberson CF, Hertel H. Chemical and morphological analyses of the *Lecidea lithophila-plana* group (Lecideaceae). Bryologist 1979; 82:189–197.
74. Culberson CF, Nash TH III, Johnson A. 3-α-Hydroxybarbatic acid, a new depside in chemosyndromes of some Xanthoparmeliae with β-orcinol depsides. Bryologist 1979; 82:154–161.
75. Lumbsch HT, Elix JA. A new species of the lichen genus *Diploschistes* from Australia. Pl Syst Evol 1985; 150:275–279.
76. Elix JA, Jenie UA, Arvidsson L, Jorgensen PM, James PW. New depsidones from the lichen genus *Erioderma*. Aust J Chem 1986; 39:719–722.
77. Strack D, Feige GB, Kroll R. Screening of aromatic secondary lichen substances by high performance liquid chromatography. Z Naturforsch Part C 1979; 34:695–698.
78. Huovinen K, Hiltunen R, von Schantz M. A high performance liquid chromatographic method for the analysis of lichen compounds from the genera *Cladina* and *Cladonia*. Acta Pharm Fenn 1985; 94:99–112.
79. Feige GB, Lumbsch HT, Huneck S, Elix JA. Identification of lichen substances by a standardized high-performance liquid chromatographic method. J Chromatogr 1993; 646:417–427.
80. Huneck S, Feige GB, Lumbsch HT. High performance liquid chromatographic analysis of aliphatic lichen acids. Phytochem Anal 1994; 5:57–60.

81. Yoshimura I, Kinoshita Y, Yamamoto Y, Huneck S, Yamada Y. Analysis of secondary metabolites from lichen by high performance liquid chromatography with a photodiode array detector. Phytochem Anal 1994; 5:195–205.
82. Santesson J. Chemical studies on lichens. 10. Mass spectrometry of lichens. Ark Kemi 1969; 30:363–377.
83. Elix JA, Johnston J, Parker JL. A computer program for the rapid identification of lichen substances. Mycotaxon 1988; 31:89–99.
84. Mietzsch E, Lumbsch HT, Elix JA. Wintabolites. Published by the authors, Essen, Germany, 1992.
85. Mietzsch E, Lumbsch HT, Elix JA. Notice: a new computer program for the identification of lichen substances. Mycotaxon 1993; 47:475–479.
86. Leighton WA. Notulae lichenologicae. XI. On the examination and rearangement of the Cladoniei, as tested by hydrate of potash. Ann Mag Nat Hist Ser 3 1866; 18: 405–420.
87. Lindsay WL. On chemical reaction as a specific character in lichens. J Linn Soc Bot 1869; 11:36–63.
88. Gyelnik V. *Alectoria* Studien. Nyt Mag Naturvid 1932; 70:35–62.
89. Erichsen CFE. Beiträge zur Kenntnis der Flechtengattung *Pertusaria*. Repert Spec Nov Regni Veg 1936; 41:77–101.
90. Degelius G. Lichens from southern Alaska and the Aleutian Islands, collected by Dr. E. Hultén. Acta Horti Gothob 1937; 12:105–144.
91. Almborn O. The species concept in lichen taxonomy. Bot Not 1965; 118:454–457.
92. Asahina Y. Über den taxonomischen Wert der Flechtenstoffe. Bot Mag Tokyo 1937; 51:759–764.
93. Asahina Y. Lichens of Japan. Vol. III. Genus *Usnea*. Tokyo: Research Institute for Natural Resources, 1956.
94. Tétényi P. Proposition à propos de la nomenclature des races chimiques. Taxon 1958; 7:40–41.
95. Targe A, Lambinon J. Étude chimiotaxonomique du groupe de *Parmelia borreri* (Sm.) Turn. en Europe occidentale. Bull Soc R Bot Belg 1965; 98:295–306.
96. Esslinger TL. A chemosystematic revision of the brown *Parmelia*. J Hattori Bot Lab 1977; 42:1–211.
97. Esslinger TL. Systematics of *Oropogon* (Alectoriaceae) in the New World. Syst Bot Monogr 1989; 28:1–111.
98. Imshaug I, Brodo I. Biosystematic studies on *Lecanora pallida* and some related lichens in the Americas. Nova Hedwigia 1966; 12:1–59.
99. Leuckert C, Poelt J. Studien über die *Lecanora rupicola*-Gruppe in Europa (Lecanoraceae). Nova Hedwigia 1989; 49:121–167.
100. Lumbsch HT, Feige GB, Elix JA. Chemical variation in two species of the *Lecanora subfusca* group (*Lecanoraceae*, lichenized *Ascomycotina*). Pl Syst Evol 1994; 191: 227–236.
101. Nimis PL, Poelt J, Puntillo D. *Candelariella faginea* spec. nov. (Lichenes, Candelariaceae) eine bemerkenswerte neue Art einer schwierigen Gattung aus Südeuropa. Nova Hedwigia 1989; 49:274–280.
102. Dickhäuser A, Lumbsch HT, Feige GB. A synopsis of the *Lecanora subcarnea* group. Mycotaxon 1995; 56:303–323.

103. Gilbert OL, Henderson A, James PW. Citrine-green taxa in the genus *Candelariella*. Lichenologist 1981; 13:249–251.
104. Clayden SR. Chemical divergence of eastern North American and European populations of *Arctoparmelia centrifuga* and their sympatric usnic acid-deficient chemotypes. Bryologist 1992; 95:1–4.
105. Hale ME Jr. *Arctoparmelia*, a new genus in the Parmeliaceae (Ascomycotina). Mycotaxon 1986; 25:251–254.
106. Follmann G. Schulz M. Stress-induced changes in the secondary products of lichen thalli. Bibl Lichenol 1993; 53:75–86.
107. Feige GB, Lumbsch HT, Mies B. Morphological and chemical changes in *Rocella* thalli infected by *Lecanactis grumulosa* (lichenized Ascomycetes, Opegraphales). Crypt Bot 1993; 3:101–107.
108. Hawksworth DL, Paterson RRM, Vote N. An investigation into the occurrence of metabolites in obligately lichenicolous fungi from thirty genera. Bibl Lichenol 1993; 53:101–108.
109. Lumbsch HT. Die *Lecanora subfusca*-Gruppe in Australasien. J Hattori Bot Lab 1994; 77:1–175.
110. Poelt J, Leuckert C. Die Arten der *Lecanora dispersa*-Gruppe (Lichenes, Lecanoraceae) auf kalkreichen Gesteinen im Bereich der Ostalpen—Eine Vorstudie. Bibl Lichenol 1995; 58:289–333.
111. Paus S, Daniels FJA, Lumbsch HT. Chemical and ecological studies in the *Cladonia subulata* complex in northern Germany (Cladoniaceae, lichenised Ascomycotina). Bibl Lichenol 1993; 53:191–200.
112. Brodo IM. The North American species of *Lecanora subfusca* group. Beih Nova Hedwigia 1984; 79:63–185.
113. Kalb K. Pyxine species from Australia. Herzogia 1994; 10:61–69.
114. Poelt J. Das Konzept der Artenpaare bei den Flechten. Vortr Gesamtg Bot NF (Dtsch Bot Ges) 1970; 4:187–198.
115. Tehler A. The species pair concept in lichenology. Taxon 1982; 31:708–717.
116. Mattsson JE, Lumbsch HT. The use of the species pair concept in lichen taxonomy. Taxon 1989; 38:238–241.
117. Poelt J. Flechtenflora und Eiszeit in Europa. Phyton Ann Rei Bot 1963; 10:206–215.
118. Poelt J. Die taxonomische Behandlung von Artenpaaren bei den Flechten. Bot Not 1972; 125:77–81.
119. Culberson W. The *Parmelia perforata* group: niche characteristics of chemical races, speciation by parallel evolution, and a new taxonomy. Bryologist 1973; 76: 20–29.
120. Culberson W, Culberson C. Parallel evolution in the lichen-forming fungi. Science 1973; 180:196–198.
121. Culberson W. Norstictic acid as a hymenial constituent of *Letharia*. Mycologia 1969; 61:731–736.
122. Galloway DJ. Studies in *Pseudocyphellaria* (lichens). I. The New Zealand species. Bull Br Mus (Nat Hist) Bot Ser 1988; 17:1–267.
123. Rikkinen J. What's behind the pretty colours? A study on the photobiology of lichens. Bryobrothera 1996; 4:1–239.
124. Culberson W, Culberson C. The lichen genera *Cetrelia* and *Platismatia* (Parmeliaceae). Contrib US Natl Herb 1968; 34:449–558.

125. Hale M Jr. Chemical strains of the lichen *Parmelia furfuracea*. Am J Bot 1956; 43: 456–459.
126. Hale M Jr. A synopsis of the lichen genus *Pseudevernia*. Bryologist 1968; 71: 1–11.
127. Hawksworth D, Chapman D. *Pseudevernia furfuracea* (L.) Zopf and its chemical races in the British Isles. Lichenologist 1971; 5:51–58.
128. Culberson C. A note on the chemical strains of *Parmelia furfuracea*. Bryologist 1965; 68:435–439.
129. Halvorsen R, Bendiksen E. The chemical variation of *Pseudevernia furfuracea* in Norway. Nord J Bot 1982; 2:371–380.
130. Culberson W. Analysis of chemical and morphological variation in the *Ramalina siliquosa* complex. Brittonia 1967; 19:333–352.
131. Culberson W. The behavior of the species of the *Ramalina siliquosa* group in Portugal. J Öster Bot Z 1969; 116:85–94.
132. Culberson W, Culberson C. Habitat selection by chemically differentiated races of lichens. Science 1967; 158:1195–1197.
133. Culberson WL, Culberson CF. Chemosyndromic variation in lichens. Syst Bot 1977; 1:325–339.
134. Culberson CF, Culberson WL, Esslinger TL. Chemosyndromic variation in the *Parmelia pulla* group. Bryologist 1977; 80:125–135.
135. Leuckert C, Knoph J-G. Secondary compounds as taxonomic characters in the genus *Lecidella* (Lecanoraceae, Lecanorales). Bibl Lichenol 1993; 53:161–171.
136. Gowan SP. A character analysis of the secondary products of the Porpidiaceae (lichenized Ascomycotina). Syst Bot 1989; 14:77–90.
137. Zeybek U, Lumbsch HT, Feige GB, Elix JA, John V. Chemosyndromic variation in *Hypogymnia* species, mainly from Turkey (lichenized Ascomycotina). Crypt Bot 1993; 3:260–263.
138. Culberson CF. Biogenetic relationships of the lichen substances in the framework of systematics. Bryologist 1986; 89:91–98.
139. Wirth V. Checkliste der Flechten und flechtenbewohnenden Pilze Deutschlands—eine Arbeitshilfe. Stuttg Beitr Naturk Ser A 1994; 517:1–63.
140. Culberson CF, Culberson WL, Johnson A. Gene flow in lichens. Am J Bot 1988; 75: 1135–1139.
141. DePriest PT. Phylogenetic analyses of the variable ribosomal DNA of the *Cladonia chlorophaea* complex. Crypt Bot 1995; 5:60–70.
142. DePriest PT. Molecular innovations in lichen systematics: the use of ribosomal and intron nucleotide sequences in the *Cladonia chlorophaea* complex. Bryologist 1993; 96:314–325.
143. DePriest PT. Variation in the *Cladonia chlorophaea* complex II: Ribosomal DNA variation in a Southern Appalachian population. Bryologist 1994; 97:117–126.
144. Culberson WL, Culberson CF, Johnson A. Speciation in lichens of the *Ramalina siliquosa* complex (Ascomycotina, Ramalinaceae): gene flow and reproductive isolation. Am J Bot 1993; 80:1472–1481.
145. Archer AW. A chemical and morphological arrangement of the lichen genus *Pertusaria*. Bibl Lichenol 1993 53:1–17.
146. Moberg R. The lichen genus *Physcia* and allied genera in Fennoscandia. Symb Bot Upsal 1977; 22:1–108.

147. Clauzade G, Roux C. Likenoj de Okcidenta Europo. Ilustrita Determinlibro. Bull Soc Bot Centre-Ouest Nouv Ser Num Spec 1985; 7:1–893.
148. Eriksson O, Hawksworth DL. Outline of the ascomycetes—1986. Syst Ascomycetum 1986; 5:185–324.
149. Poelt J, Vezda A. Bestimmungsschlüssel europäischer Flechten. Ergänzungsheft II. Bibl Lichenol 1981; 16:1–390.
150. Kärnefelt I, Mattsson J-E, Thell A. Evolution and phylogeny of cetrarioid lichens. Pl Syst Evol 1992; 183:113–160.
151. Kärnefelt I, Mattsson J-E, Thell A. The lichen genera *Arctocetraria, Cetraria*, and *Cetrariella* (Parmeliaceae) and their presumed evolutionary affinities. Bryologist 1993; 96:394–404.
152. Kärnefelt I, Thell A, Randlane T, Saag A. The genus *Flavocetraria* Kärnefelt & Thell (Parmeliaceae, Ascomycotina) and its affinities. Acta Bot Fenn 1994; 150: 79–86.
153. Randlane T, Saag A, Thell A, Kärnefelt I. The lichen genus *Tuckneraria* Randlane & Thell—a new segregate in the Parmeliaceae. Acta Bot Fenn 1994; 150:143–151.
154. Kärnefelt I, Thell A. Chemical evolution in cetrarioid lichens. Bibl Lichenol 1993; 53:115–127.
155. Wilkins AL. The distribution, biosynthetic and possible evolutionary inter-relationship of triterpenoids in New Zealand *Pseudocyphellaria* species. Bibl Lichenol 1993; 53:277–288.
156. Hertel H, Rambold G. *Miriquidica* genus novum Lecanoracearum (Ascomycetes lichenisati). Mitt Bot Staats München 1987; 23:377–392.
157. Leuckert C, Poelt J, Hahnel G. Zur Chemotaxonomie der eurasischen Arten der Flechtengattung *Rhizoplaca*. Nova Hedwigia 1977; 28:71–129.
158. McCune B. Distribution of chemotypes of *Rhizoplaca* in North America. Bryologist 1987; 90:6–14.
159. Lumbsch HT. Die holarktischen Vertreter der Flechtengattung *Diploschistes* (Thelotremataceae). J Hattori Bot Lab 1989; 66:133–196.
160. Brodo IM. Lichenes Canadenses Exsiccati: Fascicle III. Bryologist 1984; 87:97–111.
161. Eigler G. Studien zur Gliederung der Flechtengattung *Lecanora*. Diss Bot 1969; 4:1–195.
162. Brodo IM, Elix JA. *Lecanora jamesii* and the relationship between *Lecanora* s. str. and *Straminella*. Bibl Lichenol 1993; 53: 19–25.
163. Lumbsch HT, Feige GB, Elix JA. A revision of the usnic acid containing taxa belonging to *Lecanora sensu stricto* (Lecanorales; lichenized Ascomycotina). Bryologist 1995; 98:561–577.
164. Lumbsch HT. A new species in the *Lecanora subfusca* group containing usnic acid in addition to atranorin. Lichenologist 1995; 27:161–167.
165. Almborn O. Distribution and Ecology of Some South Scandinavian Lichens. Lund: Gleerup, 1948.
166. Hanko B. Die Chemotypen der Flechtengattung *Pertusaria* in Europa. Bibl Lichenol 1983; 19:1–296.
167. Schmitz KE, Lumbsch HT, Feige GB. Systematic studies in the Pertusariales II. The generic concept in the Pertusariaceae (lichenized Ascomycotina). Acta Bot Fenn 1994; 150:153–160.

168. Culberson W, Culberson C. A phylogenetic view of chemical evolution in the lichens. Bryologist 73:1–31.
169. Hafellner J. Studien in Richtung einer natürlicheren Gliederung der Sammelfamilien Lecanoraceae und Lecideaceae. Beih Nova Hedwigia 1984; 79:241–371.
170. Henssen A. Studies in the developmental morphology of lichenized ascomycetes. In: Brown DH, Hawksworth DL, Bailey RH, eds. Lichenology: Progress and Problems. London: Academic Press, 1976:107–138.
171. Lunke T, Lumbsch HT, Feige GB. Anatomical and ontogenetical studies on the Schaereriaceae (Agyriineae, Lecanorales). Bryologist 1995; 99:53–63.
172. Rogers RW, Hafellner J. *Haematomma* and *Ophioparma*: two superficially similar genera of lichenized fungi. Lichenologist 1988; 20:167–174.
173. Bruun T, Lamvik A. Haemaventosin. Acta Chem Scand 1971; 25:483–486.
174. Huneck S, Culberson CF, Culberson WL, Elix JA. Haematommone, a red pigment from apothecia of *Haematomma puniceum*. Phytochemistry 1991; 30:706–707.
175. Lumbsch HT, Rambold G, Elix JA. *Ramalinora*—a new lichen genus from Australia. Aust Syst Bot 1995; 8:521–530.
176. Elix JA, Crook CE, Lumbsch HT. The chemistry of foliicolous lichens. 1. Constituents of *Sporopodium vezdeanum* and *S. xantholeucum*. Mycotaxon 1992; 44:409–415.
177. Lücking R, Lumbsch HT, Elix JA. Chemistry, anatomy and morphology of foliicolous species of *Fellhanera* and *Badimia* (lichenized Ascomycotina: Lecanorales). Bot Acta 1994; 107:393–401.
178. Elix JA, Lumbsch HT, Lücking R. The chemistry of foliicolous lichens. 2. Constituents of some *Byssoloma* and *Sporopodium* species. Bibl Lichenol 1995; 58:81–96.
179. Vezda A. Neue Gattungen der Familie Lecideaceae s. lat. (Lichenes). Folia Geobot Phytotax 1986; 21:199–219.
180. Serusiaux E. The nature and origin of campylidia in lichenized fungi. Lichenologist 1986; 18:1–35.
181. Hakulinen R. Dir Flechtengattung *Candelariella* Müll. Argoviensis, mit besonderer Berücksichtigung ihres Auftretens und ihrer Verbreitung in Fennoskandien. Ann Bot Soc Zool Bot Fenn 'Vanamo' 1954; 27:1–127.
182. Poelt J. Zur Kenntnis der Flechtenfamilie Candelariaceae. Ein Beitrag mit besonderer Berücksichtigung einiger südamerikanischer Arten. Phyton (Aust) 1974; 16: 189–210.
183. Lumbsch HT. Calycin in *Lecanora fulvastra*. Lichenologist 1994; 26:94–96.
184. Egan RS. Correlations and non-correlations of chemical variation patterns with lichen morphology and geography. Bryologist 1986; 89:99–110.
185. Krog H. *Punctelia*, a new lichen genus in the Parmeliaceae. Nord J Bot 1982; 2: 287–292.
186. Hale ME Jr. *Flavopunctelia*, a new genus in the Parmeliaceae (Ascomycotina). Mycotaxon 1984; 20:681–682.
187. Elix JA. The lichen genus *Relicina* in Australasia. In: Galloway DJ, ed. The Systematics Association Special Volume, No. 43: Tropical Lichens: Their Systematics, Conservation and Ecology. Oxford: Clarendon Press, 1991:17–34.

Index